SELF-TESTING VLSI DESIGN

V.N. YARMOLIK
I.V. KACHAN
Computer Department
Minsk Radio Engineering Institute
Minsk, Byellorussia

1993
ELSEVIER
Amsterdam – London – New York – Tokyo

ELSEVIER SCIENCE PUBLISHERS
Sara Burgerhartstraat 25
P.O. Box 211, 1000 AE Amsterdam, The Netherlands

Library of Congress Cataloging-in-Publication Data

IArmolik, V. N. (Vīacheslav Nikolaevich)
   Self-testing VLSI design / V.N. Yarmolik, I.V. Kachan.
     p.   cm.
   Includes bibliographical references and index.
   ISBN 0-444-89640-6 (alk. paper)
   1. Integrated circuits--Very large scale integration--Testing.
 2. Integrated circuits--Very large scale integration--Design.
 I. Kachan, I. V. II. Title.
 TK7874.I16 1993
 621.39'5--dc20   92-46538
                                                                 CIP

ISBN: 0 444 89640 6

© 1993 Elsevier Science Publishers B.V. All rights reserved.

No part of this publication may be reproduced, stored in a retrieval system or transmitted in any form or by any means, electronic, mechanical, photocopying, recording or otherwise, without the prior written permission of the publisher, Elsevier Science Publishers B.V., Copyright & Permissions Department, P.O. Box 521,1000 AM Amsterdam, The Netherlands.

Special regulations for readers in the U.S.A. This publication has been registered with the Copyright Clearance Center Inc. (CCC), Salem, Massachusetts. Information can be obtained from the CCC about conditions under which photocopies of parts of this publication may be made in the U.S.A. All other copyright questions,including photocopying outside of the U.S.A., should be referred to the publisher.

No responsibility is assumed by the publisher for any injury and/or damage to persons or property as a matter of products liability, negligence or otherwise, or from any use or operation of any methods, products,instructions or ideas contained in the materials herein.

This book is printed on acid-free paper.

Printed in The Netherlands.

P R E F A C E

A distinctive feature of the present stage in computer equipment development is the continuous increase in functionality and complexity of computer components. As a result of these advances, very large-scale integration (VLSI) circuits have found extensive application in manufacture of computer products, personal computers included. Recently, among a variety of VLSI design technologies the self-test VLSI design for testability techniques have evolved. The present study deals with the problems of self-test VLSI design.

This is a summary of the basic facts about self-test VLSI design that have been obtained both by the scientists in the leading scientific centers in the CIS and by the scientists of industrially-developed countries in the world. The study gives prominence to theoretic fundamentals of designing self-test VLSI building blocks such as built-in test generators and output response analyzers. It has been shown here that the design-for- testability techniques amployed in self-test VLSI design provide for the most natural self-test concept implementation.

Among the results obtained by the authors of the present study, most significant are: structural design of self-test VLSI circuits; design of universal modules for self-test VLSI circuits; examination the VLSI circuits for signature testability.

Sections 13, 14, 15 and subsections 10.2 and 18.4 have been written together with I.V.Kachan, Candidate of Technical Science, Assistant Professor.

The author is sincerely grateful to V.G.Mikityuk, post- graduate student, and A.I.Berezhnaya, translator, for their assistance in the preparation of the manuscript. He is also indebted to his collegues from the Computer Department of the Minsk Radio Engineering Institute for their contributions and support.

V.N.Yarmolik

CONTENTS

Preface

Chapter 1
VLSI DESIGN 1
    1.1. Achievements in Microelectronics 1
    1.2. VLSI Design Procedure 2
    1.3. Self-Testing VLSI Chips 4

Chapter 2
VLSI TESTABILITY DESIGN APPROACHES 7
    2.1. VLSI Testability Design Feature 7
    2.2. Storage Element State Scan Technique 8
    2.3. Level Sensitive Scan Design (LSSD) 13
    2.4. Random Access Scan Technique 18
    2.5. VLSI Design by Boundary-Scan Technique 20

Chapter 3
SELF-TESTING VLSI STRUCTURED DESIGN 24
    3.1. Self-testing VLSI Using Scan Path 24
    3.2. Ring VLSI Self-Test 28
    3.3. Use of General-Purpose Modules for Self-Test VLSI Design 30
    3.4. Self-Testing Microcomputers 36

Chapter 4
PSEUDORANDOM TEST PATTERN GENERATORS 38
    4.1. Pseudorandom Sequences 38
    4.2. Generators of Uniformly Distributed Pseudorandom Test Sequences 46
    4.3. Pseudorandom Test Pattern Generators 54

Chapter 5
PSEUDORANDOM AND RELATED SEQUENCE GENERATORS 65
    5.1. Design of Pseudorandom Test Sequence Generator 65
    5.2. Generators of Test Sequences Related to Pseudorandom Sequences 69
    5.3. Generation of Weighted Pseudorandom Test Sequences 78

Chapter 6
RANDOM TESTING 83
    6.1. Concept of Random Testing 83
    6.2. Probabilistic Analysis Techniques for Digital Circuits 86
    6.3. Fault Detection Probability Estimation 91
    6.4. Test Sequence Length Calculation 96
    6.5. Methods for Optimal Selection of Input Variable Probabilities 101
    6.6. Automatic Search for Optimum Probability 106

Chapter 7
PSEUDORANDOM TESTING 110
    7.1. Pseudorandom Sequences as the Test Sequences of the Circuit 110
    7.2. Estimating Fault Coverage for Random and Pseudorandom Tests 114
    7.3. Pseudorandom Test Length Calculation 118
    7.4. Structured VLSI Design with Built-In Random and Pseudorandom Tests 120

Chapter 8
EXHAUSTIVE TESTING 123
    8.1. Exhaustive Testing Principles 123

8.2. Complexity of Exhaustive Test Sequence
Generator Design 128
8.3. Exhaustive Testing by Preweighted Vectors 132
8.4. Iterative Algorithm for Exhaustive Test
Generation 136

Chapter 9
SIGNATURE ANALYSIS 141
9.1. Signature Analysis as a Binary Polynomial
Division Algorithm 141
9.2. Structured Design of Signature Analyzers 147
9.3. Quadratic Signature Analyzer Property 152
9.4. Multifunctional Signature Analyzer 156

Chapter 10
SIGNATURE ANALYSIS EFFICIENCY 163
10.1. Estimation of Signature Analysis Efficiency .. 163
10.2. Examination of Error Occurrence in Output
Responses of Digital Circuits 171
10.3. Signature Analysis Efficiency Estimation
Techniques 177

Chapter 11
EVALUATION TECHNIQUES FOR REGULAR BINARY
SEQUENCE SIGNATURES 187
11.1. Evaluation of Regular Sequence Signatures 187
11.2. A Technique for Calculating Periodic Sequence
Signatures 194
11.3. Fast Signature Calculation Algorithm 198

Chapter 12
MULTI-LINE COMPRESSION SCHEMES 205
12.1. Design of Parallel Signature Analyzers 205
12.2. Compression in Space and Time 206
12.3. State Count Testing 209

12.4. Universal Module BILBO 212

Chapter 13
ANALYSIS OF BILBO-PSA EFFICIENCY 214
13.1. Equivalent PSA Circuits with Internal and External XOR Gates 214
13.2. Estimating PSA Detectability on a Two-Stage Equivalent Circuit 221
13.3. Examining Equivalent PSA Circuits with Internal XOR gates 223
13.4. PSA Efficiency Testing by Software Simulation 233

Chapter 14
PSA DESIGN FOR VLSI SELF TEST 241
14.1. An Efficient Parallel Signature Analyzer 241
14.2. An Efficient Two-Stage Parallel Signature Analyzer 245
14.3. Efficiency of a Parallel Signature Analyzer . . . 250
14.4. Two-Stage PSA Design 256

Chapter 15
PSA with T-flip-flops Design and Analysis 260
15.1. Introduction to PSA with T-flip-flops Design . . 260
15.2. Parallel Signature Analyzer with T-flip-flops as Storage Elements 264
15.3. Efficiency of PSA with T-flip-flops as Storage Elements 266
15.4. Simulation of Parallel Signature Analyzers with T-flip-flops 270

Chapter 16
SIGNATURE TESTABILITY 273
16.1. Signature Testability Analysis 273

16.2. Signature as a Function of Multiple
Variables 276
16.3. Investigation of Signature Properties 279
16.4. Generalized Condition for Signature
Testability 286

Chapter 17
EVALUATION OF SIGNATURES 288
17.1. Analytical Evaluation of Boolean Function
Signatures 288
17.2. Calculation of Signatures for Identical
Generating Polynomials 294
17.3. Finding Signatures for Reciprocal
Polynomials 299

CHAPTER 18
SIGNATURE TESTABILITY OF VLSI CHIPS 306
18.1. Signature Testability of a Circuit 306
18.2. Signature Testability of Faults at Primary
Nodes of Combinational Circuit 310
18.3. Signature Testability of Two-Level
Combinational Circuits 312
18.4. Analytic Approach to Evaluating Signatures
and Signature Testability of faults by PSA. 314

Appendix 1. 321
Appendix 2. 327
References 329
Index 342

Chapter 1

VLSI DESIGN

1.1. Achievements in Microelectronics

Modern microelectronic products which are widely used in national economy originates from the transistor invented in 1947. Later on in 50s planar silicon technology had been introduced which allows to reproduce active and passive devices on a silicon wafer. Since then considerable advances were made in the area of manufacturing silicon devices with multiple components whose number have been almost doubled each year by Moore's law. Table 1.1 shows the evolution of silicon integration scale.

The speedy development of microelectronics resulted in application of its products not only in specialized equipment, but in such devices as calculators, digital clocks, videogames, photographic equipment, washing machines, etc. widely used in everyday life. The major field of application for microelectronic products is modern computers whose size and cost steadily decrease with the increase of integration scale. Ever-growing increase in semiconductor memory size allows modern data bases to be created to hold large amounts of information. The introduction of microprocessors provided the basis for creating a new industry, i.e. personal computer manufacture. Microelectronics undoubtly offers great potential and its products will find extensive application in various field.

There are different technologies used in chip manufacture. One of the first technologies for large-scale IC production is MOS technology. It had been widely used at the early stages of microelectronics development in spite of high power consumption and relatively large area occupied by circuitry on a semiconductor chip. These drawbacks were relieved by introducing complementary MOS (CMOS) technology which had been based on complimentary transistor couples. That technology provided for low power consumption and large-scale integration for ICs, although, as opposed to MOS technology, reguired more processes for IC production. In addition to the above technologies, it is worth noting bipolar technology which had been used to create both discrete elements and the first samples of integrated microcircuits. There exist different modifications of that technology: resistor-transistor logic (RTL), transistor-transistor logic (TTL), emitter-coupled logic (ECL), I^2 logic and etc.

Continuous development of integrated circuit processes resulted in creation and wide application of large (LSI) and very large (VLSI) integrated circuits.

Table 1.1

Circuit	Number of gates	When introduced
Transistor		1947
Integrated circuit		late 50s
Small-scale integrated circuit		60s
Medium-scale integrated circuit	up to 200	late 60s
Lage-scale integrated circuit	from 200 to 5000	70s
Very lage-scale integrated circuit	from 5000 to 10000	80s
Very high-speed integrated circuit	from 10000 to 100000	80s
Ultrahigh-speed integrated circuit	over 1000000	mid 80s

Most popular VLSIs are those whose characteristics make them practicable. Firstly, VLSI comprise tens and hundreds of gates thereby allowing many elements to be replaced with a single VLSI, secondly, they are compact, reliable, consume low power and relatively low-cost. However, VLSI and tleir application require further development.

1.2. VLSI Design Procedure

Design levels. Ideally, a VLSI design procedure starts with definition of functions to be performed and ends with definition of its modes of operation.

Between the start and end of VLSI design, the use is made of many definitions (specifications) both for VLSI performance and its implementation.

The entire design procedure is divided into some levels. At the first level architectural design takes place, i.e. a VLSI block diagram is set up. The stage consists in specifying both hardware (RAM size, bit capacity, etc) and sofware (instruction set, interrupt system, etc) features as well as their interaction.

Next stage is logical design which basically consists in practical definition of various logical specifications. The simplest example of such design is implementation of the set of Boolean functions by gates in the specified basis. Logical design results in a VLSI schematic. Library elements contained in computer-aided design systems are used to prepare it. During this design stage architectural solutions obtained at the previous stage may be refined or modified.

Note that at any design level there is an opportunity to return back to an earlier VLSI development stage to radically change the design.

At the VLSI logic design step, a VLSI chip is simulated to test the obtained solution against pre-specified requirements. The process is often referred to as design verification. Besides, we can define time and precision responses of VLSI by simulation. Apart from simulation, the logical design stage is associated with testing the designed circuits.

Test design. A logical model is used to generate a test for probable VLSI faults. There exist two approaches to test generation. The first is to generate a random input test used to check for detectabity of VLSI faults which have been successively introduced in the VLSI program model. A particular fault is detected by variations at the VLSI outputs. No variations for current input test is insensitive to the introduced fault.

The second approach to test generation consists in generating a test pattern for each fault. For the purpose computer-aided test design systems are used which implement different techniques for their construction. Among them the most prominent are one-dimentional and multiple-path sensitization, boolean differences, equivalent normal form and etc. A more widely used technique is the multiple-path sensitization technique which is referred to as d-algorithm or Roth's method. There are many software modifications of the technique used for VLSI test generation.

A VLSI test can only be generated in compliance with the design for testability requirements and if VLSI functionality is low. Such requirement can be exemplified by limiting the use of synchronous VLSI implementation or the number of clocks required for a modified input to appear at the VLSI outputs. Testability design techniques considerably simplify test generation but involve certain problems due to the need to create complex test equipment for test experiment. These problems can be solved by designing self-testing VLSI chips which determine their state in the testing mode by extra built-in hardware and software. If a fault occurs, the self-testing VLSI chip will detect the fault and inform the system about its inoperable state.

The first stage of VLSI design ends by solving the test problems. It deals mainly with logical conversion and to a lesser degree depends on the technology of VLSI fabrication. The second design stage is oriented to specific technology. Here we must solve VLSI chip design problems, arrange components on a chip, create photolithographic masks, create test structures for parameter monitoring as will as develop specific technologies for VLSI production.

Computer-Aided Design. Designing modern VLSI chips with dozens and hundreds of gates is a rather complex and time-consuming task which can not be solved without special facilities. For this purpose, computer-aidid design (CAD) systems implemented on high-performance computers are widely used.

Modern CAD systems implement the following VLSI design steps:

- circuit behavior simulation,

- testing circuit performance against the design requirements for its time characteristics,

- test generation, VLSI testability estimation and analysis of test/selftest results,

- circuit decomposition into segments that can be readily implemented on a chip,

- arrangement of segments on a chip,

- component layout,

- layout design.

The VLSI design accelerates significantly by using special CADs which are called silicon compilers. Silicon compilation is a VLSI design methodology which provides information required for a designer to make correct decisions on all design steps. The silicon compiler relies on using design rules which help it to generate complex VLSI units by their functional description.

1.3. Self-Testing VLSI Chips

General. The advances in computing and microelectronics in the past 15-20 years gave rise to a versatility of LSI and VLSI and as a result of further development of their architectural and circuitry solutions to superchips which implement complex data processing functions. For example, a family of TRW superchips, each of which comprises more than 10^6 discrete components, consists of six chips being currently on trial. Four superchipes measure 35.6x 35.6 sq.mm, the remaining two measure 35.6x47.7 sq.mm. The greatest chip in the family comprises 34.7 mln. transistors. The above figures are revolutionary enough because they allow us to clarify many aspects of modern VLSI fabrication and use. Three underlying concepts have been used by TRW to improve VLSI reliability during superchip development. Among them, the key concept is self testing. Here built-in seft test is taken to mean absolutely autonomous built-in diagnostic test techniques when test slimuli generation and test response analysis are carried out separatily without any external support.

Self-testing is traditionally based on the following principles:

1. Test sequences are generated directly on a VLSI chip.

2. Output test responses or their compact characteristics are also stored on a VLSI chip.

3. VLSI self test requires only test initialization and response analysis.

4. Self-test implementation assumes that the number of extra primary VLSI nodes and hardware overhead be minimum.

Examples of Self-Test VLSI. Although limitations steming from the above principles are severe enough, self-testing finds ever-widening application in advanced VLSI design. For example, such Motorola products as 32-bit microprocessor MC 68020, its auxiliary arithmetic coprocessor MC 68881 and the Intel's 80386 microprocessor are provided with self-test facilities. Self-test facilities in the above mentioned chips are mainly used for off-line testing of programmable logic arrays (PLAs) employed to implement microprocessor control logic.

The designers from AMd developed a special microcircuit AM29818, a sequential shadow register, to expand self-testing facilities of microprocessor devices. The use of this microcircuit allows implementation of microcomputer diagnostic by SSR technique.

Self-testing is also widely used in the gate-array based products of Motorola, National Semiconductor, VTC, etc. Hardware to be built into the gate array is arranged on the chip boundary. It consists of a test generator and output response analyser. This hardware occupies about 12% of chip area in the National Semiconductor SCX6260 chip. Honeywell has produced a gate-array chip HC 20000 where self-test hardware occupies about 6% of the total area of the array, i.e. 2000 of 20000 logic gates.

Although extra hardware occupies about 20% of a VLSI chip area in average, self-testing techniques have become firmly established in complex computing structure design practices. Motorola has played a dominant role in bringing these techniques to a commercial level. Among the early self-test products of this company is microcompiuter 6804P2 and the central processing unit MC 68HC11 which were the first from the devices oriented to the use of signature analysis. Since the very first Motorola products, signature analysis has become practically imperative for all designs with self-test facilities.

The idea behind signature analysis is to generate short keywords (signatures) in pre-specified nodes of a digital circuit and compare them against the signatures obtained for a known-good state. As a test stimuli sequence generator, a pseudorandom number generator formed on the basis of a linear feedback shift register is used. The built-in self-test circuits employing signature analysis and pseudorandom test sequences require an increase of chip area by 12% in average in the majority of gate arraays. For other design for testability techniques, however, such as level sensitive scan design, the figure is about 15-20%. Extra expenses for a custom IC with specialized circuits are completely justified by lower design, fabrication and diagnostic costs as compared to commercial IC.

Self-testing concepts have been practically embodied in self-test microcomputers which employ commercial large-scale and middle-scale integration circuits. Most representative among such techniques is the Microbit method which employs special circuits functioning as pattern generators and signature analyzers. In the opinion of specialists, such circuits is practically the only way to decrease the complexity of self-test techniques at this stage. As a universal structure which has become a standard of such circuits, we can consider a structure called BILBO (Built-In Logic Block Observer). This is a 16- bit register which can opperate in any of the four modes depending on a control signal:

- as 16 separate flip-flops,

- as a 16-bit series-parallel shift register,

- as a 16-bit signature analyzer,

- as a 16-bit pseudorandom test sequence generator. The BILBO design is mainly improved by increasing its functionality. An example is the STR design and the first native design of such class TEDI 1 (Express diagnostic test LSI).

Chapter 2

VLSI TESTABILITY DESIGN APPROACHES

2.1. VLSI Testability Design Feature

VLSI testability design is based on the use of built-in facilities for test pattern generation and the analysis of responses produced at intermediate and output nodes. Analysis is carried out by compressing the VLSI responses into short keywords with their successive comparison against the reference values. The idea of self-testing is based on the concepts (refer to Chapter 1) which specify whether self-test feature are self-contained and set bounds on the number of additional nodes in a VLSI.

At present there exist two basic approaches to VLSI self-test design. The first of them is based on the use of universal modules for building test generators and response analyzers. Most often a Built-In Logic Block Observer (BILBO) is used as a universal module. In such case, a pseudorandom test sequence generator (PRSG) arranged by the BILBO in one of the modes serves as a test generator.

The PRSG of the BILBO consists of a shift register and a small number of extra elements. Connections between the shift register stages and extra elements are defined by an irreducible primitive polynomial $\phi(x)$ and are stable for the specific BILBO. The PRSG generates pseudorandom test patterns applied to the inputs of a VLSI chip with the test responses produced at its output and intermediate nodes. It is evedent that the universal module BILBO provides for generation of one and the same test sequence for all self-testing VLSIs irrespective of their architecture. By increasing the number of test generator modes, we can slightly improve the functionality of BILBO but its implementation complexity increases.

To obtain compact estimates of self-test result, the BILBO compresses output responses into signatures. In this case, the BILBO is transformed into a parallel signature analyzer (PSA), which is described by the primitive polynomial $\phi(x)$ the same as PRSG. A mismatch between the signature and its expected value indicates to a fault in the VLSI; a match with the pre-specified signature indicates that the VLSI is most probable fault-free. Similar to the PRSG, the PSA structure is rigorous for specific BILBO thereby decreasing its efficiency.

A major disadvantage of the first approach is low efficiency of VLSI self-test due to the use of universal units. The second approach which is based on synthesizing a PRSG and PSA for each self-test VLSI has gained wide acceptance. The synthesis procedure for PRSG and PSA is based on detailed description of a VLSI and taking full account of its facilities thereby it becomes more time-consuming. However this approach makes self-test more efficient due to the maximum fault coverage in VLSI.

The problem of building a self-test VLSI chips covers both the inclusion of on-chip test generators and output response analyzers as well as VLSI design for off-line testing. For the purpose testability design techniques are of considerable current use in building modern digital devices. Among the versatility of testability design techniques, the most generally employed are those for designing sequential circuits based on scanning the states of storage elements. This allows to use a PRSG and PSA in a self-testing VLSI chip most efficiently. Their implementation requires less extra facilities, since storage elements of a single shift register in the scanning path can be used as storage elements of PRSG and PSA.

2.2. Storage Element State Scan Technique

Concept. The reports devoted to estimating the efficiency of test generation methods draw attention to the fact that major difficulties arise at testing sequential networks which make up the current digital units, including VLSIs. Therefore, remembering that test pattern generation for complex VLSIs is impracticable, we must modify them so as to simplify both test generation and test experiment realization. The simplest way of modifying a sequential network is to divide it into two components, combinational and memorizing, to be further tested individually.

One of the first implementations of the above approach is the scan-path technique, which is often referred to as the shift-register modification approach. The most established term for single-shift-register VLSI design techniques is Stanford Scan-Path Design. By this technique the VLSI structure is added with multiplexers for switching over input storage elements of VLSI. Thus the multiplexers connect all VLSI storage elements into a single shift register at testing thereby creating extra facilities for VLSI testing. All storage elements in a VLSI are tested isolated from the combinational portion of the network. Any successive state of the network can be set without regard to its current state. Furthermore, inputs connected from the VLSI combinational portion to storage elements can be readily observed by scanning the states of storage elements.

Implementation. Fig.2.1 shows the block-diagram of a VLSI circuit modified by the Stanford Scan-Path design. In the general case, the VLSI will comprise a combinational circuit (CC), l storage elements $S_1, S_2, S_3,...,S_l$ and l two input multiplexers. The input lines of VLSI are applied with variables $X_1, X_2, X_3, ..., X_n$, and the values $Y_1, Y_2, Y_3, ..., Y_m$ are produced on its output lines. Also, the VLSI will comprise a system clock input C_1, a shift register mode input C_2, a shift-register data input D and its output Q.

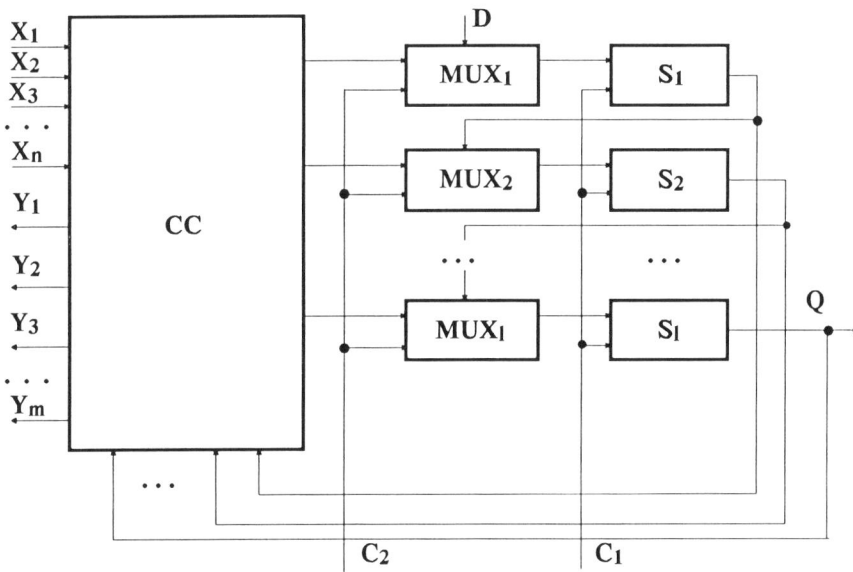

Fig. 2.1. VLSI modified by the Stanford Scan-path design

Suppose that with $C_2 = 0$ the multiplexers connect the combinational portion output to the inputs of storage elements. Then the VLSI assumes its normal configuration and is used for its proper purpose. Accordingly, with $C_2 = 1$, the external input D is connected via the first multiplexer to the input of the first storage element, whose output is connected to the input to the second element, the second-element output is connected to the input to the third element, and etc. The last storage element output is connected to the external output Q. Thus all storage elements of VLSI form a single shift register with external input D, external output Q, and input C_1 where to control signals are applied for a shift microoperation.

Scan-implementing VLSI test. The procedure of testing a VLSI circuit that has been implemented by scan design consists of the following steps.

1. Set up a shift register of VLSI storage elements by specifying $C_2=1$.

2. Test the status and operation of each storage element in a shift register:

(a) All storage elements are initialized to 0 by applying the sequence $000...0$ to input D and the sequence of clocks that is greater or equal to l to input C_1.

(b) The sequence $100...0$ is applied to input D and clocks are applied to input C_1. As a result, each storage element is successively set to 1.

(c) Steps (a) and (b) are repeated with a single 0 flushed through a background of 1s.

This test (steps (a) through (c)) for output responses at shift register output Q checks the ability of each storage element to assume a steady-state 0 and a steady-state 1.

(d) Shift test is performed by applying the sequence $00110011...$ to input D. In this case, each storage element flushes through all combinations of present and future state.

3. Storage elements in the shift register are initialized to 0.

4. Specify the test pattern by applying $X_1, X_2, X_3, ..., X_n$ to input lines.

5. Specify the test pattern to be supplied to the CC inputs from storage elements in the circuit; for the purpose, apply successively the values for S_l, $S_{l-1}, S_{l-2}, S_{l-3}, ..., S_1$ to the shift register input D.

6. The value C_2 is set to 0 causing the circuit to return to its normal configuration.

7. The response to the test pattern is observable at the outputs that are the outputs of the circuit $(Y_1, Y_2, Y_3, ..., Y_m)$.

8. A clock C_1 is generated to modify the states of the CC outputs connected to the storage elements under the effect of test response.

9. Specify $C_2=1$ to form a shift register.

10. By successively applying pulses C_1, observe the responses of CC outputs loaded to storage elements.

11. The procedure is repeated for any successive test pattern starting from step 4.

Advantages and modifications of technique. The most prominent advantage of the discussed technique is that a sequential circuit is converted into a combinational circuit at testing, thus making test generation easier. This advantage contributed greatly to the wide spread of the concept of scanning storage element states. Thus the use of two-input storage elements permitted to exclude multiplexers from VLSI design, thereby increasing VLSI performance.

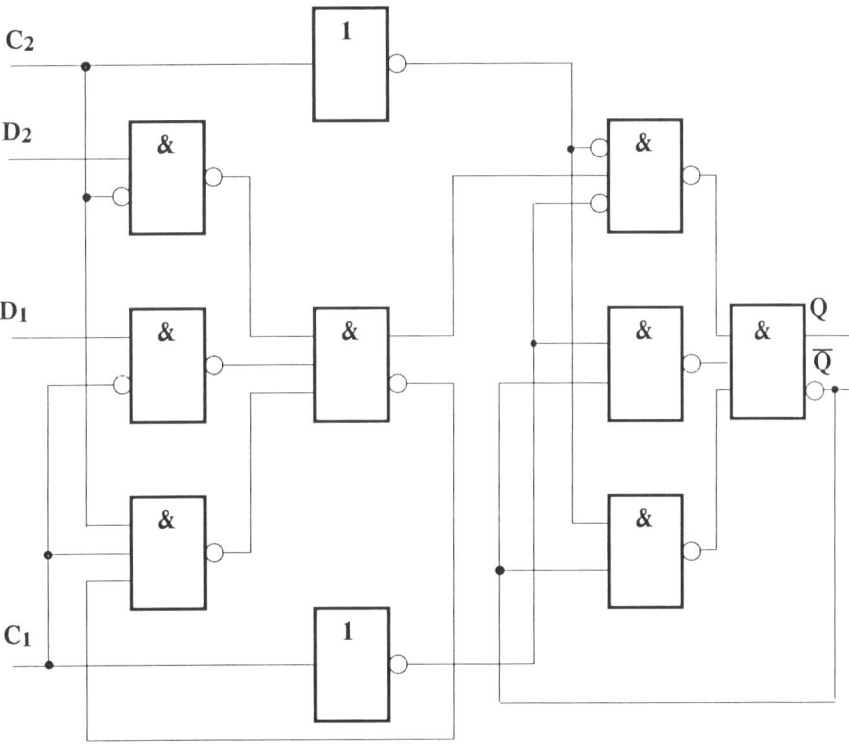

Fig. 2.2. Two-input storage element

A typical structure of a two-input storage element is shown in Fig.2.2. A two-stage flip-flop is controlled by clocks applied to input C_1 or C_2. When the flip-flop is used for its designated purpose, $C_2=1$. By a clock applied to input C_1 (low), the value applied to the data input D_1 is stored in the first stage of the flip-flop. After sufficient time for the signal to propagate through the feedback circuit of the first flip-flop stage passes C_1 becomes 1, and the first stage stores input data. Besides, with C_1, being one, the contents of the first flip-flop stage can be stored in its second stage. The flip-flop operates similarly for clock C_2, with $C_1=1$.

A shift register controlled by clock C_2 is formed by connecting output Q from one flip-flop to input D_2 to a successive flip-flop. With the whole set of storage elements thus formed, a VLSI chip can be represented as a collection of registers each of which is controlled by a clock C_2 wherever an enabling signal appears at the first 2-input NAND date (Fig.2.3). The enabling signal concurrently appears at the input to the second 2-input NAND gate thereby connecting the register output through the OR gate to the sequential test data output *(TDO)* in the VLSI chip. The outputs from all registers are ORed and connected to the *TDO*. The inputs to the VLSI registers are connected to the test data input *(TDI)* in the VLSI chip thereby providing for storing data required in any register. To store or read through the *TDO/TDI* inputs to a VLSI, specific register is selected by a one signal generated at its input Z.

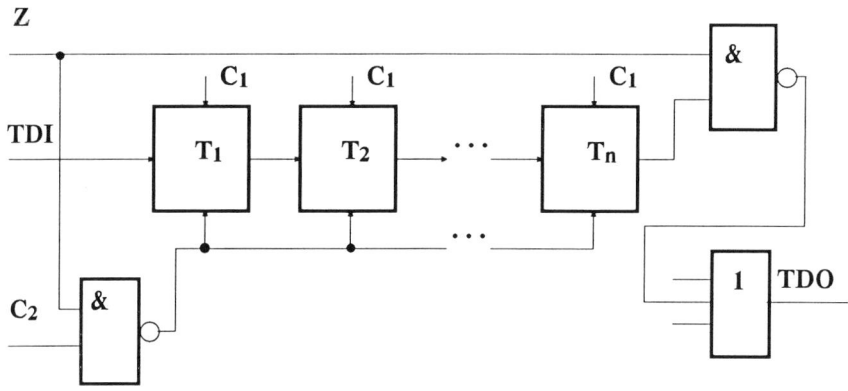

Fig. 2.3. VLSI modified by Scan-path technique

The use of special storage elements or multiplexers has noticeably degraded the performance of a VLSI designed. Therefore, the idea of scanning the states of storage elements evolved into its truncated forms. The scan-set technique has gained wide acceptance. The idea behind this technique is to build an extra shift register with an external output and controlled by independent clocks.

The original VLSI structure remains unchanged. The inputs to the shift register are connected to the test points in the VLSI, which are unaccessible at its testing and the outputs are connected to the most significant storage elements in the VLSI. As indicated in the block diagram (Fig.2.4), the additional shift register can perform shifting without affecting the original VLSI performance. Besides, the scan-test technique provides for parallel data loading into the shift register. The data can be shifted sequentially to the *TDO* terminal. In the set mode, the shift register contents are shifted to the inputs to the storage elements in the original VLSI chip.

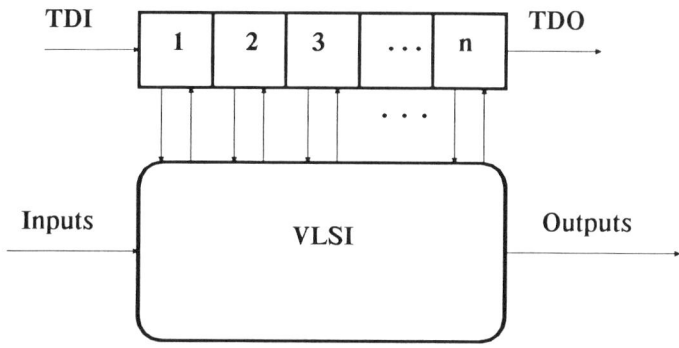

Fig. 2.4. VLSI modified by Scan-set technique

The scanning technique has been further improved by the development of a VLSI design with scanning elements on an external register placed between the VLSI chip and its input and output buses (Boundary Scan). This technique has become the IEEE Boundary-Scan Standard No 1149.1-1989/D4 and is currently being used practically by all key manufacturers in the world.

2.3. Level Sensitive Scan Design (LSSD)

Concept. The level sensitive scan design (LSSD) is the standard design technique in current use at IBM. By this technique, each system storage element is replaced by a two-port latch L1 and a one-port latch L2, which form a single shift register in the test mode. Latch L2 is used only for testing, and latch L1 is used both for on-line operation and for testing procedure.

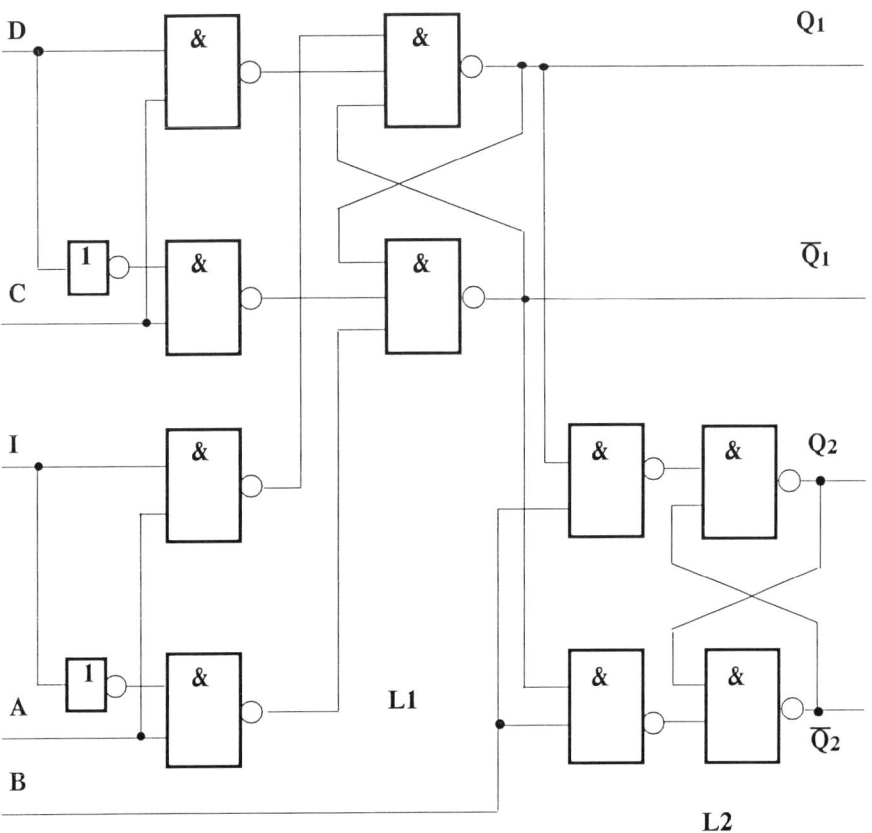

Fig. 2.5. Storage element for LSSD technique

The functional diagram of storage element formed of latches L1 and L2 is shown in Fig.2.5. Latch L2 connected to L1 output has a single clock input B and outputs Q_2 and $\overline{Q}_2$. Latch L1 has a data input D used to store system data with a clock applied to input C and a data input I, which is used only for testing. Data applied to input I are stored by input clock A. Outputs Q_1 and $\overline{Q}_1$ are used as specified by the function implemented by the storage element.

The output Q_2 of latch L2 in one of the storage elements is connected to input I of L1 in another storage element. Inputs A and B of all latches are interconnected (Fig.2.6) whereas clocks are applied sequentially first to input A and, when clock A changes, to input B. Input I of first storage element is an external *TDI* input and output Q_2 of the last storage element is an external *TDI* output. To implement the LSSD technique for a VLSI chip we must introduce for extra external terminals: two terminals for *TDI* and *TDO*, and the other two for nonoverlapping sequences of clocks A and B. The number of external VLSI terminals can be increased, for example for forming more than one independent shift register at testing. Thereby, time required for sequential test data storing and sensing can be substantially decreased.

When using the LSSD technique to design a VLSI chip, meet the following two requirements.

1) Any change in a digital circuit state involves the changing of synchronization level rather than its edge (level-sensitive) to prevent from races.

2) At testing, all storage elements in the circuit are interconnected into a shift register (registers) to which any VLSI state can be written to be used for analyzing the response of its combinational part (scan design).

To implement the above technique at VLSI design, we need that a VLSI LSSD structure employs either one storage element latch or two latches.

Fig.2.7 shows a typical structure of VLSI LSSD with only one storage element latch. A VLSI circuit contains combinational circuits CC_1 and CC_2. Besides, there must be two system clocks C_1 and C_2 to control the VLSI. This prevents from critical races and ensures stable operation.

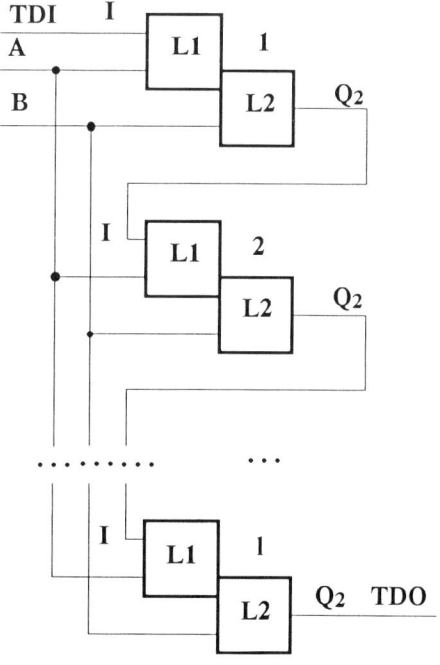

Fig. 2.6. LSSD storage elements interconnection

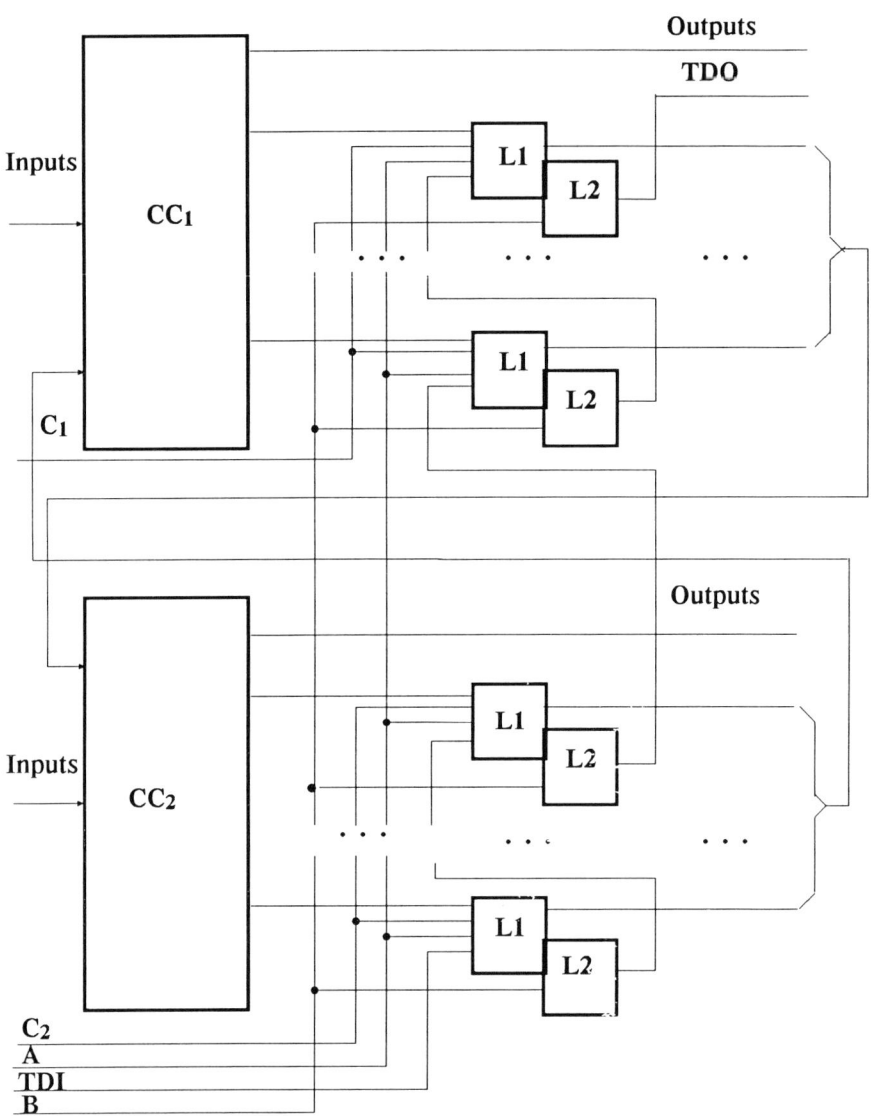

Fig. 2.7. VLSI LSSD structure

When two storage element latches L1 and L2 are used as system latches, the clock sequence B is used for shifting in the test mode, and system clock is used when the VLSI performs as specified.

When designing a VLSI circuit by one of the above techniques, adhere to the following rules.

1. All storage elements in a sequential circuit must be implemented as two-stage shift registers employing level-gated latches L1 and L2.

2. Storage elements must be controlled by two or more non-overlapping clock sequences. This is required because the two storage elements that are connected in series, change their state at different times.

3. All clock inputs must be controlled independently such that it is possible to set any of the clock inputs to the ON state, when all the remaining inputs are OFF.

4. Clock primary inputs may be connected only to clock inputs of storage elements.

5. All storage elements must be interconnected into a single shift register whose primary input and output are controllable by shift clocks at inputs A and B. Storage elements may be also interconnected into several shift registers.

6. All clock sequences other than those produced at inputs A and B must be OFF at testing.

7. Any clock produced at input A must be time-independent of that produced at input B and vice versa.

Only a digital circuit designed in compliance with these rules can meet the requirements of the LSSD technique. The resulting digital circuit or VLSI chip should have the following properties

1) design testability should be applied in the combinational section of a circuit or VLSI chip only. Therefore any classical test generation approach can be used.

2) stable operation does not depend on such characteristics as pulse edge rise time and fall time, races, etc.

Implementation. The use of VLSI circuits implemented by LSSD for designing digital devices and systems allows one to enforce unified diagnostic testing at all levels: a VLSI chip, a module comprising lots of chips, and the whole device or system. It should be noted, however, that the advantages of the LSSD technique are achieved with the hardware overhead of 20% and the use of four additional primary inputs/outputs in each VLSI circuit. Besides, in LSSD VLSI exhibits some reduction in its speed due to its synchronous operation defined by the design rules. The time required for testing is defined by sequential data store/read for the register.

The shift register length is the key value characterizing the complexity of circuit testing. To make testing easier, the original shift register is subdivided into shorter ones. Thereat, time required for storing test data or reading a CC response is specified by bit capacity of the longest among them. However, this increases the number of primary inputs/outputs, which is most unfavourable for VLSI circuits.

2.4. Random Access Scan Technique

A random access scan technique is an alternative approach to the scan-path technique, which allows each storage element to be readily set to the required state and its output value observed at the VLSI primary terminals. This technique is distinguished for the use of randomly addressed storage elements in order that each their state can be independently set, reset or analyzed. Fig.2.8 illustrates an addressable storage element.

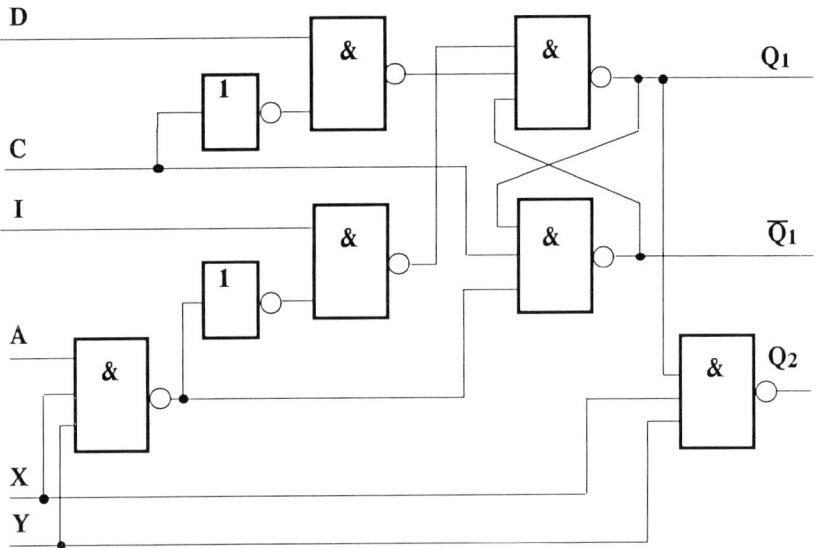

Fig. 2.8. Addressable storage element

When the storage element is used for its designated purpose data applied to input is stored by a clock applied to input C. The value stored at the latch can be taken from output Q_1 for further use in the system. In the diagnostic testing mode, test data applied to input I is stored by clock A. Data loading to the latch in the test mode and its reading from output Q_2 can only be possible with X and Y being one on address lines to specific storage element.

Outputs Q_2 of all storage elements are interconnected at the multi-input AND gate whose output is connected to the primary VLSI node. Thus the state of any storage element can be observed at the primary VLSI node at a time. The storage element number is selected by specifying ones at the appropriate address buses X and Y. Note that the state of a storage element can be observed also when the VLSI circuit is used for its designated purpose.

The addressable storage elements in the VLSI circuit are used together with X and Y address decoders to increase observability and controllability for a VLSI circuit.

Fig.2.9 shows a typical VLSI structure designed by random access scan technique. Sometimes the VLSI structure can be added with address registers X and Y, which are not generally necessary. Thus, if a VLSI circuit is not

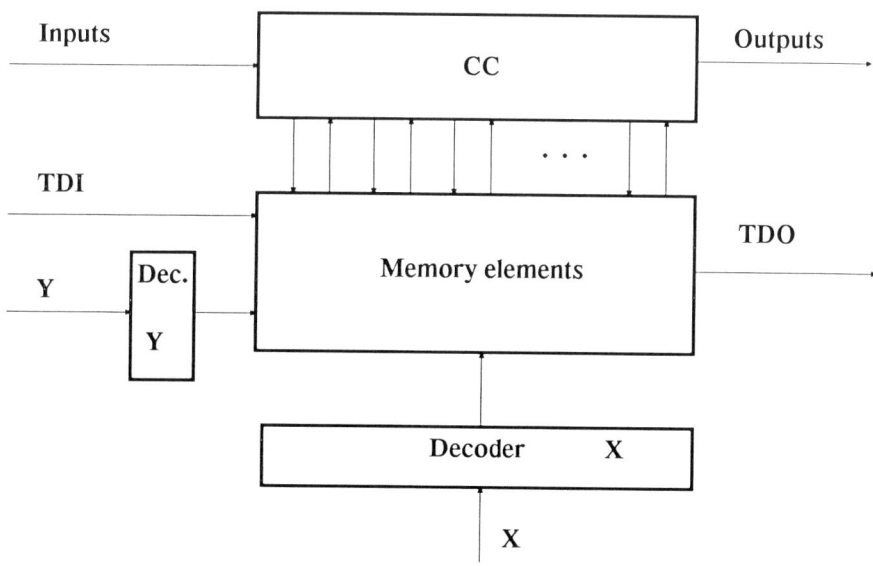

Fig. 2.9. VLSI designed by random access scan techique

critical to extra inputs, it may contain no address registers and the selected storage element address will be specified on its primary inputs.

The above discussed random access scan technique requires three or four extra gates per storage element as well as extra addressing circuits. Circuit redundancy and the number of extra inputs are determined the same as for level-sensitive scan design by the total number of storage elements in a VLSI circuit.

2.5. VLSI Design by Boundary-Scan Technique

Concept. As noted earlier, another alternative to the scan-path technique is boundary-scan architecture. The idea behind this technique is in forming a shift register out of VLSI storage elements located between its primary nodes and the functional portion in the testing mode (Fig.2.10).

All storage elements (SE) connected with the input node X_1, X_2, X_3, ..., X_n and with the output nodes Y_1, Y_2, Y_3, ..., Y_m form a single shift register. The shift register scan-path exists only in the test mode. Its input is *TDI* and its output is *TDO*. To create a shift register for shift microoperation we must add two extra inputs and a special test port to a VLSI circuit. The existence of a scan path provides for sequential storing and reading test data in the test mode.

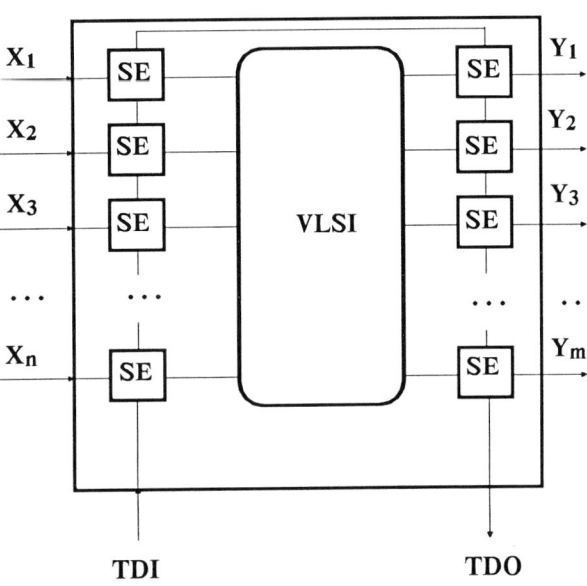

Fig.2.10. VLSI with boundary-scan arhitecture

The use of boundary scan technique consists in implementing the JTAG architecture, which allows the VLSI test support facilities to be allowed and controlled to access and control the VLSI test support. The key element in the

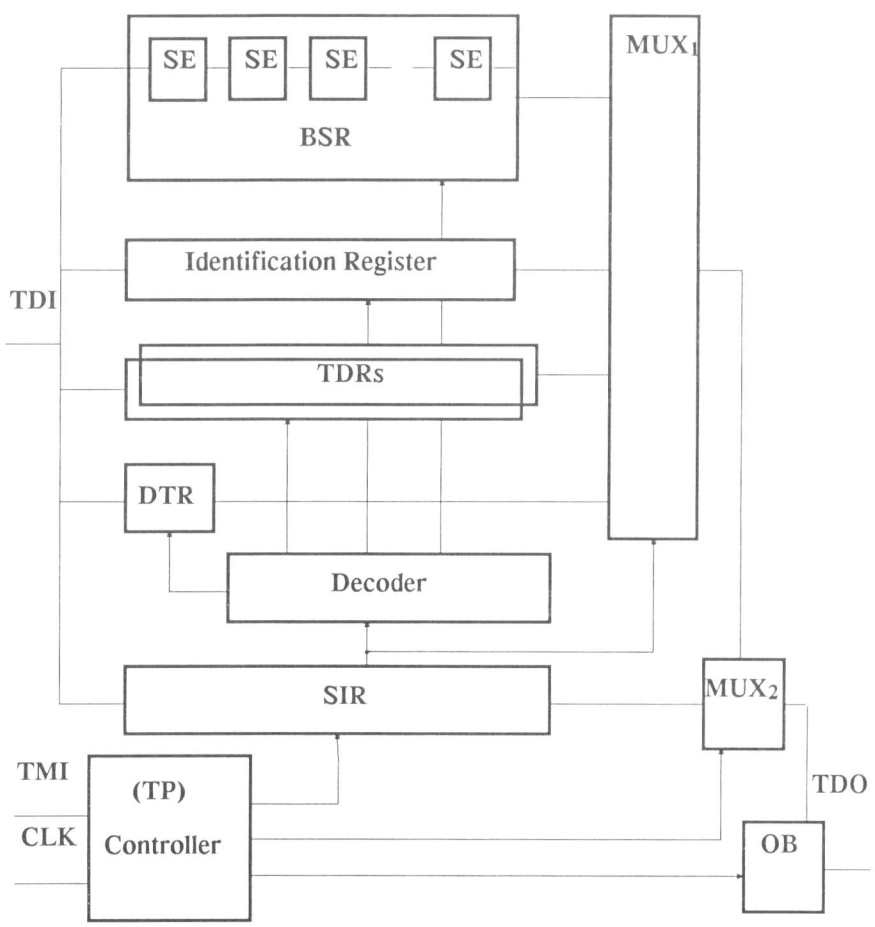

Fig. 2.11. JTAG architecture

above architecture is a test port *(TP)* which consists of a controller, a scan instruction register *(SIR)*, and four primary nodes (Fig.2.11).

Each *TP* must contain at least two scan test-data registers TDR_1 and TDR_2, and a boundary-scan register *(BSR)* formed by storage elements located between the primary nodes and the basic functional portion in the VLSI circuit. Sometimes the *TP* comprises a device identification register *(DIR)*, which has sequential input and output as well. To provide for selecting one of the registers in the test mode, The *TP* contains extra secondary units such as multiplexer *1* (MUX_1), multiplexer 2 (MUX_2), output buffer *(OB)*, decoder, and data transfer register *(DTR)* for translating test data through a VLSI circuit.

A *TP* has three external inputs and one output. The *CLK* input is applied with clock pulses whose leading edge causes data applied through the remaining two inputs to be stored in TDR_1- TDR_2 or *SIR*. By a trailing edge applied to the *CLK* input, data is produced at the *TP* output. The test data input *(TDI)* is applied with test information (command or data) which is stored at the selected data register by a *CLK*. The contents of a register are sequentially read through the sequential data output *(TDO)*. The test mode input *(TMI)* is applied with information, which specifies the *TP* operation mode for LSI diagnostic testing.

<u>Application of Basic TP Units.</u> Let us consider each of the *TP* registers in detail.

SIR is a scan instruction register with a serial input and parallel output. Instructions are loaded to the SIR in the serial mode though the *TDI* input. They specify a test mode or a register which is made accessible through nodes *TDI* and *TDO*.

DTR is a single-bit shift register providing for serial data transfer through the *TP* between *TDI* and *TDO*. This mode is used to transfer test information through the VLSI circuit to other circuits without affecting its current state. In this case, the VLSI circuit functions in its usual mode.

TDR_1 is a test sequence generator which produces test sequences in accordance with the linear shift register scheme described by a primitive polynomial. It generates pseudorandom test sequences to implement self-test VLSI.

TDR_2 is a test data register used for analyzing output test responses. Typically, TDR_2 implements a single-line signature analyzer and is used in self-test mode. Initially, the TDR_2 state is loaded though *TDI* similarly to TDR_1 and the final signature value is read through *TDO*.

DIR is an auxiliary register which allows for finding out the VLSI identification number, its modification, etc. This register can be loaded with any information identifying a specific VLSI.

BSR is a register formed of storage elements connected to input VLSI nodes. It makes it possible to sequentially address the VLSI inputs and outputs, to apply test sequences, and analyze output responses.

The SIR contents specify the modes of VLSI operation. Below are the basic of them:

1. Mode of VLSI use for its designated purpose. When power is on, the *SIR* is always set to this mode.

2. External testing. In this mode, test stimuli are applied to the BSR via *TDI*, input and output responses are analyzed via *TDO* output.

3. Self-testing. The VLSI implements the built-in self test procedure by using a test generator formed by TDR_1 and a signature analyzer formed by TDR_2.

4. Serial scanning. This mode is used to transfer test information through the group of VLSI circuits that form a single register, to a target VLSI.

At present there appeared the pioneering designs based on the boundary-scan technique, although this technique requires more than 7% of extra VLSI chip area for its implementation. The above discussed structural VLSI design techniques allow to make test organization and self-test procedure as simple as possible. It should be noted, however, that the main problem involved in diagnosing is generation of test sequences and organization of a test procedure. Below we shall successively discuss the basic approaches used both for VLSI self-test organization and for test generation.

Chapter 3

SELF-TESTING VLSI STRUCTURED DESIGN

3.1. Self-testing VLSI Using Scan Path

<u>Self Test Using Scan Path and Signature Analysis.</u> At present there are some practical techniques of VLSI design. Multiplicity of VLSI design techniques can by attributed to structure and design features of integrated circuits that present from using a unique approach to VLSI design. It should be noted, however, that the techniques based on scanning storage element states are most popular in self-test VLSI design. A practical design approach making use of signature analysis is S^3 (Self-Test Using Scan Path and Signature Analysis). By this approach, self-testing VLSI chip is represented in testing mode as a combinational section and storage elements combined in a single shift register (Fig.3.1). A pseudorandom test pattern generator (PRTG) producing probabilistic, pseudo-random or exhaustive test is used to generate test stimuli. The proce-

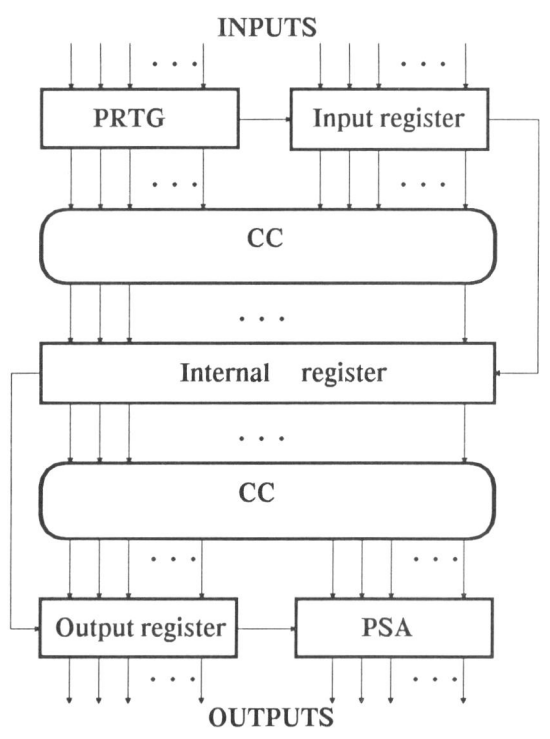

Fig. 3.1. VLSI chip

dure for producing a successive test pattern consists in sequential modifying the single shift register content in the VLSI chip. The VLSI chip response to a test stimulus is compressed into a signature by a parallel signature analyzer (PSA) whose inputs are connected to the VLSI combinational section outputs that are VLSI outputs as well. Besides, a VLSI scan path output is applied to one of the PSA inputs.

A S^3-technique modification is the LOCST technique that has been designed by IBM for LSSD-VLSI circuits. Fig. 3.2 shows the LOCST implementation block-diagram. The self-test procedure is run by the self-test control unit which initiates the testing procedure and analyses the results. If an actual signature mismatches the reference one, the VLSI circuit signals its faulty condition through the self-test control unit. Compared to S^3- technique, the main difference of the LOCST is that extra storage elements have been introduced in the VLSI design to create a PRTG and a signature analyzer (SA).

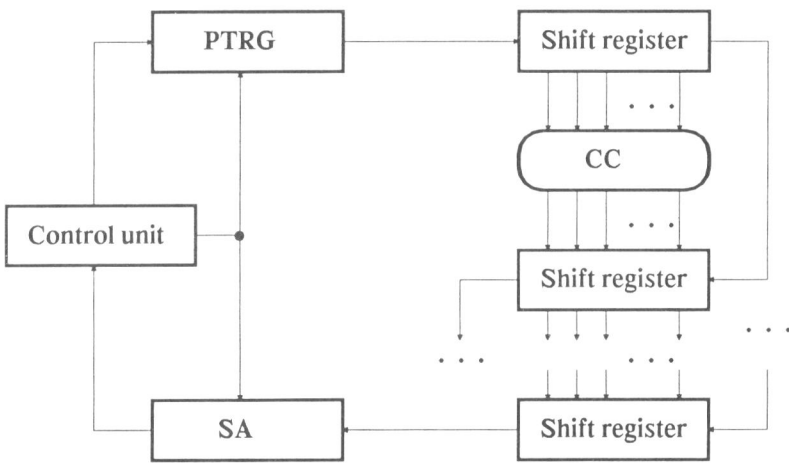

Fig. 3.2. LOCST implementation of VLSI

<u>Self-Test Using Modified Two-stage Flip-flop.</u> To implement a LSSD-VLSI self test, it is necessary to modify the structure of a basic storage element. A storage element is normally represented as a classic two-stage circuit with two extra elements: a two-input modulo-2 adder and a 2-input AND gate (Fig.3.3). If a VLSI circuit is in the function mode *(T=0)*, the storage element is controlled by system clocks C as a system flip-flop whose D-input is applied with input data or as shift register element that is controlled by clocks applied to inputs A and B. When in self-test mode *(T=1)*, clock inputs A and B are only used. The

module-2 sum of binary variables applied to inputs D and I is placed on L1 flip-flop by clock A and on L2 by clock B.

Self-testing of a VLSI circuit with storage elements (Fig.3.3) starts with placing an original test pattern in a single shift register ($T=0$). Then T is set to 1. In this mode, storage elements forming a single shift register perform the functions of the test pattern generator and the VLSI output response compactor alternately. Data applied to D inputs is compressed with that of modulo-2 adders and placed on the primary L1 stages in storage elements by clock A. Note that in this case the secondary L2 stages produce test stimuli and remain unchanged on compression of output responses. They may be changed by a clock applied to input B. Thus, in the self-test mode, the VLSI storage elements compress the VLSI responses and produce a successive test pattern alternately. With the specified number of cycles completed, the value of T is set to 0, and the content of storage elements recorded by clock A will be the signature to be sequentially read and compared with the reference signature.

A major advantage of using storage elements of Fig.3.3 is the maximum throughput at VLSI self-test. Only one shift is to be performed in the shift register to produce a successive test pattern. The above discussed S^3 and LOCST techniques required several shifts to produce a successive test stimulus. As the number of shifts is mainly determined by the shift register length the self-test time increases noticeably.

However, the above discussed flip-flops may be responsible for some drawbacks. In the first place, they prevent from testing the behavior of VLSI logic located between external VLSI pins (input, output) and the storage elements. To provide for testing the logic, the VLSI modification needs.

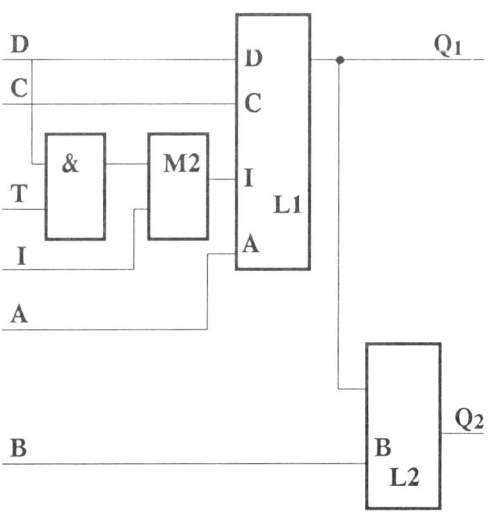

Fig. 3.3. Storage element

Chapter 3

In the second place, data paths used in the self-test mode are not the same as used for data transfer in the VLSI normal mode. Therefore, extra test stimuli are to be introduced to test system data path and system clock inputs in VLSI storage elements.

In the third place, the properties of test stimuli produced at self-testing time depend to a great extent on sequences to be compressed and eventually on the VLSI scheme. Tests produced by storage elements (Fig.3.3) are usually characterized by poor quality.

Self Test Using Three-Stage Storage Elements. To improve the quality of VLSI response compaction and test stimuli generation, the classical two-stage storage elements used in LSSD-structures would require further modification. An example of such modification is a flip-flop of Fig.3.4 which has been extended to include: a two-input modulo-2 adder, a data input I of flip-flop L2 controlled by clocks F, and a flip-flop L3 whose input is connected to L2 output. The state of L3 is changed by clocks applied to input P. Inputs D, C, I, A and B are used as designed according to the classical two-stage flip-flop (refer to Fig.2.5). Output Q_2 is connected to input I of the next storage element and input J is connected to output Q_3 of the preceding storage element.

In the normal mode, the system input D and the system clock C or the input I and clock inputs A and B are used to form a shift register element.

In the self-test mode, all VLSI storage elements are connected to form two shift registers described by primitive polynomials. Both registers are active at a time, with one behaving as a PRTG and the other as a PSA. Storage element clocks are applied in the following sequence: P,B,A,F. A clock applied to input P shifts each L2 flip-flop into the appropriate L3 flip-flop. Therefore, flip-flops L3 store the current PSA state. Then the PRTG state held

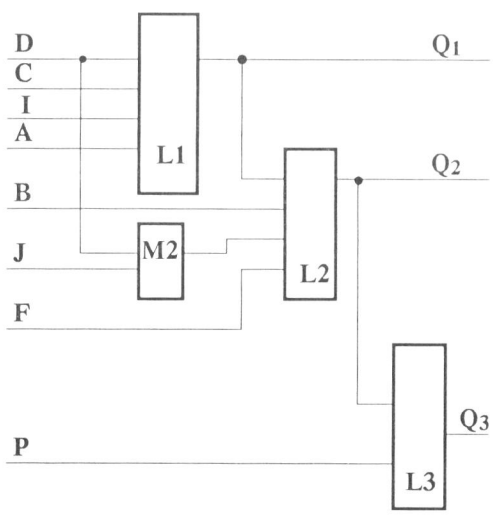

Fig. 3.4. Storage element

by L1 flip-flop is shifted into L2 flip-flops. On arrival of clock A, the contents of L2 flip-flops is changed in accordance with the primitive generating polynomial and placed on L1 flip-flop. A successive test pattern produced as a result is applied through outputs Q_I to the VLSI combinational section whose response appears on D-input to storage elements and the primary inputs to a 2-input modulo-2 adders. The secondary inputs to the adders are applied with the previous PSA state. By clock F a new value of PSA in produced as the content of flip-flops L2.

The above sequence of modifying the content of flip-flop for all VLSI storage elements is repeated the required number of cycles as determined by the test length. Then the resulting signature is read from storage elements and compared with the reference value.

Thus the storage element modification (Fig.3.4) provides for generation of test stimuli described by a primitive polynomial. Note that in this case any VLSI storage element performs as a PRTG (L1 and L2) under control of clocks B and A, and as a PSA (L2 and L3) under control of clocks F and P.

3.2. Ring VLSI Self-Test

Concept. An attempt to use one and the same test structure both as a pseudorandom sequence generator and PSA for an arbitrary VLSI structure has given rise to the so called ring testing. The simplest form of ring VLSI self-test implementation is based on the use of a HILDO register which functions in the following modes:

1) Normal mode where the HILDO storage elements are used for their designated purpose.

2) Self-test mode where the HILDO is reconfigured into a parallel signature analyzer with the output nodes of VLSI connected to its inputs.

3) Shift-register mode which allows the resulting signature to be read successively and the new HILDO state be written.

4) Initialization mode which uses inputs R and S of the HILDO flip-flops for setting the register into the required initial state.

The shift register length is specified by the number of inputs to a self-testing VLSI circuit. If there is a mismatch between VLSI inputs and outputs, which is rather common, the outputs are united by modulo-2 adders to equate the number of inputs and outputs in the HILDO register.

In the self-testing mode, the register content is applied to the VLSI inputs as a test stimulus. The VLSI response to the stimulus is compressed in the HILDO, that is a PSA, thereby producing a successive test pattern. Self-testing proceeds until the test stimuli produce the test of the desired length. It is evident that the number of test patterns in this case is completely determined by the VLSI structure, since the VLSI circuit and the HILDO register form a separate circuit in the self-test mode. To provide for test coverage at self-testing, we must increase the length of test which limiting value is 2^m, where m is the number of storage elements in the HILDO.

<u>Method implementation.</u> The most widely used structure for ring self-test implementation is the one where the VLSI circuit under test in conjunction with extra logic implements linear feedback for the shift register. Consider ring self-testing for an elementary circuit that implements a Boolean function $f = x_1 \oplus x_2 \oplus x_4$ (Fig.3.5). For the purpose, let us use the register described by polynomial $\varphi(z) = 1 \oplus z^1 \oplus z^4$. In order to obtain the maximum test length, introduce extra function $f' = x_2$. Then the modulo-2 sum of f and f' will be defined by the expression $f \oplus f' = x_1 \oplus x_4$, which complies with the selected primitive polynomial $\phi(x)$.

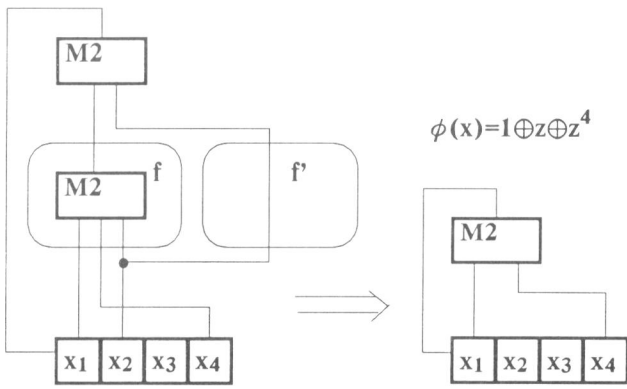

Fig. 3.5. Elementary circuit

The test coverage for the circuit of Fig.3.5 that characterizes the self-test quality as well as the appropriate characteristic for the general case of self-test are determined by the following factors:

1) High degrees of primitive polynomial $\phi(x)$, which describes the test sequence generator.

2) Layout of connections between the HILDO register positions and the inputs to the self-testing VLSI fragment.

3) Initial code of HILDO register.

4) The number of steps in the ring testing procedure.

The above considerations are closely related. Thus with the number of steps being 2^m-1, where m is the high degree of generating polynomial $\phi(x)$, the initial setting of HILDO register may assume any nonzero value. However, with fewer than 2^m-1 steps, the initial setting becomes important. Therefore, in the general case, the problem of defining the major ring self-testing parameters is rather complex.

3.3. Use of General-Purpose Modules for Self-Test VLSI Design

General. The use of both level-sensitive scan design and the built-in logic block observers allows us to implement efficient self-test structures. In this case, the self-test property can be expanded to higher levels in a computing environment: from the VLSI level up to the board, block, device or system level.

The most popular self-test VLSI design method is based on LSSD structures and the universal self-test facilities such as a pseudorandom test-pattern generator and a signature analyzer. Besides, the underlying idea of the method in question is boundary scan when the scan path is formed of only VLSI input/output storage elements. Similar to the classic LSSD scheme, every VLSI storage element is a two-stage flip-flop consisting of the basic L1 stage and the extra L2 stage used in the testing mode. The storage elements connecting input and output VLSI pins (Fig.3.6) are only somewhat different. The function of input/output storage elements corresponds to their description of subsection 2.2. The difference consists in the use of inputs D and outputs Q_1. Thus for input VLSI pins that are directly connected to the D-inputs of input storage elements delays of signals on latch L1 are often undesirable in the normal mode. To reduce the delay one can use multiplexers connecting the value of D input or that produced at L1 outputs to the output of storage element Q_1. Control input P of multiplexer normally changes state when switched between normal and test modes. Also, output storage elements are similarly switched in synchrony. In this case output VLSI values produced on D-inputs of storage elements located immediately between the VLSI chip and its output pins can be passed undelayed to the external environment.

According to the LSSD design technique the entire set of VLSI storage elements are combined in the test mode into three basic groups of registers: internal, input and output. In its simplest form the VLSI chip contains three registers. The first register R1 is formed by input storage elements and extra modulo-2 adders thereby acting as a pseudorandom test pattern generator. The second register R2 combines all internal storage elements of VLSI, the third register R3 consisting of output storage elements forms a parallel signature analyzer employing extra modulo-2 adders. The signature analyzer structure corresponds to that of a universal self-test element BILBO which has four modes of operation: data read-in mode when data applied to D inputs are read in by clocks C; shift under control of clocks applied to inputs A and B; single and parallel signature analyzer modes. Register R1 much as register R3 can function in several modes, of which the more important is the test generator mode. Multiple modes of operation for R1 and R3 allow different procedures to be organized at VLSI self-testing.

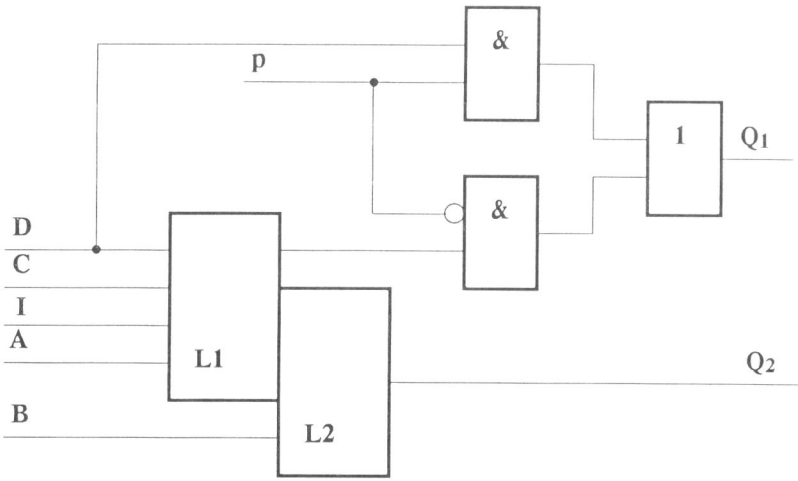

Fig. 3.6. The storage element

Self-testing VLSI structure. Fig.3.7 shows a general scheme of a self-testing VLSI. It incorporates 4 multiplexers MUX1, MUX2, MUX3 and MUX4 to implement different VLSI test modes. The *TDI* input is an input for test data applied from external environment and *TDO* is a sequential data output. We should also note that register R3, which is a BILBO structure with linear feedback, switches to any mode required under external control signals. At the same time, the test stimuli generator is formed of register R1 and a multiple-input modulo-2 adder both by control signals and the multiplexer MUX1.

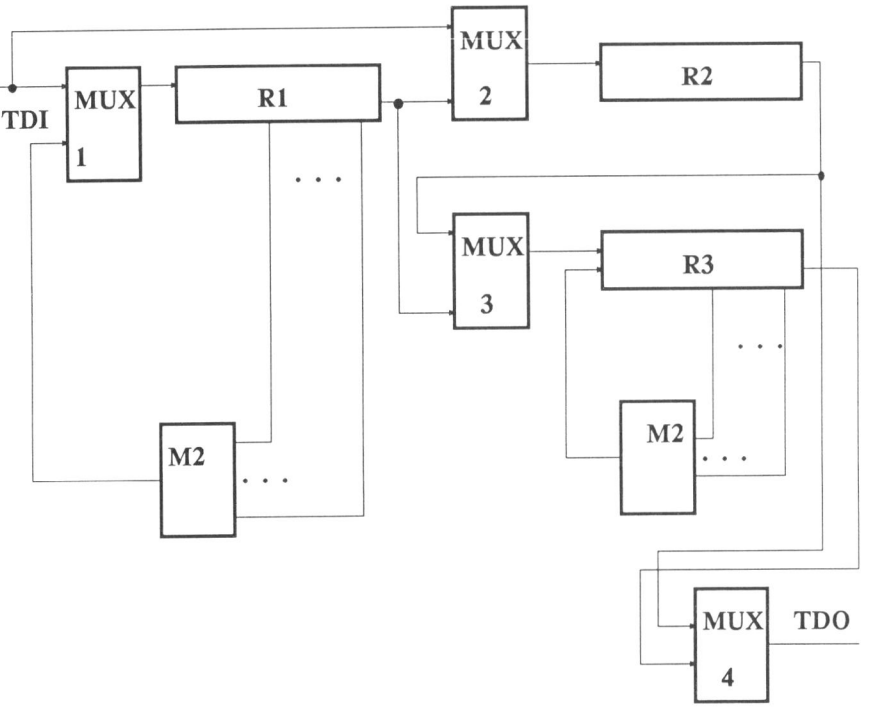

Fig. 3.7. Self-testing VLSI

To implement a general VLSI self-testing procedure, extra eight pins are to be introduced. The purpose of them: shift clocking-inputs A and B, parallel read-in control for R1 and R3 (inputs C_1 and C_2), *TDI* input, *TDO* output and two inputs for specifying self-test mode. Below are the possible VLSI self-test modes and their respective configurations.

Let us consider the VLSI behavior in the above modes (Table 3.1).

In the first mode, any test data can be placed on VLSI register R2 via *TDI* and then sequentially read out via *TDO* as in a normal LSSD architecture. By clocks applied to inputs C_1 and C_2, respectively, registers R1 and R3 can be loaded with data. If no clocks are applied, input/output elements do not affect testing.

In the second mode, registers R1 and R3 form a single scan path that is independent of internal storage elements (register R2). Test stimuli that have been sequentially loaded to output storage elements of a VLSI are applied to VLSI outputs connected to the inputs of other VLSIs. By clock C_1, test stimuli are registered on input storage elements of VLSI. Therefore, this mode tests interconnections between the complex environment elements representing self-test VLSIs.

Table 3.1

N	Mode	VLSI configuration according to Fig. 3.6 and Fig. 3.7
1.	LSSD test	*TDI* is connected to R2 input via MUX2. R2 output is connected to *TDO* via MUX4. For input/output storage elements, $p=1$. Input *A* and *B* of register R2 are enabled.
2.	VLSI layout test	*TDI* is connected to R1 input via MUX1. R1 output is connected to R3 input via MUX3. R3 output is connected to *TDO* via MUX4. Input *A* and *B* are enabled in registers R1 and R3. $p=X$, $X \in \{0,1\}$ for input storage elements. $p=0$ for output storage elements. R3 is the shift register.
3.	Ring scan	Controls are the same as or mode 2, except that $p=1$ for output storage elements.
4.	Self-test	R1 forms pseudorandom test-pattern generator. R1 output is connected to R2 input via MUX2. R2 output is connected to R3 input via MUX3. Inputs *A* and *B* are enabled for all storage elements. Register R3 is changed over to parallel signature analyzer.

In the third mode, test output hold on registers R1 and R3 can be read. In particular, this mode is used to analyze the results of testing VLSI interconnections.

In the fourth mode, the VLSI self-test procedure is performed. Registers R1, R2 and R3 are combined into a single scan path. The first bits in the scan path form a pseudorandom test pattern generator whereas the last bits form a parallel signature analyzer. The only thing required to implement this mode are external clocks A and B. The self-test result is produced on R3 and can be scanned in the third mode.

Technique implementation. Fig.3.8 demonstrates a self-test VLSI block-diagram consisting of four combinational subcircuits CC_1, CC_2, CC_3 and C_4, and storage elements controlled by system clocks C_1', C_2' and C_3'. In the self-test mode, internal storage elements are combined into the three shift registers $R2_1$, $R2_2$ and $R2_3$ each of which is controlled by a sequence of system clocks C_1', C_2' and C_3', respectively. All inputs of the VLSI circuit are separated from its functional part by input storage elements that are represented by two registers $R1_1$ and $R1_2$. In the self-test mode, register $R1_2$ changes to the pseudorandom test pattern generator mode. Output storage elements combined into two registers $R3_1$ and $R3_2$, where $R3_1$ is a parallel signature analyzer, are inserted between the VLSI outputs and its external pins. Registers $R1_1$ and $R1_2$ are controlled by clocks C_1, registers $R3_1$ and $R3_2$ are controlled by clock C_2. In the self-test mode, all the VLSI registers form a single shift register (scan path) with the *TDI* input and *TDO* output controlled by clocks A and B. The self-test mode is set through inputs R_1 and R_2.

Prior to self-testing, the test generator (register $R1_2$) and the parallel signature analyzer must be initialized to predefined states. For the purpose, inputs R_1 and R_2 are applied with the code of interconnection test mode when registers $R1_1$, $R1_2$, $R3_1$ and $R3_2$ are combined into a single scan path. All elements of the registers in question are sequentially initialized through the *TDI* input. The contents of $R1_1$, $R3_1$ and $R3_2$ are normally zeroed, and register $R1_2$ is set to a nonzero code *10...0*.

To make the self-testing procedures reproducible, the internal storage elements of VLSI must be also preset (Fig.3.8). The procedure is performed in the LSSD test mode. As a result, all internal storage elements $R2_1$, $R2_2$, and $R2_3$ are zeroed. With all VLSI registers initialized, the self-test mode is set through inputs R_1 and R_2. Then the pseudorandom test pattern generator (register $R1_2$), when strobed by clocks A and B, produces the first test pattern causing the states of registers $R1_1$, $R2_1$, $R2_2$ and $R2_3$ to change. The number of shifts is determined by the total width of the said registers. In the general case, clocks A and B are to be produced only to update registers $R2_1$, $R2_2$ and $R2_3$ in order to obtain a successive test stimulus. Therefore, a successive test

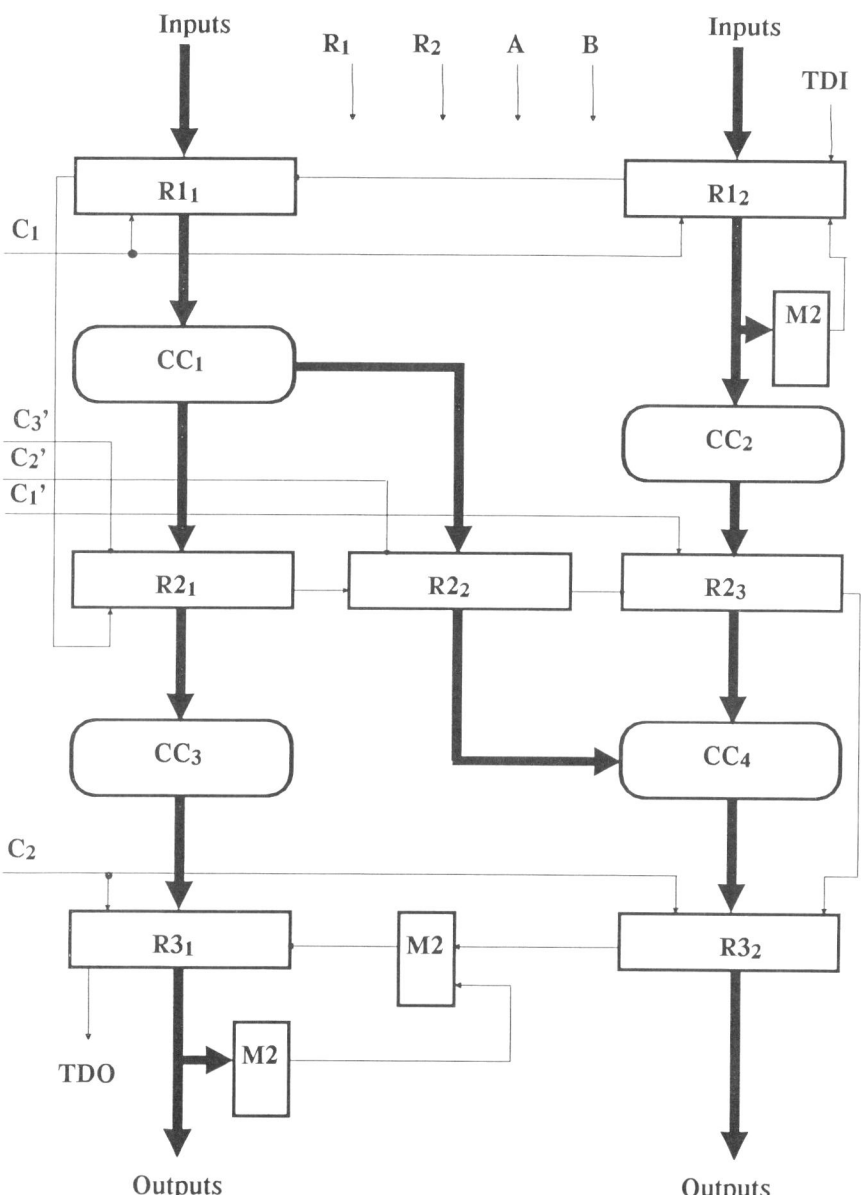

Fig. 3.8. A self-testing VLSI block-diagram

pattern will be represented by the contents of registers $R1_1$, $R1_2$, $R2_1$, $R2_2$, and $R2_3$. With the stimuli applied to the outputs of VLSI combinational subcircuits CC_1, CC_2, CC_3, and CC_4, output responses are produced which can be stored on registers $R2_1$, $R2_2$, $R2_3$, $R3_1$, and $R3_2$ under control of clocks C_1', C_2', C_3', and C_2. In the simplest form, all the clocks in question are applied at a time.

If parallel implementation of VLSI response store is impossible, the self-test procedure becomes more complicated.

Suppose that a VLSI response to a test stimulus is stored in parallel. In this case, the response is fixed on registers $R2_1$, $R2_2$, $R2_3$, $R3_1$, and $R3_2$ where $R3_1$ is a parallel signature analyzer. For the purpose, the obtained response must be compressed into a short signature by applying clocks A and B whose number is defined by the length of registers $R2_1$, $R2_2$, $R2_3$, and $R3_2$. Signature generation is accompanied with generation of a successive test stimulus. Then the procedure repeats for a next stimulus.

Thus, the complete set of test pattern generated produces the value of an actual signature to be compared with the reference signature for decision making about the self-testing VLSI status.

3.4. Self-Testing Microcomputers

<u>General.</u> Self-testing microcomputers as well as self-testing VLSI chips are based on the use of BILBOs. Among the most popular self-test microcomputer design approaches in the Microbit.

The approach consists in introducing extra hardware and software facilities in a microcomputer design to provide for self-testing at the functional level. The extra facilities used are a pseudorandom test pattern generator, a signature analyzer, a control unit, and an initialization circuit. This circuit is a primary control input whereto a signal initiating microcomputer self-test is applied. Standard BILBOs are used to implement both the test-pattern generator and the signature analyzer. The width of both units is determined by that of the microcomputer. Thus for an 8-bit microprocessor, the test generator and the output response compressor are also 8-bit. The basic function of control unit is to provide for the required clock sequences for self-test implementation. Self-testing is controlled by software consisting of a master program and test programs for individual microcomputer blocks. Software is pre-stored in the ROM together with the reference signatures.

After power-on, all storage elements in the microcomputer are set to known specified states and system initialization occurs. Then self-testing starts.

Microcomputer Self Test Implementation. The first test is the so called nucleus test that is run to make sure of microcomputer viability and the lack of evident faults. It consists in performing an elementary addition of two operands *10101010* and *01010101* and comparing the results with code *11111111*.

If the result matches the reference value, the pseudorandom test pattern generator will be tested. The test pattern generator testing consists in placing an 8-bit non-zero initial code (for an 8-bit microprocessor) into its storage elements and running 256 or other number of cycles. Then the contents of storage elements is compared with the precomputed value. If the values match the signature analyzer test will be run which is practically identical with that of the test generator since both devices are designed as a standard BILBO. The difference is in the fact that the analyzer inputs are applied with fixed values which are generally zero sequences.

At the next stage ROM test is run by compressing sequentially the content of all its locations in the signature analyzer. The signature thus obtained is compared with the ROM pre-loaded value.

Random Access Memory (RAM) Test. RAM test is performed by pages. All location in the current RAM page are pre-loaded with pseudorandom patterns produced by the test generator. For each page, there is an actual signature produced to be compared with the pre-loaded value. If the RAM test results coincide with the reference values, the microprocessor will be tested exhaustively. During the test different microprocessor instructions are executed, and the intermediate results are compared with the reference values. If at least one intermediate results mismatches the expected value, the test run is unsuccessful, and the microprocessor is considered to be faulty.

If the microprocessor test run is successful, the I/O ports will be tested in the microcomputer. First, each port is checked for data output and then for input. As the test patterns covering the entire set of stuck-at faults, we use two patterns, *00000000* and *11111111*. The test used for detection of stuck-at and bridge faults consists of the following patterns:

```
1 1 1 1 0 0 0 0
1 1 0 0 1 1 0 0
1 0 1 0 1 0 1 0
0 0 0 0 1 1 1 1
```

The last test run is the microcomputer timer.

The Microbit technique has been used originally to construct a self-testing microcomputer employing an 8-bit microprocessor 8085 which is a classic architecture.

Chapter 4

PSEUDORANDOM TEST PATTERN GENERATORS

4.1. Pseudorandom Sequences

Congruential pseudorandom sequences. Two methods are most commonly used for generating pseudorandom sequences. The first method, which forms the basis of a major proportion of present-day program sensors, makes use of the recurrence relation.

$$X_k = AX_{k-1} + B \pmod{M}, \quad k = 1,2,3,\dots \quad (4.1)$$

where A, B, M are constants and $X_0 > 0$, $A > 0$, $B \geq 0$, $M > X_0$, $M > A$, $M > B$. This method has been called multiplicative congruential (with $B=0$) or mixed congruential (with $B \neq 0$), so a sequence generated by the method is called linear congruential. With an appropriate choice of the value of $M = r^m$, where r is the basis number that is used to represent the values of m-digit pseudorandom numbers X_k, equation (4.1) may be rearranged to give

$$X_k = AX_{k-1} + B \pmod{r^m}, \quad k = 1,2,3,\dots \;.$$

Not all sets of values X_0, A, B, r and m give rise to sequences whose properties are close to those of uniformly distributed random sequences. For example, with $X_0=110$, $A=011$, $B=001$, $r=2$ and $m=3$, the generated number sequence has the form: $X_0=110$, $X_1=011$, $X_2=010$, $X_3=111$, $X_4=110$. As is evident, the values of the low-order digit in the pseudorandom number sequence $\{X_k\}$ ($\{X_k\} = X_0, X_1, X_2, \dots$) have period 2. For the high-order digit and the digit generated in the second bit, the periods are 4 and 1, respectively. The period L_γ for a γth digit, $\gamma \in \{1,2,3,\dots,m\}$, in the pseudorandom number sequence $\{X_k\}$ can be estimated by

$$\max L_\gamma = r^{\gamma-1}(r-1) \quad (4.2)$$

with $B=0$ or by

$$\max L_y = r^{2\gamma - 1}(r - 1) \quad (4.3)$$

with $B \neq 0$. Thus, for binary representation $(r=2)$, we find that the period L_1, of the low-order digit is I by (4.2). Hence the low-order digit in a random sequence $\{X_k\}$ is invariable and equal to I. The second-digit period L_2 thereby denoting that its value is either invariable or reversible in each cycle.

By using equation (4.1) with $B \neq 0$, the period for digits in the numbers can be slightly increased as follows from equation (4.3). However, the above periodicity cannot be radically eliminated, thereby affecting the quality of pseudorandom number sequence. As applied to the problem of digital circuit testing, periodicity can significantly degrade the test coverage. Thus, for example, by applying a constant logical signal to a definite input of digital circuit it is not possible to detect even a stuck-at fault associated with the generated signal value at the tested input node. What is more, inplementation of the discussed pseudorandom number generation method and its modifications, in particular, is very complicated due to multiplications involved. Because the above method exhibits the mentioned deficiencies, the second method is more popular in generating pseudorandom sequences by means of the relation

$$a_k = \sum_{i=1}^{m} {}^{\oplus} \alpha_i a_{k-1}, \quad k=0,1,2,\ldots \quad (4.4)$$

where k is the clock number, $a_k \in \{0,1\}$ are sequence digits, $\alpha_i \in \{0,1\}$ are constant coefficients and $\sum^{\oplus}$ represents modulo-2 addition of m logic variables.

With an appropriate choice of the α_i coefficients, the bit sequence $\{a_k\}$ has maximal length 2^m-1 by the characteristic polynomial $\phi(x) = 1 \oplus \alpha_1 x^1 \oplus \alpha_2 x^2 \oplus \ldots \oplus \alpha_{m-1} x^{m-1} \oplus \alpha_m x^m$, which must be primitive. Such a sequence is called a M-sequence.

For a specific generating polynomial there exists a M-sequence generator structure to generate a maximal-length sequence. As an example for $\phi(x) = 1 \oplus x^3 \oplus x^4$, Fig.4.1 shows its block diagram with the storage elements loaded with values of Table 4.1. The table shows that the period for the ith digit, $i = \overline{1,4}$, of bit $a_i(k)$ of M-sequence generator is $2^4 - 1$.

Table 4.1

k	$a_1(k)$	$a_2(k)$	$a_3(k)$	$a_4(k)$	k	$a_1(k)$	$a_2(k)$	$a_3(k)$	$a_4(k)$
0	1	0	0	0	9	1	1	0	1
1	0	1	0	0	10	1	1	1	0
2	0	0	1	0	11	1	1	1	1
3	1	0	0	1	12	0	1	1	1
4	1	1	0	0	13	0	0	1	1
5	0	1	1	0	14	0	0	0	1
6	1	0	1	1	15	1	0	0	0
7	0	1	0	1	16	0	1	0	0
8	1	0	1	0	17	...	...	...	...

In synthesizing a M-sequence generator a major problem is to find a polynomial $\phi(x)$ that meets the conditions for maximal length sequence generation. It is known that for a given $m = deg\phi(x)$ there exist $\Phi(L=2^m-1)/m$ distinct primitive polynomials where $\Phi(L)$ is the Euler function.

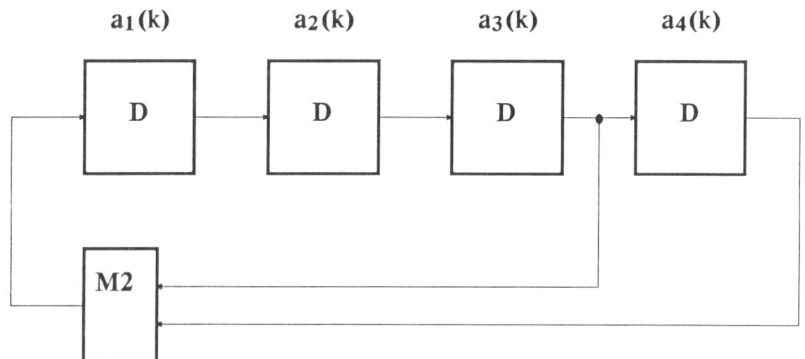

Fig. 4.1. M-sequence generator

Since $\Phi(L)$ increases quickly with m, the number of polynomials $\phi(x)$ of degree m that generate M-sequences will accordingly increase. Among the set of polynomials $\phi(x)$ with identical high degree m there may exist polynomials with the least number of unit coefficients α_i, $i=\overline{1,m}$. This condition is characteristic for the structures based on M-sequences and most simple in implementation. Table 4.2 shows polynomials with a minimal number of non-zero coefficients for all $m \leq 40$.

Table 4.2

m	$\phi(x)$	m	$\phi(x)$
1	$1+x$	21	$1+x^2+x^{21}$
2	$1+x+x^2$	22	$1+x+x^{22}$
3	$1+x+x^3$	23	$1+x^5+x^{23}$
4	$1+x+x^4$	24	$1+x+x^3+x^4+x^{24}$
5	$1+x^2+x^5$	25	$1+x^3+x^{25}$
6	$1+x+x^6$	26	$1+x+x^7+x^8+x^{26}$
7	$1+x+x^7$	27	$1+x+x^7+x^8+x^{27}$
8	$1+x+x^5+x^6+x^8$	28	$1+x^3+x^{28}$
9	$1+x^4+x^9$	29	$1+x^2+x^{29}$
10	$1+x^3+x^{10}$	30	$1+x+x^{15}+x^{16}+x^{30}$
11	$1+x^2+x^{11}$	31	$1+x^3+x^{31}$
12	$1+x^3+x^4+x^7+x^{12}$	32	$1+x+x^{27}+x^{28}+x^{32}$
13	$1+x+x^3+x^4+x^{13}$	33	$1+x^{13}+x^{33}$
14	$1+x+x^{11}+x^{12}+x^{14}$	34	$1+x+x^{14}+x^{15}+x^{34}$
15	$1+x+x^{15}$	35	$1+x^2+x^{35}$
16	$1+x^2+x^3+x^5+x^{16}$	36	$1+x^{11}+x^{36}$
17	$1+x^3+x^{17}$	37	$1+x^2+x^{10}+x^{12}+x^{37}$
18	$1+x^7+x^{18}$	38	$1+x+x^5+x^6+x^{38}$
19	$1+x+x^5+x^6+x^{19}$	39	$1+x^4+x^{39}$
20	$1+x^3+x^{20}$	40	$1+x^2+x^{19}+x^{21}+x^{40}$

Note that for generating M-sequences a polynomial $\phi(x)^{-1} = x^m \phi(x^{-1})$, which is the reverse of the primitive polynomial $\phi(x)$, may be used also. The resulting maximal-length sequence will be the reverse of the sequence generated by $\phi(x)$. For example, polynomial $\phi(x) = 1 \oplus x^1 \oplus x^4$ is the reverse of the polynomial $\psi(x) = 1 \oplus x^3 \oplus x^4$.

A major advantage of the pseudorandom sequence generation method described by (4.4) is simplicity of its implementation, hardware implementation included. The M-sequence generator that functions in accordance with expression (4.4) comprises only a m-bit shift register (SR) and the set of modulo - 2 adders in the feedback circuit. In generator operation the SR stores m-bits of the M-sequence and shifts the m-bit code one bit to the right. The modulo - 2 adders in the feedback circuit compute the values of successive bits that are sequentially written to the left-most SR position. The SR states can be represented as a sequence of m-dimensional vectors $A(k) = a_1(k)\, a_2(k)\, a_3(k) \ldots a_m(k)$, $k = 0,1,2,\ldots$, where $a_i(k) \in \{0,1\}$. For the vectors the following relation applies

$$\begin{vmatrix} a_1(k) \\ a_2(k) \\ a_3(k) \\ \ldots \\ a_m(k) \end{vmatrix} = \begin{vmatrix} \alpha_1\ \alpha_2\ \alpha_3 \ldots \alpha_{m-1}\ \alpha_m \\ 1\ \ 0\ \ 0\ \ldots\ \ 0\ \ \ 0 \\ 0\ \ 1\ \ 0\ \ldots\ \ 0\ \ \ 0 \\ \ldots\ldots\ldots\ldots\ldots \\ 0\ \ 0\ \ 0\ \ldots\ \ 1\ \ \ 0 \end{vmatrix} \cdot \begin{vmatrix} a_1(k-1) \\ a_2(k-1) \\ a_3(k-1) \\ \ldots \\ a_m(k-1) \end{vmatrix}$$

More compactly, $A(k) = V^1 A(k-1)$, from whence the relation

$$A(k+S) = V^{S+1} A(k-1)$$

or

$$A(k+S) = V^S A(k) \qquad (4.5)$$

where

$$V = \begin{vmatrix} \alpha_1 & \alpha_2 & \alpha_3 & \ldots & \alpha_{m-1} & \alpha_m \\ 1 & 0 & 0 & \ldots & 0 & 0 \\ 0 & 1 & 0 & \ldots & 0 & 0 \\ \ldots & \ldots & \ldots & \ldots & \ldots & \ldots \\ 0 & 0 & 0 & \ldots & 1 & 0 \end{vmatrix}$$

holds for any S.

For a specific M-sequence defined by polynomial $\phi(x) = 1 \oplus x^3 \oplus x^4$, the matrix V has the form

$$V = \begin{vmatrix} 0 & 0 & 1 & 1 \\ 1 & 0 & 0 & 0 \\ 0 & 1 & 0 & 0 \\ 0 & 0 & 1 & 0 \end{vmatrix}$$

By sequentially applying equation (4.4) or (4.5) for $S=0$, we may generate single or multiple-bit pseudorandom number sequences that have some statistical properties.

<u>M-sequence properties.</u> Consider now the properties of maximal-length sequences (M-sequences).

1. The period of a M-sequence generated by (4.4) is described by the high degree of generating polynomial $\phi(x)$ and is $L=2^m-1$.

2. For the specified polynomial $\phi(x)$, there exist L distinct M-sequences that differ by phase shifts. Thus 15 M-sequences (Table 4.3) are associated with the polynomial $\phi(x)=1 \oplus x^3 \oplus x^4$.

3. In a M-sequence a_k, $k=0,1,2,\ldots,L-1$, there are ones and zeros. Their probabilities of occurrence is given by:

$$p(a_k=1) = \frac{2^{m-1}}{2^m-1} = \frac{1}{2} + \frac{1}{2^{m+1}-2},$$

$$p(a_k=0) = \frac{2^{m-1}-1}{2^m-1} = \frac{1}{2} - \frac{1}{2^{m+1}-2}$$

and can take on any values, however close to *1/2*, as *m* increases.

Table 4.3

							{a_i}							
0	0	0	1	0	0	1	1	0	1	0	1	1	1	1
0	0	1	0	0	1	1	0	1	0	1	1	1	1	0
0	1	0	0	1	1	0	1	0	1	1	1	1	0	0
1	0	0	1	1	0	1	0	1	1	1	1	0	0	0
0	0	1	1	0	1	0	1	1	1	1	0	0	0	1
0	1	1	0	1	0	1	1	1	1	0	0	0	1	0
1	1	0	1	0	1	1	1	1	0	0	0	1	0	0
1	0	1	0	1	1	1	1	0	0	0	1	0	0	1
0	1	0	1	1	1	1	0	0	0	1	0	0	1	1
1	0	1	1	1	1	0	0	0	1	0	0	1	1	0
0	1	1	1	1	0	0	0	1	0	0	1	1	0	1
1	1	1	1	0	0	0	1	0	0	1	1	0	1	0
1	1	1	0	0	0	1	0	0	1	1	0	1	0	1
1	1	0	0	0	1	0	0	1	1	0	1	0	1	1
1	0	0	0	1	0	0	1	1	0	1	0	1	1	1

4. In a M-sequence, the probability of runs of l, $l \in \{1,2,...,m-1\}$, identical digits (*1*s or *0*s) closely approximates that of the random sequence.

5. For any s, ($1 \leq s \leq L$) there exists an $r \neq s$ ($1 \leq r \leq L$) such that $\{a_k\} \oplus \{a_{k-s}\}$ = $\{a_{k-r}\}$. This is commonly called the 'shift-and-add' property.

6. The M-sequence autocorrelation function is defined by

$$R(\tau) = \begin{cases} 1 & \text{for } \tau = 0 \ (mod\ L) \\ \dfrac{1}{L} & \text{for } \tau \neq 0 \ (mod\ L) \end{cases}$$

7. Among the l non-zero M-sequence formed by the polynomial $\phi(x)$ there exists a single sequence with the property $a_k=a_{2k}$, $k=0,1,2,...$ which is called characteristic and defined as follows. For the specified polynomial $\phi(x)$, the system of linear equations

$$a_i = a_{2i}, \quad i = \overline{0, m-1}. \quad (4.6)$$

is written in accordance with the recurrence relation (4.4). The unique non-zero solution of the system (4.6) will be the characteristic sequence associated with the given $\phi(x)$.

By way of example, we shall find a characteristic sequence for the generating polynomial $\phi(x)= 1 \oplus x^3 \oplus x^4$. The system (4.6) takes the form

$$a_0 = a_0,$$
$$a_1 = a_2,$$
$$a_2 = a_4 = a_0 \oplus a_1,$$
$$a_3 = a_6 = a_2 \oplus a_3.$$

By solving the system of equations, we shall find that the initial values of the characteristic sequence $\{a_k\}^*$ associated with the generating polynomial $\phi(x)= 1 \oplus x^3 \oplus x^4$ are defined by the relation $a_0\, a_1\, a_2\, a_3 = 1000$.

Decimation of the M-sequence $\{a_i\}$ by $q \in (q=1,2,3,...)$ means the generation of another sequence $\{b_k\}$ of qth elements of $\{a_k\}$, i.e. $b_k = a_{kq}$. If $\{b_k\}$ is a non-zero sequence it is generated by polynomial $\phi(x)$ whose roots are the qth degrees of the original polynomial $\phi(x)$ roots and its period is $L/(L,q)$, where (L,q) is the least common divisor of L and q. With $(L,q) = 1$, the period of $\{b_k\}$ is $2^m - 1$, where $m = \deg \phi(x)$, and decimation is said to be proper or normal. Any proper decimation results in a M-sequence generated by a polynomial $\phi(x)$. In this case, decimation performed on a sequence that has been shifted by s clocks relative to an original $\{a_k\}$ will result in a sequence shifted by several clocks relative to $\{b_k\}$. In other words, irrespective of the shift value that has been chosen for $\{a_k\}$, decimation will always result in a M-sequence generated by polynomial $\phi(x)$.

4.2. Generators of Uniformly Distributed Pseudorandom Test Sequences

<u>PRSG with external adders</u>. A key relation used for building pseudorandom test sequence generators is that of (4.5) which takes on the form

$$\begin{vmatrix} a_1(k+1) \\ a_2(k+1) \\ a_3(k+1) \\ \ldots \\ a_m(k+1) \end{vmatrix} = \begin{vmatrix} \alpha_1 \, \alpha_2 \, \alpha_3 \ldots \alpha_{m-1} \, \alpha_m \\ 1 \; 0 \; 0 \; \ldots \; 0 \; \; 0 \\ 0 \; 1 \; 0 \; \ldots \; 0 \; \; 0 \\ \ldots \ldots \ldots \ldots \ldots \\ 0 \; 0 \; 0 \; \ldots \; 1 \; \; 0 \end{vmatrix} \cdot \begin{vmatrix} a_1(k) \\ a_2(k) \\ a_3(k) \\ \ldots \\ a_m(k) \end{vmatrix} \quad (4.7)$$

where $k=0,1,2,\ldots$, and $\alpha_i \in \{0,1\}$ is the coefficient of generating polynomial $\phi(x)$ whose high degree $(deg\ \phi(x))$ equals to m. The necessary condition for generating pseudorandom test sequences is meeting the following inequality

$$a_1(-1)\ a_2(-1)\ a_3(-1)\ \ldots\ a_m(-1) \neq 0\ 0\ 0\ \ldots 0 \quad (4.8)$$

In this case, the states of storage elements in PRSG can assume any value but zero. When the condition of (4.8) is not met, i.e. with $a_1(-1)a_2(-1)a_3(-1) \ldots a_m(-1)=000\ldots0$, the states of storage elements in PRSG are invariably zero.

By using the equation (4.7) we can obtain the system of logical equations that determine the basic functional connections of PRSG. For the general case, such system will appear as

$$\begin{aligned} a_1(k) &= \sum_{i=1}^{m} {}^{\oplus} \alpha_i\, a_i(k-1), \\ a_j(k) &= a_{j-1}(k-1), \quad j=\overline{2,m},\ \ k=0,1,2,\ldots. \end{aligned} \quad (4.9)$$

Fig.4.2 shows the functional diagram of PRSG by (4.9). A PRSG consists of m storage elements which are D flip-flops connected into a shift register to perform a one-bit-to-right shift microoperation. The PRSG feedback loop includes two-input modulo-2 adders whose connections with the shift register stages are specified by coefficients $\{\alpha_j\}$ of the generating polynomial. If $\alpha_j=1$,

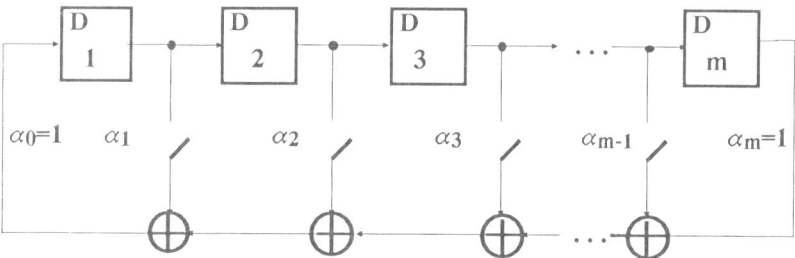

Fig. 4.2. Functional diagram of PRSG

$j \in \{1,2,...,m-1\}$, then the output from the j-th shift register stage must be connected with an input to the modulo-2 adder. Otherwise $(j=0)$, no connection exists.

Fig.4.3 shows the schematic diagram of a generator for generating polynomial $\phi(x) = 1 \oplus x \oplus x^4$. The system of logical equations which describes the generator connection appear as

$$\begin{aligned}
a_1(k) &= a_1(k-1) \oplus a_4(k-1), \\
a_2(k) &= a_1(k-1), \\
a_3(k) &= a_2(k-1), \\
a_4(k) &= a_3(k-1).
\end{aligned} \quad (4.9)$$

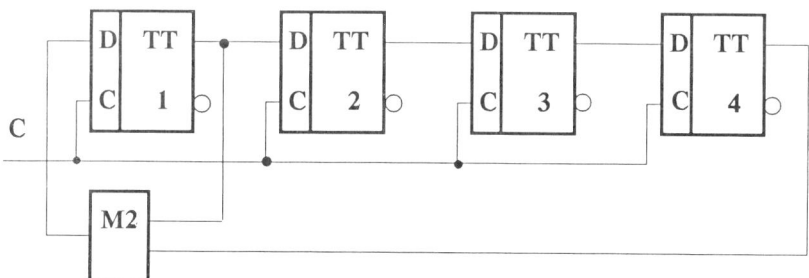

Fig. 4.3. Schematic diagram of PRSG

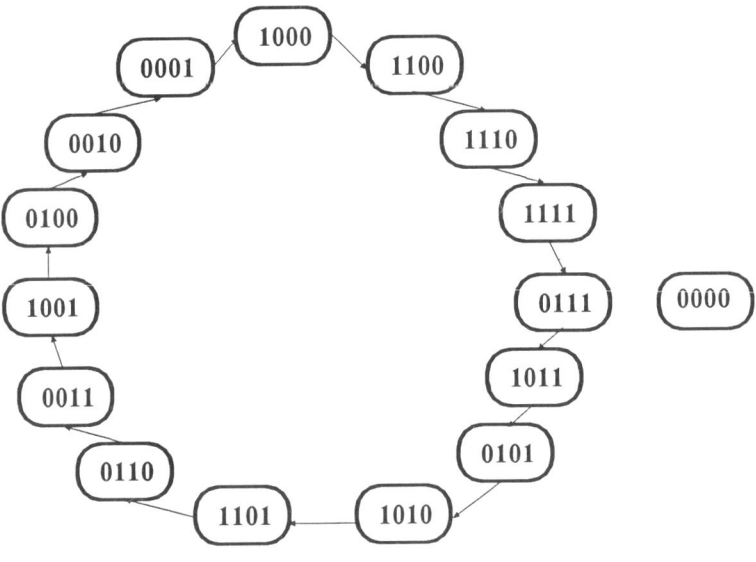

Fig. 4.4. PRSG states

When a successive sync pulse is applied to C inputs of D flip-flops, the latter change their states according to (4.10). Fig.4.4 shows the chart of PRSG states for $a_1(-1)a_2(-1)a_3(-1)a_4(-1) \neq 0000$. If the condition of (4.8) has not been met, storage elements in PRSG remain unchanged.

A major advantage of PRSGs built in accordance with expression (4.7) is the simplicity of their hardware implementation. To implement a PRSG as a stand-alone unit, we can use only a m-stage shift register and some two-input modulo-2 adders. The number of modulo-2 adders is determined by the number of nonzero coefficients of the generating polynomial $\phi(x)$. However, when the generating polynomial $\phi(x)$ is inadequate, the number of adders may become significant. This causes the PRSG complexity to increase with the decrease of its speed due to signal propagation delay in a sequence of series-connected modulo-2 adders. The complete list of polynomials with the minimal number of nonzero coefficients for m, is given in Appendix I.

<u>PRSG with internal adders.</u> As opposed to a PRSG scheme with external modulo-2 adders, an alternative scheme with internal modulo-2 adders has the maximum speed. A basic relationship, which determines the functional connection of such PRSG, is

$$\begin{vmatrix} a_1(k) \\ a_2(k) \\ a_3(k) \\ \ldots \\ a_m(k) \end{vmatrix} = \begin{vmatrix} 0 & 0 & 0 & \ldots & 0 & 1 \\ 1 & 0 & 0 & \ldots & 0 & \alpha_1 \\ 0 & 1 & 0 & \ldots & 0 & \alpha_2 \\ \ldots & \ldots & \ldots & \ldots & \ldots & \ldots \\ 0 & 0 & 0 & \ldots & 1 & \alpha_{m-1} \end{vmatrix} \cdot \begin{vmatrix} a_1(k-1) \\ a_2(k-1) \\ a_3(k-1) \\ \ldots \\ a_m(k-1) \end{vmatrix} \quad (4.11)$$

Based on the relation (4.11), we can build the set of equations to describe functional connections of PRSG. For the general case it assumes the form

$$a_1(k) = a_m(k-1),$$
$$a_j(k) = \alpha_{j-1} a_m(k-1) \oplus a_{j-1}(k-1), \quad j = \overline{2,m}, \quad k = 0,1,2,\ldots.$$

Fig.4.5 shows the functional diagram of the PRSG.

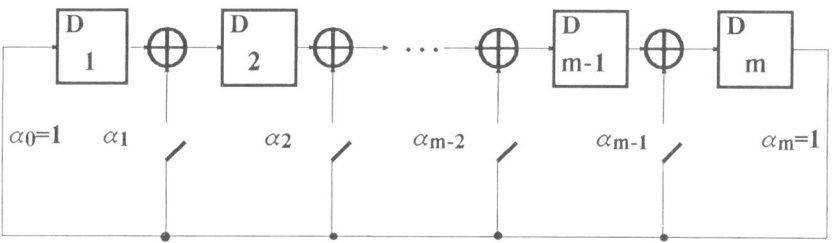

Fig. 4.5. Functional diagram of the PRSG

Hence the relations (4.7) and (4.11) allow us to build PRSGs with external or internal modulo-2 adders which differ in time responses, as noted above. From the maximum speed standpoint, a PRSG scheme with internal adders is more preferable. However, its implementation as a stand-alone device is based on the use of individual storage elements which makes the PRSG scheme more complicated. A simpler implementation is evidently a device based on a single shift register (it may be one or more ICs) and modulo-2 adders.

PRSG with external and internal adders. An ingenious method for building a PRSG is the one using both internal and external modulo-2 adders. A matrix relating the current generator state with its preceding state and similar to relations (4.7) and (4.11) has the following form

$$\begin{vmatrix} a_1(k) \\ a_2(k) \\ a_3(k) \\ \cdots \\ a_j(k) \\ a_{j+1}(k) \\ a_{j+2}(k) \\ \cdots \\ a_m(k) \end{vmatrix} = \begin{vmatrix} \alpha_1 \alpha_2 \ldots \alpha_{j-1} & 0 & \ldots & 0 & 1 \\ 1 & 0 & \ldots & 0 & 0 & \ldots & 0 & 0 \\ 0 & 1 & \ldots & 0 & 0 & \ldots & 0 & 0 \\ \cdots & \cdots & \cdots & \cdots & \cdots & \cdots \\ 0 & 0 & \ldots & 1 & 0 & \ldots & 0 & 0 \\ 0 & 0 & \ldots & 0 & 0 & \ldots & 0 & \beta_j \\ 0 & 0 & \ldots & 0 & 1 & \ldots & 0 & \beta_{j+1} \\ \cdots & \cdots & \cdots & \cdots & \cdots & \cdots \\ 0 & 0 & \ldots & 0 & 0 & \ldots & 1 & \beta_{m-1} \end{vmatrix} \cdot \begin{vmatrix} a_1(k-1) \\ a_2(k-1) \\ a_3(k-1) \\ \cdots \\ a_j(k-1) \\ a_{j+1}(k-1) \\ a_{j+2}(k-1) \\ \cdots \\ a_m(k-1) \end{vmatrix} \quad (4.12)$$

where $\beta_j=1$ and $\alpha_r, \beta_w \in \{0,1\}$ take the value 0 or 1 for specific maximal-length sequence ($r=\overline{1,j-1}$, $w=\overline{j+1, m-1}$). The above relation combines the preceding two therefore the set of equations which describes the functional connections of PRSG built in accordance with the expression (4.12), copies the appropriate equations for a PRSG with external and internal modulo-2 adders. Actually, by analyzing the relation (4.12), we obtain:

$$a_1(k) = a_m(k-1) \oplus \sum_{r=1}^{j-1} {}^{\oplus} \alpha_r a_r(k-1),$$
$$a_i(k) = a_{i-1}(k-1), \quad i = \overline{2,j}, \quad (4.13)$$
$$a_{j+1}(k) = a_j(k-1) \oplus a_m(k-1),$$
$$a_v(k) = a_{v-1}(k-1) \oplus \beta_{v-1} a_m(k-1), v = \overline{j+2, m}, k = 0,1,2,\ldots.$$

The functional diagram of a PRSG built in accordance with the set of equations (4.13) is shown in Fig.4.6. Note that coefficients define connections between the first $j-1$ bits of PRSG and external modulo-2 adders, whereas β_j and β_w define connections between the last bit and internal modulo-2 adders cut in between the ending bits of the generator. Considering the differences between coefficients α_r and β_w, we can write the generating polynomial for a PRSG as

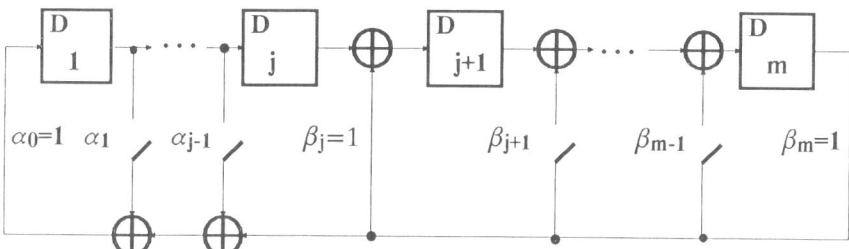

Fig. 4.6. Functional diagram of PRSG

$$\varphi^*(x) = 1 \oplus \alpha_1 x^1 \oplus \alpha_2 x^2 \oplus \ldots \oplus \alpha_{j-1} x^{j-1} \oplus \beta_j x^j \oplus \\ \oplus \beta_{j+1} x^{j+1} \oplus \ldots \oplus \beta_{m-1} x^{m-1} \oplus x^m \quad (4.14)$$

A major request for a PRSG is the use of maximal-length sequences (M-sequences) for their generation. Recall that M-sequences are described by irreducible primitive polynomials. Hence the basic theoretic outcome used for synthesizing PRSGs with internal and external modulo-2 adders is the following theorem.

Theorem 4.1. If the primitive generating polynomial

$$\varphi(x) = 1 \oplus h_1 x^1 \oplus h_2 x^2 \oplus \ldots \oplus h_{m-1} x^{m-1} \oplus h_m x^m$$

can be transformed to

$$\varphi(x) = 1 \oplus (\alpha_1 x^{j-1} \oplus \alpha_2 x^{j-2} \oplus \ldots \oplus \alpha_{j-1} x^1 \oplus x^j)(1 \oplus \\ \oplus \beta_{j+1} x^1 \oplus \ldots \oplus \beta_{m-1} x^{m-j-1} \oplus x^{m-1}),$$

where h_i, α_r and β_w are coefficients, which take the value 0 or 1 ($i=\overline{1, m}$, $r=\overline{1, j-1}$, $w=\overline{j+1, m-1}$) a synchronous automaton created by the generating polynomial

$$\varphi^*(x) = 1 \oplus \alpha_1 x^1 \oplus \alpha_2 x^2 \oplus \ldots \oplus \alpha_{j-1} x^{j-1} \oplus \beta_j x^j \oplus$$
$$\oplus \beta_{j+1} x^{j+1} \oplus \ldots \oplus \beta_{m-1} x^{m-1} \oplus x^m,$$

where $\beta_j = 1$ and coefficient α_r relates the rth automaton's bit with an external modulo-2 adders (Fig.4.6) will produce a maximum-length sequence.

The reverse is also true.

As an example let us consider synthesizing a PRSG for a primitive polynomial $\phi(x) = 1 \oplus x^2 \oplus x^3 \oplus x^4 \oplus x^5$. This polynomial can be transformed into

$$\varphi(x) = 1 \oplus (x^{4-2} \oplus x^4)(1 \oplus x) \qquad (4.15)$$

which meets the condition of theorem 4.1. Then by using expression (4.15) we can transform the primitive polynomial $\phi(x)$ into a generating polynomial $\phi^*(x)$ in accordance with expression (4.14) which for the case appears as

$$\varphi(x) = 1 \oplus \alpha_2 x^2 \oplus \beta_4 x^4 \oplus x^5$$

A schematic diagram for the PRSG built in accordance with the latter generating polynomial is shown in Fig.4.7. It is evident from the PRSG diagram that the number of extra modulo-2 adders required for its implementation is less than for the cases when only internal or only external modulo-2 adders are used for PRSG implementation. However, the speed of such PRSGs may desrease due to signal delay in series-connected modulo-2 adders in the external circuit as compared to the scheme with only internal adders. The necessity of testing the theorem (4.1) condition for the primitive polynomial $\phi(x)$ is somewhat inconvenient.

Appendix 2 contains the outcome of analyzing the limited number of primitive polynomials whose high degree is less then *100*. Each m here is associated with a generating polynomial which allows synthesizing a PRSG with a single internal and a single external modulo-2 adder. Similar to primitive polynomials, each polynomial of Appendix 2 is associated with an inverse polynomial, which can be obtained by the formula

$$\psi*(x, \alpha, \beta) = x^m \varphi*(x^{-1}, \alpha, \beta).$$

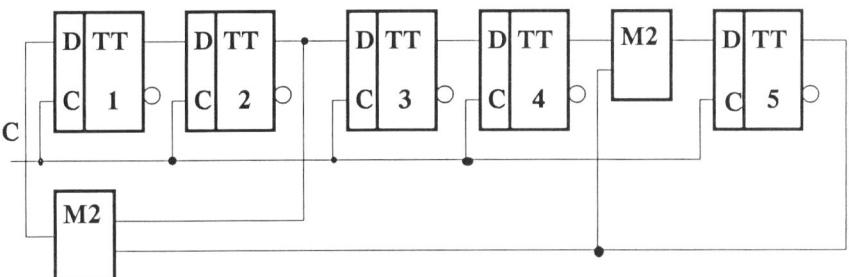

Fig. 4.7. Schematic diagram of PRSG

Thus for

$$\varphi*(x) = 1 \oplus \alpha_2 x^2 \oplus \beta_4 x^4 \oplus x^5$$

we obtain

$$\psi*(x) = x^5 (1 \oplus \alpha_2 x^{-2} \oplus \beta_4 x^{-4} \oplus x^{-5}) = 1 \oplus \beta_4 x^1 \oplus \alpha_2 x^3 \oplus x^5.$$

By associating the indices of coefficients α and β with the degrees of variable x, we can finally obtain

$$\psi*(x) = 1 \oplus \beta_1 x^1 \oplus \alpha_3 x^3 \oplus x^5$$

where, the same as for $\phi^*(x)$, coefficients β define connection with internal modulo-2 adders whereas coefficients α with external modulo-2 adders.

4.3. Pseudorandom Test Pattern Generators

Sequential PRPG. The above discussed methods of building PRPG allow us to create a device to generate sequences of uncorrelated r-bit test patterns. In practice, the so-called sequential test-pattern generation concept is used for the purpose. The idea behind this concept is generation of any successive pattern of r bits sequentially produced by a structure of Fig.4.2, 4.3 or 4.5. A pseudorandom test pattern is produced on the outputs from r shift register stages in the PRPG every sr cycles of its operation. The latter inequality is the condition for statistical independence of adjacent test patterns. Any test pattern from the sequence $\{A_k\} = A_0, A_1, A_2, ... A_k$ can be described as

$$A_k = a_1(ks)a_1(ks-1)a_1(ks-2) \ldots a_1(ks-r+1) \qquad (4.16)$$

where $a_1(ks)$ is the contents of the first register bit in the ksth cycle of generator operation.

Considering that sequences $\{a_1(ks)\}, \{a_1(ks-1)\}, ..., \{a_1(ks-r+1)\}$, $k=0,1,2,...$, are periodic, we can readily prove that $\{A_k\}$ will be also periodic for a value obtained by dividing 2^m-1 ($m = deg\phi(x)$), $\phi(x)$ is a PRPG generating polyno-

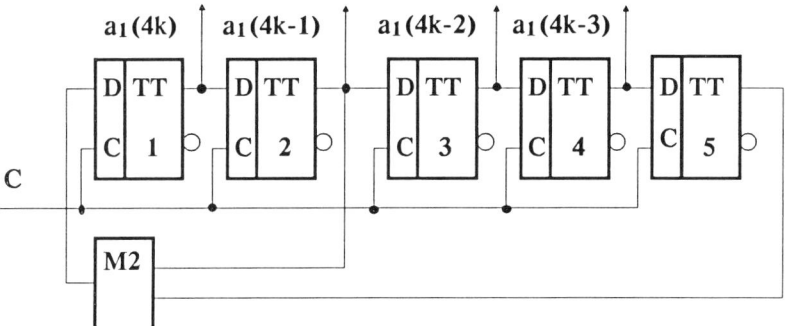

Fig. 4.8. Sequental PRSG

mial) by the greatest common divisor of 2^m-1 and $s - (2^m-1, s)$. With $(2^m-1, s) = 1$, the generator produces $(2^m-1, s)$ different test sequences whose form and probability responses depend on the initial state of the shift register in PRPG. When s is mutually prime with $L = 2^m-1$, the period of $\{A_k\}$ is L, and its characteristics are independent of the initial state of the shift register. Therefore, to select a proper value of s, we must factor 2^m-1 into components.

By way of example, we consider creation of a sequential PRPG for $r=4$ and a generating polynomial $\phi(x)=1\oplus x^2 \oplus x^5$ (Fig.4.8, Table 4.4). Since $L=2^m-1=31$ is a Mersen number, we can select $s=4$ to define the number of shifts.

Undoubtedly the major advantage of a serial PRPG is the ease of its hardware implementation and the major disadvantage is its low speed. Actually, to produce a r-bit pseudorandom test pattern the contents of PRPG must be shifted s times, where $s \geq r$.

PRPG based on multi-bit shift. One way of speeding up the PRPG operation is to use the multi-place shift concept. The solution is based on implementation of the relation (4.5) for $S \neq 1$. According to equation (3.5), for $S>1$, the M-sequence will be shifted by S stages within a cycle. The functional connections in the generator will then be defined by matrix V^S. Consider as an example the construction of a sequential 4-bit PRPG for polynomial $\phi(x)=1\oplus x^2 \oplus x^5$ and shift $S=4$. The matrix V^4 will have the form:

$$V^4 = \begin{vmatrix} \alpha_1 & \alpha_2 & \alpha_3 & \alpha_4 & \alpha_5 \\ 1 & 0 & 0 & 0 & 1 \\ 0 & 1 & 0 & 0 & 0 \\ 0 & 0 & 1 & 0 & 0 \\ 0 & 0 & 0 & 1 & 0 \end{vmatrix}^4 = \begin{vmatrix} 0 & 1 & 1 & 0 & 1 \\ 1 & 0 & 0 & 0 & 1 \\ 0 & 1 & 0 & 0 & 0 \\ 0 & 0 & 1 & 0 & 0 \\ 0 & 0 & 0 & 1 & 0 \end{vmatrix}^4 =$$

$$= \begin{vmatrix} 1 & 1 & 0 & 1 & 0 \\ 0 & 1 & 1 & 0 & 1 \\ 1 & 0 & 0 & 1 & 0 \\ 0 & 1 & 0 & 0 & 1 \\ 1 & 0 & 0 & 0 & 0 \end{vmatrix}.$$

The equation (4.5) may be written as

$$\begin{vmatrix} a_1(k+4) \\ a_2(k+4) \\ a_3(k+4) \\ a_4(k+4) \\ a_5(k+4) \end{vmatrix} = \begin{vmatrix} 1 & 1 & 0 & 1 & 0 \\ 0 & 1 & 1 & 0 & 1 \\ 1 & 0 & 0 & 1 & 0 \\ 0 & 1 & 0 & 0 & 1 \\ 1 & 0 & 0 & 0 & 0 \end{vmatrix} \begin{vmatrix} a_1(k) \\ a_2(k) \\ a_3(k) \\ a_4(k) \\ a_5(k) \end{vmatrix} \qquad (4.17)$$

Table 4.4

k	$a_1(k)$	$a_2(k)$	$a_3(k)$	$a_4(k)$	$a_5(k)$	$a_1(4k)$	$a_1(4k-1)$	$a_1(4k-2)$	$a_1(4k-3)$
0	1	0	0	0	0	1	0	0	0
1	0	1	0	0	0				
2	1	0	1	0	0				
3	0	1	0	1	0				
4	1	0	1	0	1	1	0	1	0
5	1	1	0	1	0				
6	1	1	1	0	1				
7	0	1	1	1	0				
8	1	0	1	1	1	1	0	1	1
9	1	1	0	1	1				
10	0	1	1	0	1				
11	0	0	1	1	0				
12	0	0	0	1	1	0	0	0	1
13	1	0	0	0	1				
14	1	1	0	0	0				
15	1	1	1	0	0				
16	1	1	1	1	0	1	1	1	1
17	1	1	1	1	1				
18	0	1	1	1	1				
19	0	0	1	1	1				
20	1	0	0	1	1	1	0	0	1
21	1	1	0	0	1				
22	0	1	1	0	0				
23	1	0	1	1	0				
24	0	1	0	1	1	0	1	0	1

Based on this we may set up a set of logical equations specifying the functional diagram of the generator:

$$a_1(k+4) = a_1(k) \oplus a_2(k) \oplus a_4(k),$$
$$a_2(k+4) = a_2(k) \oplus a_3(k) \oplus a_5(k),$$
$$a_3(k+4) = a_1(k) \oplus a_4(k),$$
$$a_4(k+4) = a_2(k) \oplus a_5(k),$$
$$a_5(k+4) = a_1(k).$$

A successive test pattern may be defined as vector $a_1(k)a_2(k)a_3(k)a_4(k)$ (Fig.4.9). Table 4.5 presents the state sequence for the PRPG storage elements under initial conditions $a_1(k)=1$, $a_j(k)=0$, $j=\overline{2, 5}$.

The identity of test pattern sequences produced by the generators shown in Figs 4.8 and 4.9 can be readily verified. The PRPG of Fig.4.9 has the higher speed and high hardware costs.

To reduce hardware cost, it is necessary to minimize the number of unities in matrix V^S.

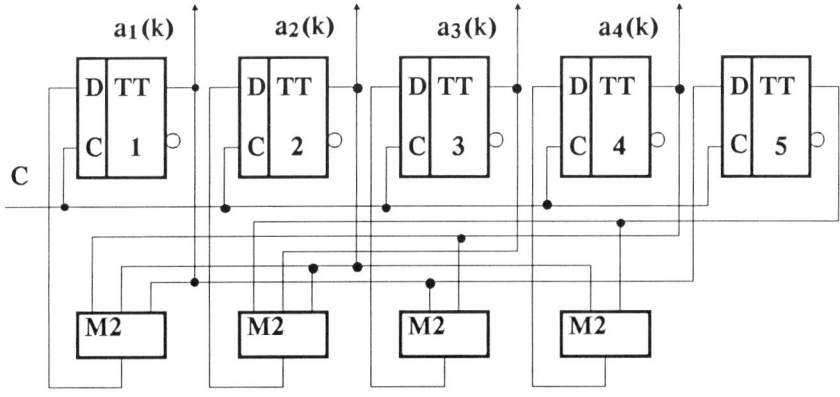

Fig. 4.9. PRPG based on multi-stage shift

<u>PRPG based on T- and D-flip-flops.</u> An ingenious way of constructing high-speed PRPGs with minimum hardware costs is based on specific feature of generating polynomials. The said method is applicable to polynomials having the form of a trinomial $\phi(x)=1\oplus x^j\oplus x^m$.

Let $S=j$, $j \geq m/2$, and $(j, 2^m-1)=1$. Then V^S will take the form:

$$V^S = \begin{vmatrix} \alpha_j & 0 & 0 & \ldots & \alpha_m & 0 & 0 & \ldots & 0 & 0 & 0 \\ 0 & \alpha_j & 0 & \ldots & 0 & \alpha_m & 0 & \ldots & 0 & 0 & 0 \\ 0 & 0 & \alpha_j & \ldots & 0 & 0 & \alpha_m & \ldots & 0 & 0 & 0 \\ \ldots & \ldots & \ldots & \ldots & \ldots & \ldots & \ldots & \ldots & \ldots & \ldots & \ldots \\ 0 & 0 & 0 & \ldots & 0 & 0 & \alpha_j & \ldots & 0 & 0 & \alpha_m \\ 1 & 0 & 0 & \ldots & 0 & 0 & 0 & \ldots & 0 & 0 & 0 \\ 0 & 1 & 0 & \ldots & 0 & 0 & 0 & \ldots & 0 & 0 & 0 \\ \ldots & \ldots & \ldots & \ldots & \ldots & \ldots & \ldots & \ldots & \ldots & \ldots & \ldots \\ 0 & 0 & 0 & \ldots & 0 & 1 & 0 & \ldots & 0 & 0 & 0 \end{vmatrix}. \quad (4.18)$$

Table 4.5

k	$a_1(k)$	$a_2(k)$	$a_3(k)$	$a_4(k)$	$a_5(k)$
0	1	0	0	0	0
1	1	0	1	0	1
2	1	0	1	1	1
3	0	0	0	1	1
4	1	1	1	1	0
5	1	0	0	1	1
6	0	1	0	1	1
7	0	0	1	0	0

Note that V^S, according to equation (4.18), has the maximum number of zero elements, with only two unities in the first j rows and a single unity in the last $(m-j)$ rows.

The set of logical relations for the states of storage elements in pseudorandom test pattern generator obtained from equation (4.5) and matrix (4.18) has the form:

$$\begin{aligned} a_i(k+j) &= a_i(k) \oplus a_{m-j+i}(k), & i &= \overline{1,j}, \\ a_i(k+j) &= a_{i-j}(k), & i &= \overline{j+1,m}. \end{aligned} \quad (4.19)$$

The PRPG implemented according to this set of equations may comprise T- and D-flip-flops. Fig.4.10 shows the functional scheme of such generator for the generating polynomial $\phi(x)=1\oplus x^3\oplus x^5$. Here the complementing flip-flop fulfill the same function as a modulo-2 adder in the feedback circuit of a sequential PRPG. The scheme provides for obtaining uncorrelated test patterns at the output of the first $j=3$ stages of the PRPG.

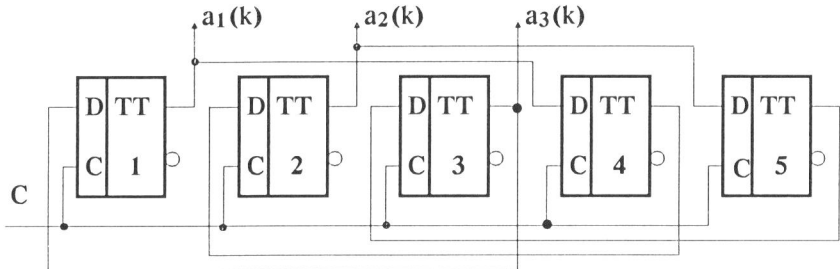

Fig. 4.10. PRPG based on T- and D-flip-flops

PRPG based on T-flip-flops. Most efficient among the pseudorandom test pattern generator schemes is the one constructed for the polynomials of the form $\phi(x)=1\oplus x^{m-1}\oplus x^m$. In this case for the maximum $r=m$ we construct V^S where S is also equal to m. For any m and the associated polynomial $\phi(x)=1\oplus x^{m-1}\oplus x^m$, the matrix takes the form:

$$V^m = \begin{vmatrix} 1 & 0 & 0 & \ldots & 0 & 1 & 1 \\ 1 & 1 & 0 & \ldots & 0 & 0 & 0 \\ 0 & 1 & 1 & \ldots & 0 & 0 & 0 \\ \ldots & \ldots & \ldots & \ldots & \ldots & \ldots & \ldots \\ 0 & 0 & 0 & \ldots & 0 & 1 & 1 \end{vmatrix}. \quad (4.20)$$

According to equation (4.20), the set of equations for V^m has the following form:

$$\begin{aligned} a_1(k+m) &= a_1(k) \oplus a_{m-1}(k) \oplus a_m(k), \\ a_j(k+m) &= a_j(k) \oplus a_{j-1}(k), \quad i = \overline{2,m}. \end{aligned} \quad (4.21)$$

The functional diagram of a PRPG that has been designed according to (4.21) consists of m sequential by connected T-flip-flops and a 2-input modulo-

2 adder (Fig.4.11). With the condition $(m, 2^{m}-1)=1$ satisfied, the diagram allows generating uncorrelated m-bit test pattern with maximum speed and minimum hardware cost. The sole limitation of the method is inability to construct a PRPG for any m since polynomials of the form $\phi(x)=1\oplus x^{m-1}\oplus x^m$ exist only for $m=2,3,4,6,7,15,22,60,63,127,153,\ldots$.

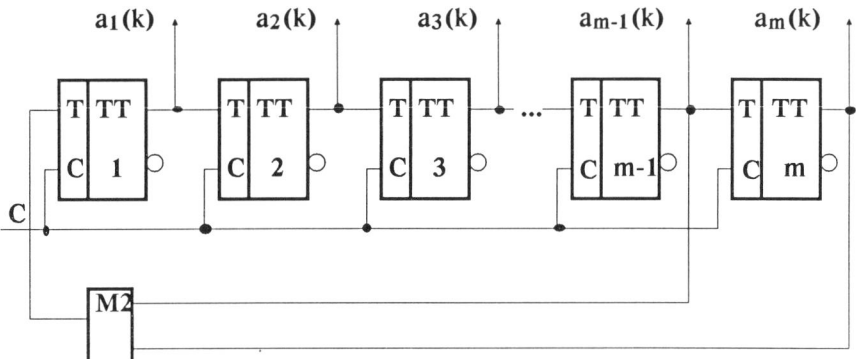

Fig. 4.11. PRPG based on T-flip-flops

Generalization of the ingenious method of designing PRPGs on the basis of T-flip-flops resulted in the design of such generators for any primitive generating polynomial. The principle of the method is that a T-flip-flop is described by polynomial $1\oplus x$ thereby allowing to change from the polynomial that describes a PRPG with D-flip-flops to a new structure. Thus the initial polynomial $\phi(x)$ may be represented as $\phi(1\oplus x)$ to define the key functional connections of a PRPG based on T-flip-flops.

Let us consider a PRPG design for the polynomial $\phi(x)=1\oplus x^2\oplus x^5$. Substituting variable x for $1\oplus x$, we obtain

$$\varphi(1\oplus x) = 1\oplus(1\oplus x)^2\oplus(1\oplus x)^5 = 1\oplus 1\oplus x^2\oplus 1\oplus x^4\oplus x\oplus x^5 = 1\oplus x\oplus x^2\oplus x^4\oplus x^5.$$

The resulting polynomial $\phi(1\oplus x)$ may take on an arbitrary form, including reducible and irreducible. Similar transformations for $\phi(x)=1\oplus x^3\oplus x^4$ result in the polynomial $\phi(1\oplus x)=1\oplus x\oplus x^2\oplus x^3\oplus x^4$. Fig.4.12 shows the functional diagram of a PRPG according to the latter polynomial. The states of storage elements in this PRPG are given in Table 4.6.

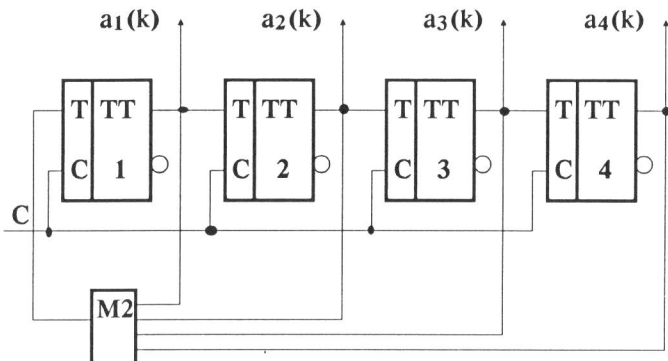

Fig. 4.12. Functional diagram of the PRPG

Analytical treatment of the state chart of PRPG storage elements reveals that their values correspond to a maximal-length sequence and with the (4.8) condition satisfied all possible binary codes other that zero are produced on the generator outputs. The M-sequence generating property of a structure based on T-flip-flops stems from their design approach and is valid for the general case.

By comparing the states of adjacent PRPG bits we can note that they are identical with the time shift. For the example of Fig.4.12 the time shift is 12 (Refer to Table 4.6). The method described below has been based on the possibility of designing a PRPG with shifted replicas of an initial M-sequence produced in its adjacent bits.

PRPG based on shifted replicas of M-sequences. The idea behind this method is to use different segments of the same M-sequence to form distinct values of bits for the patterns being generated. The advantage of the technique is the possibility of obtaining different M-sequence segments at one time including extra modulo-2 adders to a single generator. Symbols produced at the adder outputs are treated as the values of bits in pseudorandom test patterns.

Such PRPGs, called "parallel", are complicated to implement for practically used values of m and r. Since every extra r modulo-2 adders has $m/2$ inputs, on the average, the hardware cost for their implementation migth exceed by far the cost of proper M-sequence generator. The use of techniques for minimizing the number of modulo-2 adder inputs involves an increase in computation and effects no saving in hardware. Besides, the pseudorandom test patterns generated by this technique differs essentially from the former sequential construction only in its implementation. An example will make this clear.

Suppose that for the generating polynomial $\phi(x)=1\oplus x^2\oplus x^5$ a 4-bit PRPG is to be constructed. Let the value of the first bit in the test patterns making up the pseudorandom test sequence be defined by the contents of the first bit $a_1(k)$ in the M-sequence generator shift register, and those of the second, third and fourth bits by its replicas, i.e. $a_1(k+23)$, $a_1(k+15)$ and $a_1(k+7)$.

Table 4.6

k	$a_1(k)$	$a_2(k)$	$a_3(k)$	$a_4(k)$
0	1	0	0	0
1	0	1	0	0
2	1	1	1	0
3	0	0	0	1
4	1	0	0	1
5	1	1	0	1
6	0	0	1	1
7	0	0	1	0
8	1	0	1	1
9	0	1	1	0
10	0	1	0	1
11	0	1	1	1
12	1	1	0	0
13	1	0	1	0
14	1	1	1	1

To construct the functional diagram of the PRPG it is necessary to find coefficients $\{\delta_i(l)\}$, $i=\overline{1,5}$, for $l=7, 15$ and 23. Here coefficients $\delta_i(l)\in\{0,1\}$, $i=\overline{1,m}$, define the use of the values of bits in the M-sequence generator shift register for obtaining a l-shifted replica. As applied to hardware implementation, coefficients $\{\delta_i(l)\}$ specify the shift register connections for the generator with an additional modulo-2 adder producing the shifted sequence at its output. Let us use the technique for finding $\{\delta_i(l)\}$. The problem is to find the states of generator elements for $l-1$ with $l=7, 15$ and 22. As a result we obtain (Refer to Table 4.4)

$$a_1(k+6)\ a_2(k+6)\ a_3(k+6)\ a_4(k+6)\ a_5(k+6) = 1\ 1\ 1\ 0\ 1,$$
$$a_1(k+14)\ a_2(k+14)\ a_3(k+14)\ a_4(k+14)\ a_5(k+14) = 1\ 1\ 0\ 0\ 0,$$
$$a_1(k+22)\ a_2(k+22)\ a_3(k+22)\ a_4(k+22)\ a_5(k+22) = 0\ 1\ 1\ 0\ 0.$$

As a basic relation for calculating $\{\delta_i(l)\}$ we shall use:

$$\begin{vmatrix} \delta_1(l) \\ \delta_2(l) \\ \delta_3(l) \\ \dots \\ \delta_m(l) \end{vmatrix} = \begin{vmatrix} \alpha_1 & \alpha_2 & \alpha_3 & \dots & \alpha_{m-1} & \alpha_m \\ \alpha_2 & \alpha_3 & \alpha_4 & \dots & \alpha_m & 0 \\ \alpha_3 & \alpha_4 & \alpha_5 & \dots & 0 & 0 \\ \dots & \dots & \dots & \dots & \dots & \dots \\ \alpha_m & 0 & 0 & \dots & 0 & 0 \end{vmatrix} \begin{vmatrix} a_1(k+l-1) \\ a_2(k+l-1) \\ a_3(k+l-1) \\ \dots \\ a_m(k+l-1) \end{vmatrix},$$

where for polynomial $\phi(x)=1 \oplus x^2 \oplus x^5$ the matrix of coefficients $\alpha_i \in \{0,1\}$, $i=\overline{1,m}$ of the generating polynomial will appear as:

$$\begin{vmatrix} 0 & 1 & 0 & 0 & 1 \\ 1 & 0 & 0 & 1 & 0 \\ 0 & 0 & 1 & 0 & 0 \\ 0 & 1 & 0 & 0 & 0 \\ 1 & 0 & 0 & 0 & 0 \end{vmatrix}.$$

Table 4.7

l	$\delta_1(l)$	$\delta_2(l)$	$\delta_3(l)$	$\delta_4(l)$	$\delta_5(l)$
7	0	1	1	1	1
15	1	1	0	1	1
23	1	0	1	1	0

As a result we obtain the values of coefficients $\{\delta_i(l)\}$ (Table 4.7). Fig. 4.13 shows the functional scheme of the PRPG, and Table 4.8 gives the sequence of states for the storage elements in the generator and test pattern $\{A_k\}=\{a_1(k)a_1(k+23)a_1(k+15)a_1(k+7)\}$ generated under initial conditions $a_1(k)a_2(k)a_3(k)a_4(k)a_5(k) = 10101$, $k=0$.

Table 4.8

k	$a_1(k)$	$a_2(k)$	$a_3(k)$	$a_4(k)$	$a_5(k)$	A_k
0	1	0	1	0	1	1000
1	1	1	0	1	0	1010
2	1	1	1	0	1	1011
3	0	1	1	1	0	0001
4	1	0	1	1	1	1111
5	1	1	0	1	1	1001
6	0	1	1	0	1	0101
7	0	0	1	1	0	0010

On comparing the results presented in Tables 4.4 and 4.8, we see that the pseudorandom test sequences produced by the sequential and parallel PRPGs are absolutely identical. However, their implementations differ widely in speed and hardware requirements. Sequential PRPGs require less hardware than parallel devices but compare unfavourably in terms of speed.

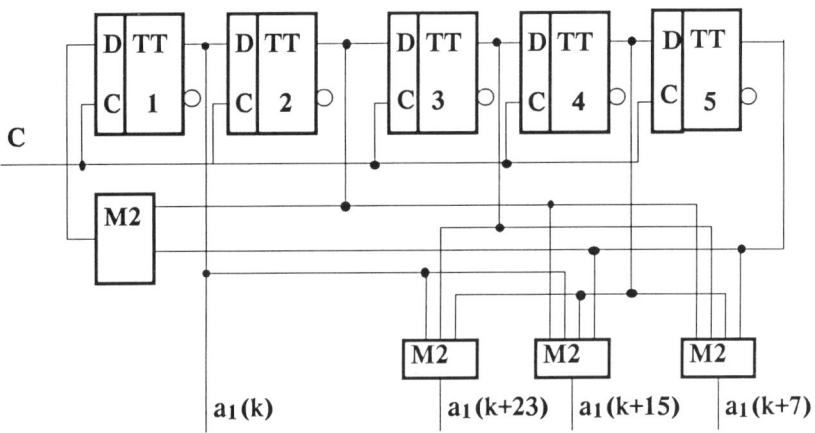

Fig. 4.13. Functional scheme of the PRPG

Chapter 5

PSEUDORANDOM AND RELATED SEQUENCE GENERATORS

5.1. Design of Pseudorandom Test Sequence Generator

<u>General.</u> The use of pseudorandom test sequences for VLSI self-testing imposes certain limitations on their properties. Thus a major requirement placed on a test sequence is maximum coverage of specified VLSI fault classes. Besides, the test sequence length is also essential for VLSI self-testing.

In the elementary case, a self-testing VLSI consists of a single shift register and a combinational portion. For such VLSI structures, the procedure of pseudorandom test sequence generator design consists in selecting the high degree of generating polynomial $\phi(x)$ such that inequality $deg\,\phi(x) > max\,n_i$ where max n_i is the maximum number of inputs to combinational subclass G_i connected to the shift register hold. This provides for the possibility to generate all possible test patterns at the inputs to any combinational subcircuit G_i. Provision for any test pattern at the inputs to combinational subcircuits means 100% coverage of all possible combinational faults at test pattern generation. Since the problem of generating the complete set of test codes is practically unsolvable, the VLSI designers are guided by the rule of "possible test coverage".

Real structures of self-testing VLSIs contain more than one shift register as an attempt to reduce testing time. In this case, the use of a PRPG may involve either structural or linear dependence of test patterns at the inputs to combinational circuits. To make clear the terms "structural" and "linear" dependencies, we shall consider a VLSI comprising three shift registers (in self-testing modes). The three VLSI registers are applied with test patterns from the outputs of PRPG described by polynomial $\phi(x) = 1 \oplus x^3 \oplus x^4$ (Fig.5.1). Each of the three registers RC1, RC2, and RC3 consist of three storage elements which outputs are connected to the inputs of combinational subcircuits G_1 and G_2. Examination of VLSI self-testing procedure reveals that all the three inputs to subcircuit G_1 will be applied with the same symbol from a M-sequence produced by the PRPG (Fig.5.1). In fact either *000* or *111* will be produced at G_1 inputs. In this case we say that structural dependence exists. For a combinational subcircuit G_2, the linear dependence condition holds. Actually, by using the add-and-shift property of a M-sequence, we can demonstrate that condition $z_1 \oplus z_2 \oplus z_3 = 0$ holds for the values z_1, z_2 and z_3 produced at G_2 inputs. Hence G_2 will be only testable for input values z_1, z_2 and z_3 that meet linear relation.

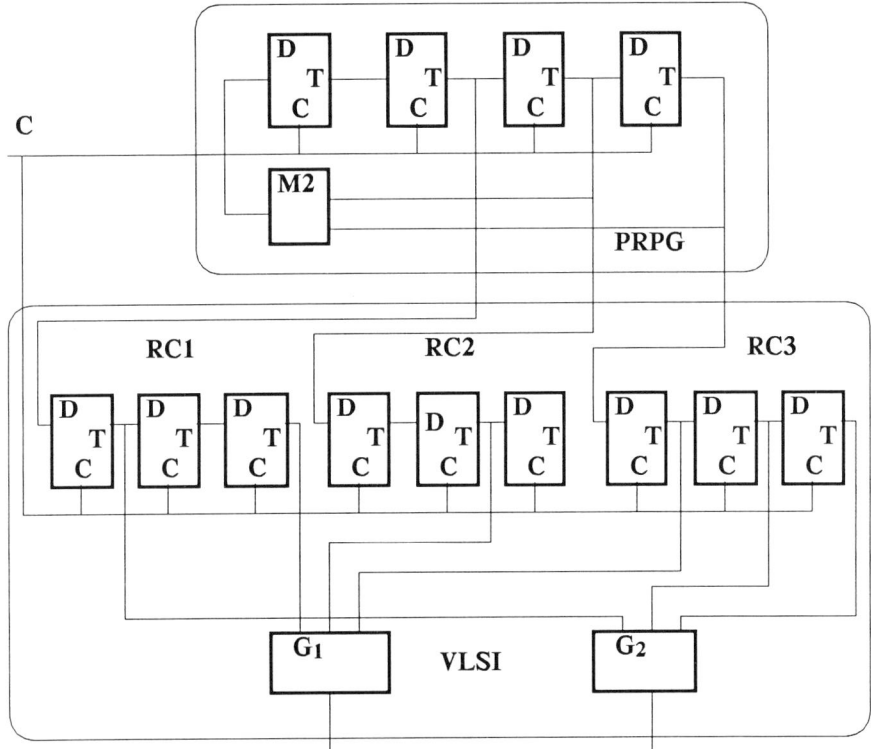

Fig. 5.1. Self-testing VLSI

The above examples suggest that linear and structural dependencies affect self-test efficiency. To avoid structural and linear dependencies at PRPG design is evidently necessary to generate test sequences with the required dimension of uniform distribution law.

Pseudorandom and Related Sequence Generators

PRPG Design. The procedure for designing PRPGs providing the required dimension of uniform distribution for test patterns and thereby full test coverage consist of the following steps.

1. For the specified VLSI, the maximum number of sequentially connected storage elements (the maximum shift register length in self-testing mode) T and the maximum number of inputs n (number of registers) are determined.

2. From the tables of primitive polynomials, a polynomial $\phi(x)$ is chosen that satisfies the condition $deg\phi(x) > (T+1)n$ and has the minimum number of non-zero coefficients $\alpha_i \in \{0,1\}$, $i=\overline{1,m}$.

3. Using the values of $\{\alpha_i\}$ a matrix is constructed for generating the elements of matrix

$$V = \begin{vmatrix} \alpha_1 & \alpha_2 & \alpha_3 & \ldots & \alpha_{m-1} & \alpha_m \\ 1 & 0 & 0 & \ldots & 0 & 0 \\ 0 & 1 & 0 & \ldots & 0 & 0 \\ \ldots & \ldots & \ldots & \ldots & \ldots & \ldots \\ 0 & 0 & 0 & \ldots & 1 & 0 \end{vmatrix},$$

4. From the relation (4.5) for $S=n$ the set of logic equations is written to associate the states of storage elements in the $(k+n)$th generator cycle with those in the kth cycle.

5. Based on the resulting set of logic equations

$$A(k+n) = V^n A(k),$$

the functional diagram of the PRPG is constructed.

By successively executing the steps of the above algorithm for PRPG design, we can construct a test sequence generator that implements a n-bit shift. For the purpose, steps 3 and 4 can be executed by n-fold application of the equation

$$a_1(k+1) = \sum_{i=1}^{m} {}^{\oplus} \alpha_i a_i(k),$$
$$a_j(k+1) = a_{j-1}(k), \quad j=\overline{2,m} \quad k=0,1,2,\ldots$$

The resulting set of logical equations

$$a_i(k+4) = F_i\ [a_1(k), a_2(k), a_3(k), ..., a_m(k)\ ], \quad i = \overline{1,m},$$

defines the functional connections for PRPG.

Consider the use of the above PRPG design procedure for the VLSI of Fig.5.1.

1. The number T of bits in the VLSI shift registers is 3 and their total number $n=3$.

2. Since $deg\ \phi(x) > (3+1)3 = 12$, we chose a polynomial $\phi(x) = x^{12} \oplus x^7 \oplus x^4 \oplus x^3 \oplus 1$ which has the minimum number of non-zero coefficients for $m=12$.

3. By using the values of coefficients $\alpha_1 = \alpha_2 = \alpha_5 = \alpha_6 = \alpha_8 = \alpha_9 = \alpha_{10} = \alpha_{11} = 0$ and $\alpha_3 = \alpha_4 = \alpha_7 = \alpha_{12} = 1$, a matrix

$$V = \begin{vmatrix} 0 & 0 & 1 & 1 & 0 & 0 & 1 & 0 & 0 & 0 & 0 & 1 \\ 1 & 0 & 0 & 0 & 0 & 0 & 0 & 0 & 0 & 0 & 0 & 0 \\ 0 & 1 & 0 & 0 & 0 & 0 & 0 & 0 & 0 & 0 & 0 & 0 \\ 0 & 0 & 1 & 0 & 0 & 0 & 0 & 0 & 0 & 0 & 0 & 0 \\ 0 & 0 & 0 & 1 & 0 & 0 & 0 & 0 & 0 & 0 & 0 & 0 \\ 0 & 0 & 0 & 0 & 1 & 0 & 0 & 0 & 0 & 0 & 0 & 0 \\ 0 & 0 & 0 & 0 & 0 & 1 & 0 & 0 & 0 & 0 & 0 & 0 \\ 0 & 0 & 0 & 0 & 0 & 0 & 1 & 0 & 0 & 0 & 0 & 0 \\ 0 & 0 & 0 & 0 & 0 & 0 & 0 & 1 & 0 & 0 & 0 & 0 \\ 0 & 0 & 0 & 0 & 0 & 0 & 0 & 0 & 1 & 0 & 0 & 0 \\ 0 & 0 & 0 & 0 & 0 & 0 & 0 & 0 & 0 & 1 & 0 & 0 \\ 0 & 0 & 0 & 0 & 0 & 0 & 0 & 0 & 0 & 0 & 1 & 0 \end{vmatrix}$$

is constructed.

The matrix, when used, results in:

$$V^3 = \begin{vmatrix} 1 & 1 & 0 & 0 & 1 & 0 & 0 & 0 & 0 & 1 & 0 & 0 \\ 0 & 1 & 1 & 0 & 0 & 1 & 0 & 0 & 0 & 0 & 1 & 0 \\ 0 & 0 & 1 & 1 & 0 & 0 & 1 & 0 & 0 & 0 & 0 & 1 \\ 1 & 0 & 0 & 0 & 0 & 0 & 0 & 0 & 0 & 0 & 0 & 0 \\ 0 & 1 & 0 & 0 & 0 & 0 & 0 & 0 & 0 & 0 & 0 & 0 \\ 0 & 0 & 1 & 0 & 0 & 0 & 0 & 0 & 0 & 0 & 0 & 0 \\ 0 & 0 & 0 & 1 & 0 & 0 & 0 & 0 & 0 & 0 & 0 & 0 \\ 0 & 0 & 0 & 0 & 1 & 0 & 0 & 0 & 0 & 0 & 0 & 0 \\ 0 & 0 & 0 & 0 & 0 & 1 & 0 & 0 & 0 & 0 & 0 & 0 \\ 0 & 0 & 0 & 0 & 0 & 0 & 1 & 0 & 0 & 0 & 0 & 0 \\ 0 & 0 & 0 & 0 & 0 & 0 & 0 & 1 & 0 & 0 & 0 & 0 \\ 0 & 0 & 0 & 0 & 0 & 0 & 0 & 0 & 1 & 0 & 0 & 0 \end{vmatrix}.$$

4. By substituting the resulting matrix into equation (4.5), we obtain the set of logic equations

$$A(k+4) = V^4 A(k).$$

In view of the single-cycle principle of PRPG operation, we may transform it to

$$\begin{aligned}
a_1(k+1) &= a_1(k) \oplus a_2(k) \oplus a_5(k) \oplus a_{10}(k), \\
a_2(k+1) &= a_2(k) \oplus a_3(k) \oplus a_6(k) \oplus a_{11}(k), \\
a_3(k+1) &= a_3(k) \oplus a_4(k) \oplus a_7(k) \oplus a_{12}(k), \\
a_4(k+1) &= a_1(k), \quad a_5(k) = a_2(k), \\
a_6(k+1) &= a_3(k), \quad a_7(k) = a_4(k), \\
a_8(k+1) &= a_5(k), \quad a_9(k) = a_6(k), \\
a_{10}(k+1) &= a_7(k), \quad a_{11}(k) = a_8(k), \\
a_{12}(k+1) &= a_9(k).
\end{aligned}$$

Fig.5.2 shows the functional diagram of the PRPG built in accordance with the above set of equations.

By using the values of any three successive PRPG bits as a test pattern, we can demonstrate that all possible combinations and their sequences are generated in three successive three-bit patterns.

5.2. Generators of Test Sequences Related to Pseudorandom Sequences

General. The preceding subsections of the present Section discussed the properties and generation techniques of only M-sequences. Prominence given to M-sequences can be explained by their wide use in various scientific and engineering applications. However, there are many sequences, including binary, that are characterized by properties close to those of M-sequences, which make hardware implementation of PRPG rather simple. Consider some classes of such sequences. From the set of binary sequences of period $L=2^m-1$, we shall separate out a subset of sequences which satisfies the equiprobability of 1s and 0s. In the repetition period such sequences contain $(L+1)/2$ ones and $(L-1)/2$ zeros. How in the set of sequences U_0 of period L we shall separate out a subset

70   Chapter 5

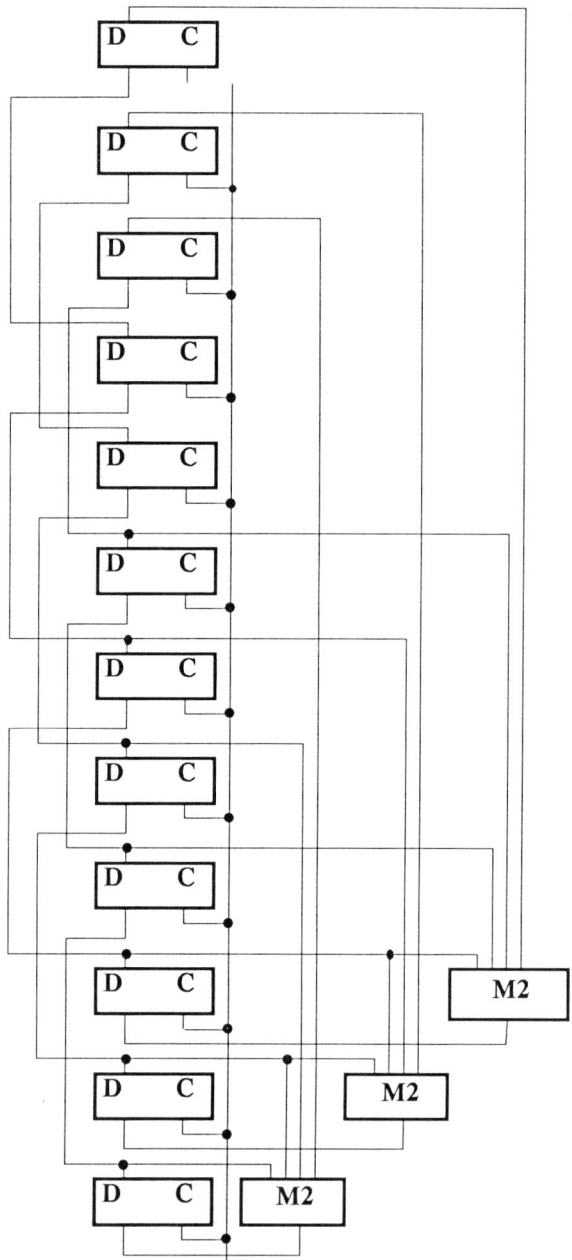

Fig. 5.2. Functional diagram of the PRPG

of sequences U_1 that are characterized by run properties. The run property consists in the fact that for period $L=2^m-1$ there exist 2^{m-2-i} runs of i ones or zeros and a single run of $(m-1)$ zeros or m ones. In its turn, subset U_2 of sequences includes sequences, all of which have 2^m-1 possible segments of m length, except for a sole m-zero segment. Thus each segment occurs only once in a repetition period. From the set of sequences U_0, subset U_3 can be separated out also for which the property of decimation by 2 holds. Subset U_3 comprises sequences U_4 that meet the property of non-correlation between cyclic shifts of the sequence over a repetition interval. With the shift value $\tau \neq 0$, the autocorrelation function value for such sequences is fixed and close to 0.

A comparison between the properties of M-sequences (refer to subsection 4.1) and those associated with U_0, U_1, U_2, U_3, and U_4 shows that the former only form subset U_5 belonging to the intersection of U_2 and U_4, as follows from the chart representing the relationship between pseudorandom sequence subsets (Fig.5.3). A class of sequences that do not belong to the subset, but are close in their properties, has been called related. Of these sequences, the most feasible are those attributed to Ford, de Bruijn, Gold and others.

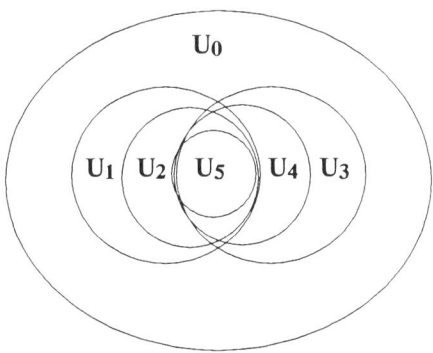

Fig. 5.3. Sequence subsets

Ford sequences. These sequences can be obtained by adding a zero to a sequence from subset U_2 such that a run of m zeros will be formed over a period of its repetition. They have a period of 2^m and are characterized by the presence of various m-length codes in the repetition period, each occurring only once. It was the latter property that attracted particular interest to the use of Ford sequences for VLSI off-line testing.

The Ford sequences have a relatively simple algorithm for their generation which is recurrent in nature. The algorithm generates any successive m-bit code $X_k(b_2,b_3,b_4,...,b_{m+1})$ on the basis of the preceding code $X_k(b_1,b_2,b_3,...,b_m)$. The value of $b_{m+1} \in \{0,1\}$ is determined as follows. A code of the form $X'_{k-1}=(b_1,b_2,b_3,...,b_m,1)$ is generated and all its cycle shifts are obtained, among which code $X''_{k-1}=(b_i,b_{i+1},...,b_m,1,b_2,...,b_{i-1})$ which is the largest m-bit number is chosen. If $b_2=...=b_{i-1}=0$ the value b_{m+1} equals $b_1 \oplus 1$, otherwise

$b_{m+1}=b_1$. Note that $X_0=(0,0,0,...,0)$ has been chosen as a starting code for sequence generation.

By way of example Table 5.1 gives generation of the Ford sequence for $n=4$.

Table 5.1

k	$X_k=(b_1,b_2,...,b_m)$	$X''_{k-1}=(b_i,...,1,b_2,...,b_{i-1})$	b_m
0	0 0 0 0	1 0 0 0	
1	0 0 0 1	1 1 0 0	$b_1 \oplus 1 = 0 \oplus 1 = 1$
2	0 0 1 1	1 1 1 0	$b_1 \oplus 1 = 0 \oplus 1 = 1$
3	0 1 1 1	1 1 1 1	$b_1 \oplus 1 = 0 \oplus 1 = 1$
4	1 1 1 1	1 1 1 1	$b_1 \oplus 1 = 0 \oplus 1 = 1$
5	1 1 1 0	1 1 1 0	$b_1 \oplus 1 = 1 \oplus 1 = 0$
6	1 1 0 1	1 1 1 0	$b_1 = 1$
7	1 0 1 1	1 1 1 0	$b_1 = 1$
8	0 1 1 0	1 1 1 0	$b_1 \oplus 1 = 1 \oplus 1 = 0$
9	1 1 0 0	1 1 0 0	$b_1 = 0$
10	1 0 0 1	1 1 0 0	$b_1 = 1$
11	0 0 1 0	1 0 1 0	$b_1 \oplus 1 = 1 \oplus 1 = 0$
12	0 1 0 1	1 1 1 0	$b_1 \oplus 1 = 0 \oplus 1 = 1$
13	1 0 1 0	1 0 1 0	$b_1 = 0$
14	0 1 0 0	1 1 0 0	$b_1 \oplus 1 = 1 \oplus 1 = 0$
15	1 0 0 0	1 0 0 0	$b_1 = 0$
16	0 0 0 0	1 0 0 0	$b_1 \oplus 1 = 1 \oplus 1 = 0$

De Bruijn sequences. These sequences can be obtained by completing an M-sequence with a zero pattern by using the *000...01* decoder. Considering the fact that the pattern of *(m-1)* zeros occurs only once, we can use *(m-1)*-input OR NOT gate for decoding the *000...01* combination. Analytical description of a de Bruijn sequence generator based on the decoder appears as:

Pseudorandom and Related Sequence Generators 73

$$a_1(k+1) = \sum_{i=1}^{m} {}^{\oplus} \alpha_i a_i(k) \oplus \bigvee_{l=1}^{m-1} \overline{a_l(k)}$$

$$a_j(k+1) = a_{j-1}(k), \quad i = \overline{2,m}, \quad k = 0,1,2,\ldots \ . \tag{5.1}$$

where $\alpha_i \in \{0,1\}$, $n=\overline{1,m}$, are constant coefficients defined by a generating polynomial $\phi(x)$; $a_1(k+1) \in \{0,1\}$ is a de Bruijn sequence element treated as the first storage elements of its generator in the $(k+1)$th cycle of its operation.

As an example we consider the case for the generating polynomial $\phi(x)=1\oplus x^2 \oplus x^5$. Then the set of logical equations that describe the sequence generator takes the form:

$$a_1(k+1) = a_2(k) \oplus a_5(k) \oplus \overline{[a_1(k) + a_2(k) + a_3(k) + a_4(k)]} \ ,$$
$$a_2(k+1) = a_1(k), \quad a_3(k+1) = a_2(k),$$
$$a_4(k+1) = a_3(k), \quad a_5(k+1) = a_4(k).$$

Fig.5.4 shows the block-diagram of de Bruijn sequence generator according to the above set of equations. Examination of the block-diagram reveals simplicity of its hardware implementation and the possibility of generating the entire set of codes.

<u>Cellular automata for pseudorandom sequence generation.</u> Regular structures represented by arrays of cellular automata can be used to implement pseudorandom test sequences for VLSI. The distinctive feature of a cellular automaton is the dependence between the current automaton state and its previous state and the previous states of the neighbouring automatous. For a one-dimensional cellular array a new value of its ith element is determined by the equation

$$a_i(k+1) = f\ [a_{i-1}(k), a_i(k), a_{i+1}(k)\ ] \tag{5.2}$$

For the cyclic boundary conditions of a one-dimensional cellular array the equation (5.2) holds for any array element, for zero boundary conditions the relations

$$a_1(k+1) = f\ [0,a_1(k),a_2(k)\ ], \quad a_m(k+1) = f\ [a_{m-1}(k),a_m(k),0\ ].$$

are used for the first and last elements in the array.

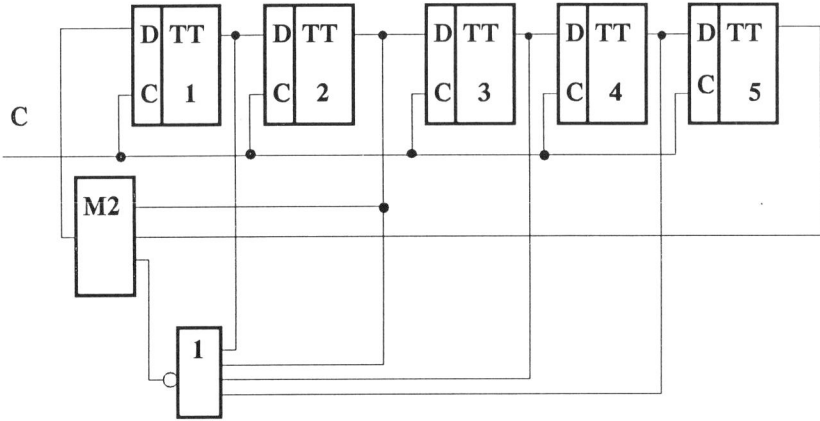

Fig. 5.4. De Bruijn sequence generator

We shall discuss here only one-dimensional arrays of binary cellular automata. However, the array can generally have higher dimensionality and the automata can process r nary data $(r>2)$. For a binary automaton, the current state depends on eight possible combinations *000, 001, ..., 111* of the previous states of the original automaton and its nearest neighbors. Such dependence is called the cellular automaton rule. For example, the value $a_i(k+1)$ can be represented as

$$a_i(k+1) = a_{i-1}(k) \oplus a_i(k) \oplus a_{i+1}(k) \qquad (5.3)$$

Depending on the rules of binary cellular automata array, their states form the sequences with distinct properties. Among them we can distinguish a subclass of stochastic sequences. Thus for a rule described by

$$a_i(k+1) = a_{i-1}(k) \oplus [a_i(k) \cup a_{i+1}(k)]$$

and with boundary conditions being cyclic, a stochastic sequence that satisfies some randomness factors is formed. Based on the rule, we can design built-in test generators (CALBO) for self-testing VLSIs. By using certain combinations of the rules, we can create built-in test generators for self-testing VLSI that offer properties close or similar to those of M-sequences. Thus by using the combination of rules (5.3) and the rule

$$a_i(k+1) = a_{i-1}(k) \oplus a_{i+1}(k) \quad (5.4)$$

with the zero boundary conditions we can build maximal-length sequence generators. The order of applying the said rules to specific cellular automata of a one-dimensional array for different values of m (m is the number of automata in an array) is defined by Table 5.2.

Thus for $m=5$ the code *11001* (refer to Table 5.2) means that the rule (5.3) is to be applied for the first, second, and fifth automaton in the array, whereas the rule (5.4) is to be applied to the third and fourth automata. Therefore a one in the code of Table 5.2 corresponds to the rule (5.3), whereas a zero to the rule (5.4).

By using Table 5.2 we can create a test sequence generator for $m=4$. Considering the zero boundary conditions we obtain

$$\begin{aligned}
a_1(k+1) &= 0 \oplus a_2(k) = a_2(k), \\
a_2(k+1) &= a_1(k) \oplus a_2(k) \oplus a_3(k), \\
a_3(k+1) &= a_2(k) \oplus a_4(k), \\
a_4(k+1) &= a_3(k) \oplus a_4(k) \oplus 0 = a_3(k) \oplus a_4(k).
\end{aligned} \quad (5.5)$$

Fig.5.5 shows the block-diagram of a test generator according to the set of equations (5.5) where each automaton is associated with the elements required for implementing the equation.

<u>Multilevel sequences of maximal length.</u> These sequences may also be classified among related sequences. In the general case, an r-level M-sequence has a period $L=r^m-1$, where m is the degree of primitive generating polynomial $\phi(x)$ over field GF(r).

The hardware implementation of an r-level M-sequence generator is an m-bit shift register whose positions contain the elements of GF(r) and whose feedback connections correspond to the form of generating polynomial $\phi(x)$.

Fig.5.6 shows the block-diagram of a three-level M-sequence generator. The sequence elements are *-1, 0* and *+1*, and a successive a_i value is determined from the recurrence relation

$$a_i = (-1 \times a_{i-1}) + (+1 \times a_{i-2}) \ [GF(3)] \qquad (5.6)$$

Table 5.2

m	Rules application order
4	0101
5	11001
6	010101
7	1101010
8	11010101
9	110010101
10	0101010101
11	11010101010
12	010101010101
13	1100101010100
14	01111101111110
15	100100010100001
16	1101010101010101
17	01111101111110011
18	010101010101010101
19	0110100110110001001
20	11110011101101111111
21	011110011000001111011
22	0101010101010101010101
23	11010111001110100011010
24	111111010010110101010110
25	1011110101010100111100100
26	01011010110100010111011000

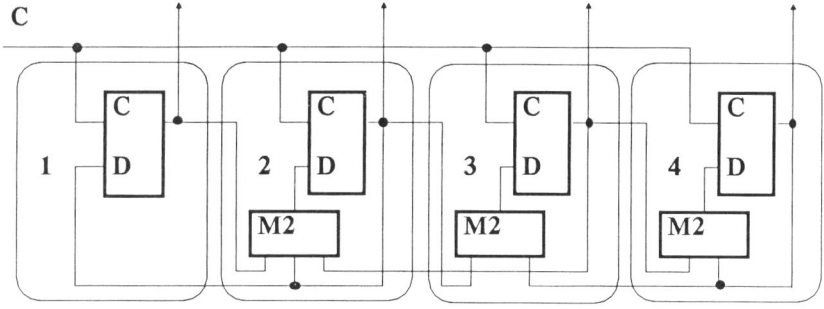

Fig. 5.5. The test generator

Table 5.3 shows the truth table for add and multiply operation over field GF(3) used in relation (5.6).

Table 5.3

a	b	S	Q
+1	+1	-1	+1
+1	0	+1	0
+1	-1	0	-1
0	+1	+1	0
0	0	0	0
0	-1	-1	0
-1	+1	0	-1
-1	0	-1	0
-1	-1	+1	+1

In the three-level M-sequence that is being generated there are 3^m-1 occurrences of +1s and -1s each and $3^{m-1}-1$ occurrences of zeros over a repetition period. Any of the possible m-bit combinations of elements 0, +1, and -1, except for an all-zero code, occurs only once over a period. The autocorrelation function of the sequence is determined by:

$$K(\tau) = \frac{1}{3^m - 1} \sum_{i=1}^{3^m-1} a_i a_{i-\tau}$$

$$= \begin{cases} \dfrac{2 \cdot 3^{m-1}}{3^m - 1} & \text{for } \tau = 0, L, 2L, \ldots, \\ -\dfrac{2 \cdot 3^{m-1}}{3^m - 1} & \text{for } \tau = \dfrac{L}{2}, \dfrac{3L}{2}, \dfrac{5L}{2}, \ldots, \\ 0 & \text{for } \textit{other value of } \tau \end{cases}$$

where $L=3^m-1$.

Regardless of the advantages of the use of multilevel M-sequence in the problems of linear/nonlinear system identification, they have not received wide acceptance due to the complexity of implementing such sequence generators.

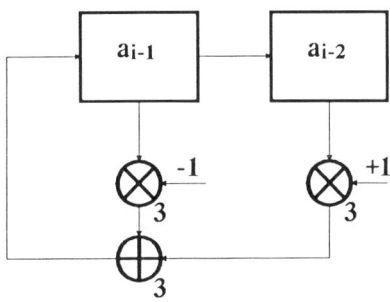

Fig. 5.6. M-sequence generator

5.3. Generation of Weighted Pseudorandom Test Sequences

General. The above discussed techniques for pseudorandom test pattern generation make it possible to produce uniformly distributed number sequences $A=a_1 a_2 a_3 \ldots a_m$ where $p(a_i=1)=p(a_i=0)=0.5$, which are used in various test types. The efficiency of testing, probabilistic testing in particular, depends on specific values of $p(a_i=1)$ and hence $p(a_i=0)=1-p(a_i=1)$. Therefor the problem of generating weighted pseudorandom test sequences arises. Let us consider the best known techniques for generating such sequences.

<u>Weighted sequences resulting from linear transformation.</u> A technique for obtaining weighted pseudorandom (random) sequences z_1 of binary characters with the probability $p(z_1=1)$ that a unity proportional to the code $B=b_1b_2b_3...b_m$ will appear is the one comparing the number B against a sequence of pseudorandom (random) uniformly distributed binary numbers $A=a_1a_2a_3...a_m$. The technique is illustrated by the block-diagram of Fig.5.7.

As may be seen from Fig.5.7, a pseudorandom binary number $A=a_1a_2a_3...a_m$ produced by a uniformly distributed pseudorandom number generator is compared against the code of deterministic number B in each cycle. With $A<B$ a 1 appears at the comparison circuit output, otherwise, a 0. Since the occurrence of any binary combination A is equiprobable, a unity value of z_1 will appear at the circuit output B times of 2^m on the average. Therefore the probability $p(z_1=1)$ is determined by the relation $B/2^m$ which will finally appear as

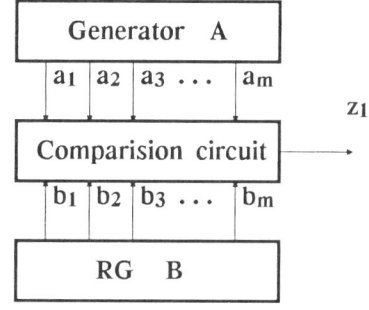

Fig. 5.7. The block-diagram

$$p(z_1 = 1) = \frac{B}{2^m} = B' = 0,b_1b_2b_3...b_m.$$

<u>Boolean functions of weighted sequence generators.</u> It has been known that the transformer producing a sequence z_2 with probability $p(z_2=1)$ which is proportional to a m-bit binary number B' can be constructed of m equiprobable binary number generators and a combinational circuit that implements a Boolean function

$$\begin{aligned}z_2 = &a_1b_1 + a_2b_2(a_1 + b_1) + a_3b_3(a_1 + b_1)(a_2 + b_2) + ... \\ &+ a_mb_m(a_1 + b_1)(a_2 + b_2)...(a_{m-1} + b_{m-1})\end{aligned} \quad (5.7)$$

where Boolean variables $a_i \in \{0,1\}$, $i=\overline{1,m}$ become one with the probability 0.5.

In the specific case of $B'=0,10101=21/32$, the expression (5.7) may be rearranged as

$$z_2 = a_1 + a_3 a_2 + a_5 a_2 a_4 = a_1 + a_2(a_3 + a_5 a_4).$$

Taking account of the fact that variables a_i and a_j, $i \neq j \in \{1,2,3,...,m\}$ are statistically independent and by using the elementary relations of the form

$$p(a_i a_j) = p(a_i)p(a_j),$$
$$p(a_i + a_j) = 1 - [1 - p(a_i)][1 - p(a_j)] = p(a_i) + p(a_j) - p(a_i)p(a_j),$$

we obtain $p(a_5 a_4) = (1/2)(1/2) = 1/4$. Then $p[a_3+(a_5 a_4)] = 1/2 + 1/4 - (1/2)(1/4) = 5/8$, $p[a_2(a_3+a_5 a_4)] = (1/2)(5/8) = 5/16$. And, finally, $p(z_2=1) = p[a_1+a_2(a_3+a_5 a_4)] = 1/2 + 5/16 - (1/2)(5/16) = 1/2 + 5/16 - 5/32 = 21/32$.

We can similarly demonstrate that the Boolean function also

$$z_3 = \overline{a_1}b_1 + \overline{a_2}b_2(\overline{a_1} + b_1) + \overline{a_3}b_3(\overline{a_1} + b_1)(\overline{a_2} + b_2) + ... \\ + \overline{a_m}b_m(\overline{a_1} + b_1)(\overline{a_2} + b_2) ... (\overline{a_{m-1}} + b_{m-1}) \quad (5.8)$$

allows to design a weighted pseudorandom test pattern generator. Actually, for the earlier discussed example with $B'=0, b_1 b_2 b_3 b_4 b_5 = 0,10101$, the relation (5.7) will take the form

$$z_3 = \overline{a_1} + \overline{a_3}\,\overline{a_2} + \overline{a_5}\,\overline{a_2}\,\overline{a_4},$$

for which we obtain $p(z_3=1)=21/32$ since $p(a_i=1)=p(a_i=0)=0.5$.

Therefore we may conclude that both relations (5.7) and (5.8) are essentially identical in terms of their final result and their implementing schemes belong to the class of weighted pseudorandom test pattern generator schemes. It should be noted that generation of pseudorandom or random patterns depends on the nature of generators producing equiprobable binary numbers $a_i \in \{0,1\}$, $i=\overline{1,m}$. Their randomness or pseudorandomness determines the form of weighted test patterns.

Therefore an entire family of Boolean functions can be used to design a weighted pseudorandom test pattern generator. An estimate of the number of such functions with the length of codes A and B equal to m takes the form

$$Q = \prod_{i=1}^{2^m-1} C_{2^m}^1.$$

From among the variety of existing function types most interesting are those allowing to create weighted test generators at minimum hardware requirements.

Among the variety of Boolean expressions in the form of a conjunction disjunction, the minimum DNF of all possible functions that describe a weighted pattern generator has the form where

$$\begin{aligned} z_{min} = a_1^{1-\alpha_1} b_1 + a_1^{\alpha_1} a_2^{1-\alpha_2} b_2 + \ldots + a_1^{\alpha_1} a_2^{\alpha_2} \ldots \\ a_{m-1}^{\alpha_{m-1}} a_m^{1-\alpha_m} b_m, \end{aligned} \quad (5.9)$$

$\alpha_i \in \{0,1\}$, $i=\overline{1,m}$, and

$$a_i^{\alpha_i} = \begin{cases} a_i & \text{for } \alpha_i = 1, \\ \overline{a_i} & \text{for } \alpha_i = 0. \end{cases}$$

The Boolean function z_{min} is represented in the orthogonal disjunctive normal form thereby greatly simplifying calculation of $p(z_{min}=1)$ due to non-concurrency of events described by the terms of (5.9). Actually, based on (5.9) we shall have for the case of $B'=0,10101$.

Whence $p(z_{min}=1) = 1/2 + (1/2)(1/2)(1/2) + (1/2)(1/2)(1/2)(1/2)(1/2) = 1/2 + 1/8 + 1/32 = 21/32$. The minimum DNF z_{min} for the functions that implement a weighted pattern generator does not produce absolutely minimum expression. The expression is obtained by collecting like terms in expression (5.9) and going to the parenthesis form. As a result we obtain

$$\begin{aligned} z'_{min} = a_1^{1-\alpha_1} b_1 + a_1^{\alpha_1}(a_2^{1-\alpha_2} b_2 + \ldots + a_{m-2}^{\alpha_{m-2}}(a_{m-1}^{\alpha_{m-1}} b_{m-1} + \\ + a_{m-1}^{\alpha_{m-1}} a_m^{1-\alpha_m} b_m)\ldots), \end{aligned} \quad (5.10)$$

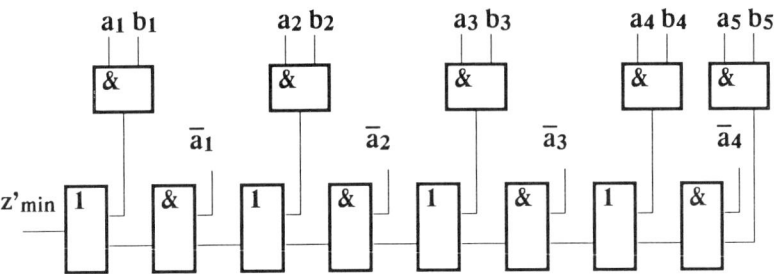

Fig. 5.8. The weighted test pattern generator

The scheme of a weighted test pattern generator constructed of AND and OR gates by formula (5.10) with the gate inputs being applied both with variables a_i, $i=\overline{1,m}$, and their negations is the minimum one among the schemes that implement the functions of class in question. The complexity of such scheme is determined by the number of two-input AND and OR gates whose total number is *3n-2*.

The functional block-diagram of a weighted test pattern generator for *m=5* is given in Fig.5.8. The diagram is associated with a family of Boolean functions that differ in the arrangement of a_i negations.

Chapter 6

RANDOM TESTING

6.1. Concept of Random Testing

A major feature of random testing is the use of random sequences of independent bits applied to the inputs of a circuit under test. Here a variable $x_i \in \{0,1\}$, $i=\overline{1,n}$, applied to the i th input is described by probability $P(x_i=1)$ of its value to be 1.

The classical random testing network consists of a random bit sequence generator, the circuit under test and a reference circuit as well as an output response comparator. The bit sequence generator is used to produce independent random numbers by the specified probability distribution $P(x_i=1)$. Binary patterns produced at the generator output are applied simultaneously to the inputs of the circuit under test and the reference circuit. Output responses of these circuits are compared in the output response comparator. When the output responses from both circuits are identical, the circuit under test is considered to be faultfree; otherwise it is assumed to be faulty and the problem of diagnosing it arises.

Publications on random testing deal with finding the reference values for the probability of 1s to appear in the intermediate and output nodes of the circuit. By the proposed techniques, the reference probability values that have been obtained analytically were compared with the actual values. Thus, the random testing procedure consists of obtaining actual probability values and their comparison with reference values. The mentioned techniques have not found wide application due to the necessity of calculating reference probability values for a 1 to appear in the intermediate and output nodes of the circuit.

Consider the relationship between output probabilities for simple logic devices and probabilities $P(x_i=1)$, $i=\overline{1,n}$, for their input variables $x_1, x_2, ..., x_n$. Considering $x_i \in \{0,1\}$, we obtain

$$P(x_i = 1) = 1 - P(x_i = 0). \qquad (6.1)$$

Then examine a NOT gate whose input variable x_i uniquely determines the output value y, where $y = \overline{x_1}$. Since the probability $P(y=1) = P(x_1=1)$, then, subject to (6.1), we obtain

$$P(y=1) = P(x_1=0) = 1 - P(x_1=1) \qquad (6.2)$$

To examine the probability relation $P(y=1)$ of the AND gate, apply sequences of random binary numbers that are distinct magnitudes to its inputs. The output value of the AND gate is 1 if and only if all input variables are 1. The probability of the event is determined as

$$P(y=1) = P(x_1=1, x_2=1, \ldots, x_n=1)$$

where $y = x_1 x_2 \ldots x_n$. Considering the events $x_1=1$, $x_2=1$, ..., $x_n=1$ to be independent, the latter relationship can be rearranged as

$$P(y=1) = \prod_{i=1}^{n} P(x_i=1). \qquad (6.3)$$

For a two-input AND gate, $P(y=1) = P(x_1=1)P(x_2=1)$. Next consider a multi-input OR gate whose output value can be obtained by the expression

$$y = x_1 + x_2 + \ldots + x_n = \overline{\overline{x_1 + x_2 + \ldots + x_n}} = \overline{\overline{x_1}\,\overline{x_2}\ldots\overline{x_n}}.$$

Applying (6.2) we obtain

$$P(y=1) = P(\overline{\overline{x_1}\,\overline{x_2}\ldots\overline{x_n}} = 1) = 1 - P(\overline{x_1}\,\overline{x_2}\ldots\overline{x_n} = 1).$$

Taking into account that the events $x_1=1$, $x_2=1$, ..., $x_n=1$ are independent and applying (6.3) we can finally obtain

$$P(y=1) = 1 - \prod_{i=1}^{n} P(\overline{x_i} = 1) = 1 - \prod_{i=1}^{n} [1 - P(x_i = 1)]. \quad (6.4)$$

From equation (6.4)

$$P(y=1) = P(x_1 = 1) + P(x_2 = 1) - P(x_1 = 1) P(x_2 = 1)$$

for the two-point OR gate.

When the events $x_i=1$ and $x_j=1$ are not simultaneous, with $i \neq j$, relation (6.4) is rearranged as

$$P(y=1) = \sum_{i=1}^{n} P(x_i = 1). \quad (6.5)$$

A more complete list of relationships between output probabilities of gates as probability functions of output variables and their correlation functions is given in some books. In these relationships the expressions for AND and OR gates can assume quite different form those of (6.3) and (6.4). Thus, for example, for a two- input AND gate with x_1 applied to both its inputs, the output probability $P(y=1)$ is given by

$$P(y=1) = P(x_1 x_1 = 1) = P(x_1 = 1).$$

We can similarly show that

$$P(x_1 \overline{x_1} = 1) = 0,$$
$$P(x_1 + \overline{x_1} = 1) = 1,$$
$$P(x_1 + x_1 = 1) = P(x_1 = 1).$$

However, in the most simple case of a combinational tree, relations (6.2)-(6.4) will be sufficient for finding probability values for intermediate and output nodes of the digital circuit. As an example, consider the circuit in Fig. 6.1 whose inputs are fed with the sequences of independent random bits with probabilities $P(x_1=1) = P(x_2=1) = P(x_3=1) = P(x_4=1) = P(x_5=1) = p$. By applying relations (6.2)-(6.4), we obtain $P(f_1=1)=p^3$ and $P(f_2=1)=2p-p^2$ respectively. For the output value of the circuit, we finally obtain

$$P(f_3=1) = P(f_1=1)P(f_2=1) = p^3(2p-p^2) = 2p^4 - p^5.$$

With $p=1/2$, the reference probability values take the form

$$P(f_1=1) = 0.125, \quad P(f_2=1) = 0.75, \quad P(f_3=1) = 0.08375.$$

When random testing is realized for the circuit under test, the actual probability values $P(f_1=1)$, $P(f_2=1)$ and $P(f_3=1)$ are compared with the reference values and then a decision on the circuit state is made. For a faulty condition, the actual probability values normally differ from their reference values. Thus, for example, for the fault $f_1 \equiv 1$, we obtain $P^*(f_1=1) = 1$, $P^*(f_2=1) = 0.75$, $P^*(f_3=1) = 0.75$ indicating that the circuit is faulty.

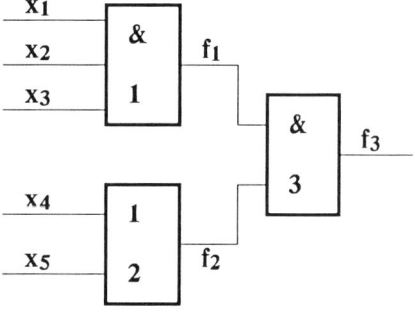

Fig. 6.1. A combinational circuit

In the general case, estimation of reference probabilities is a complex analytical task, especially for digital circuits of arbitrary form. Here special techniques are required. Let us discuss some of them.

6.2. Probabilistic Analysis Techniques for Digital Circuits

Two algorithms discussed in literature were among the first probabilistic analysis techniques for digital circuits. They make use of earlier relations (6.2)-(6.4) and are applied for combinational networks. The first algorithm has the following properties.

(i) A Boolean function that describes any node in a combinational network can be represented in the sum-of-products form of input variables and their negations.

(ii) A Boolean function that has been expressed in a sum-of-products form can be represented as a canonical sum of products:

$$f_i f_j = 0, \quad i \neq j.$$

A straightforward algorithm for finding the probability of a *1* signal to appear in any node of the network consists of the following steps.

(i) Find the analytical expression for the specified node of the network as a canonical sum of products

$$F = f_1 + f_2 + \ldots + f_k.$$

(ii) Estimate $P(f_i=1)$ for any product f_i by using relations (6.2) and (6.3).

(iii) Since the events $f_i=1$ and $f_j=1$, $i \neq j$, are not concurrent, estimate $P(F=1)$ by equation (6.5).

As an example of the algorithm discussed, let us find $P(F=1)$, where $F=(x_1+x_2)x_3+x_1x_2$, for the network of Fig.6.2.

(i) The Boolean function $F=(x_1+x_2)x_3+x_1x_2$ is represented as a canonical sum of products

$$F = x_1 x_2 x_3 + \overline{x_1} x_2 x_3 + x_1 \overline{x_2} x_3 + x_1 x_2 \overline{x_3}.$$

(ii) Find sequentially the values $P(x_1 x_2 x_3=1)$, $P(\overline{x}_1 x_2 x_3=1)$, $P(x_1 \overline{x}_2 x_3=1)$ and $P(x_1 x_2 \overline{x}_3=1)$. Then

$$\begin{aligned}
P(x_1 x_2 x_3 = 1) &= P(x_1 = 1) P(x_2 = 1) P(x_3 = 1), \\
P(\overline{x_1} x_2 x_3 = 1) &= [1 - P(x_1 = 1)] P(x_2 = 1) P(x_3 = 1), \\
P(x_1 \overline{x_2} x_3 = 1) &= P(x_1 = 1) [1 - P(x_2 = 1)] P(x_3 = 1), \\
P(x_1 x_2 \overline{x_3} = 1) &= P(x_1 = 1) P(x_2 = 1) [1 - P(x_3 = 1)].
\end{aligned}$$

(iii) According to (6.5), we find

$$P(F=1) = P(x_1x_2x_3 = 1) + P(\overline{x_1}x_2x_3 = 1) + P(x_1\overline{x_2}x_3 = 1) +$$
$$+ P(x_1x_2\overline{x_3} = 1) = P(x_1 = 1)P(x_2 = 1) + P(x_1 = 1)P(x_3 = 1) +$$
$$+ P(x_2 = 1)P(x_3 = 1) - 2P(x_1 = 1)P(x_2 = 1)P(x_3 = 1).$$

For $P(x_1=1)=P(x_2=1)=P(x_3=1)=p$, we obtain

$$P(F=1) = 3p^2 - 2p^3.$$

A major disadvantage of the technique discussed is the complexity of representing the analytical expression as a canonical sum of products. The second algorithm that allows one to obtain the probability representation of a digital circuit on the basis of its functional diagram is easier to use. This algorithm relies on the probability property of dependent events. Let us consider the said property by using a five-input AND gate as an example. Suppose three of its primary inputs are applied with variable x_1 whose probability to be 1 is $P(x_1=1)$. The fourth and fifth primary inputs are applied with x_2 and x_3 whose probabilities are $P(x_2=1)$ and $P(x_3=1)$, respectively. According to (6.3), the probability $P(F=1)$ of a 1 to appear at the output of the five-input AND gate is defined by the expression

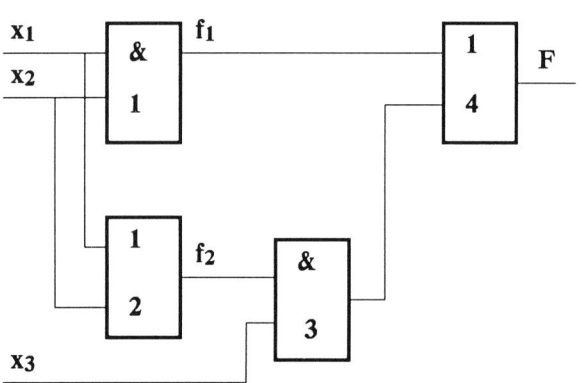

Fig. 6.2. A combinational circuit

$$P(F=1) = P^3(x_1 = 1)P(x_2 = 1)P(x_3 = 1)$$

where $P(x_1=1)$ has degree 3 in the latter expression. On the other hand, from the AND gate behaviour we may deduce that the probability of a 1 to appear at its output is characterized by the probability of concurrent events $x_1=1$, $x_2=1$ and $x_3=1$, which is calculated for independent variables x_1, x_2 and

x_3 as a product of probabilities: where all probabilities have degree *1* only. Thus, for obtaining the correct probability estimate, which describes the behaviour of a combinational circuit with reconvergent fan-out (dependent events), we must reduce the degrees of probabilities to *1* in the probability expression obtained by using (6.2.)-(6.4.).

The algorithm that makes use of the property consists of the following steps.

(i) All the nodes of the digital circuit under test are assigned different variables.

(ii) For each element of the circuit, the output probability is calculated as a function of input variable probabilities with the procedure being performed in succession from the elements to which the circuit inputs are connected to the element with the function implemented by the circuit produced on its output.

(iii) To obtain the correct probability expression for the circuit, all powers are reduced to *1*.

As an example of the above algorithm, let us find $P(F=1)$ for the circuit of Fig.6.2. By using the relations (6.3) and (6.4), we obtain

$$P(f_1 = 1) = P(x_1 = 1) P(x_2 = 1),$$
$$P(f_2 = 1) = P(x_1 = 1) + P(x_2 = 1) - P(x_1 = 1) P(x_2 = 1),$$
$$P(f_3 = 1) = P(f_2 = 1) P(x_3 = 1) = P(x_1 = 1) P(x_3 = 1) +$$
$$+ P(x_2 = 1) P(x_3 = 1) - P(x_1 = 1) P(x_2 = 1) P(x_3 = 1),$$
$$P(F = 1) = P(f_1 = 1) + P(f_3 = 1) - P(f_1 = 1) P(f_3 = 1) =$$
$$= P(x_1 = 1) P(x_2 = 1) + P(x_1 = 1) P(x_3 = 1) +$$
$$+ P(x_2 = 1) P(x_3 = 1) - P(x_1 = 1) P(x_2 = 1) P(x_3 = 1) -$$
$$- P^2(x_1 = 1) P(x_2 = 1) P(x_3 = 1) - P(x_1 = 1) P^2(x_2 = 1)$$
$$P(x_3 = 1) + P^2(x_1 = 1) P^2(x_2 = 1) P(x_3 = 1).$$

Reducing the powers to *1*, we finally have

$$P(F = 1) = P(x_1 = 1) P(x_2 = 1) + P(x_1 = 1) P(x_3 = 1) +$$
$$+ P(x_2 = 1) P(x_3 = 1) - 2P(x_1 = 1) P(x_2 = 1) P(x_3 = 1).$$

The problem of finding the probabilistic expression for a digital circuit can be substantially simplified for a circuit employing logic gates of the same type. As has been demonstrated, for a tree circuit employing n-input NAND gates with the inputs of each gate connected to the outputs of the preceding level gates, the probability expression will have the form.

$$P_l(y=0) = P_{l-1}^n(y=1) = [1 - P_{l-1}(y=0)]^n, \qquad (6.6)$$

where $P_l(y=0)$ and $P_l(y=1)$ are respectively the probabilities of a *0* and *1* to appear on the outputs of the *l*th-level NAND gates.

An original algorithm has been proposed based on transforming the source digital circuit to a tree structure. The idea behind the approach is to establish the bounds for the sought-for probability by simulating limiting probabilities in fanout points. In this case, a fanout is represented as independent inputs with the probability of a *1* to occur being in the range *0* to *1*. Thus for the example of Fig. 6.2, the probability of a *1* to appear at the input of a two-input OR gate is assumed to fall within *0* to *1*. Then $P(f_2=1)=0 \div 1$ and, hence, $P(f_3=1)=P(f_2=1)P(x_3=1)=0 \div P(x_3=1)$. The probability of a *1* to appear at the output of the single-input AND gate is $P(x_1=1)P(x_2=1)$, and the output probability is respectively $P(F=1)=P(f_1=1)+P(f_3=1)-P(f_1=1)P(f_3=1)$. By substituting the upper $(P(x_3=1))$ and lower *(0)* bounds of $P(f_3=1)$ into the latter expression, we finally obtain

$$P(F=1) = [P(x_1=1)P(x_2=1)] \div [P(x_1=1)P(x_2=1) + P(x_3=1) - P(x_1=1)P(x_2=1)P(x_3=1)].$$

For $P(x_1=1)=P(x_2=1)=P(x_3=1)=p$, we obtain

$$P(F=1) = p^2 \div (p + p^2 - p^3),$$

where p^2 and $p + p^2 - p^3$ are respectively the lower and the upper probability estimates $P(F=1)$.

More complex is the problem of probabilistic description of sequential circuits. Analytic approaches are impracticable here. The only choice is the use of statistical exppperiment. An example of the implementation of the statistical appoach for probabilistic description of digital circuits is STAFAN (STAtistical Fault ANalysis). It is a software system that allows one, in particular, to estimate the probability of a *1* to appear at any of the curcuit nodes.

A probabilistic description of a digital circuit that has been obtained either analytically or from experimental data is required only for generation of reference probabilities to be used for comparison with the actual estimates obtained by testing the digital circuit. Such probabilities are often referred to as signal probabilities. They can be used to estimate the probability of a *1* to appear at the specified nodes of the circuit under test.

The problem of estimating the probability of fault detection at the specified nodes of a digital circuit is most important since it makes it possible to evaluate the efficiency of probabilistic testing in each particular case and outline the methods of improvement. Consider some practical approaches used to evaluate the probability of specified fault detection.

6.3. Fault Detection Probability Estimation

The probability $P_d(f \equiv w)$, $w \in \{0,1\}$, of detecting a stuck-at fault $\equiv w$ on a node is a measure of random testing efficiency and characterizes the complexity of defining the specified curcuit fault. The $P_d(f \equiv w)$ is the probability of detecting the fault $\equiv w$. by random test pattern.

For the general case the probability $P_d(f \equiv w)$ is characterized by the probability of two concurrent events: fault manifestation at its point of origin and its propagation to one of the circuit outputs. The manifestation of fault $\equiv w$ can be estimated by probability $P_e(f \equiv w)$ of its manifestation at the specified node f; the fault propagation can be estimated by probability $P_t(f \equiv w)$ of its observation at any circuit output.

Let us to examine the fault detection, manifestation and propagation probabilities for the stuck-at faults $w \in \{\equiv 0, \equiv 1\}$. For $w \equiv 0$ at the node f of the VLSI the fault manifestation probability equals to the probability of the equality $f=1$, and for $w \equiv 1$ to the probability of $f=0$. For the general case, $P_e(f \equiv 0)$ and $P_e(f \equiv 1)$ may take different relationship. It depends on the complexity of the VLSI. The propagation probability of the stuck-at fault does not depend on the value of fault w. It estimates as the probability that the fault w is visible at an observable output. On the basis of $P_e(f \equiv w)$ and $P_t(f \equiv w)$ we can estimate the detection probability. For the stuck-at faults the detection probability is defined as

$$P_d(w \equiv 0) \leq \min\ [P_e(w \equiv 0), P_t(w \equiv 0)\ ],$$
$$P_d(w \equiv 1) \leq \min\ [P_e(w \equiv 1), P_t(w \equiv 0)\ ],$$

where $P_t(f \equiv 0) = P_t(f \equiv 1)$. For the general case $P_d(\equiv w)$ has very complex dependence of $P_e(f \equiv w)$ and $P_t(f \equiv w)$.

For the case of independent events only, the probability P_d may be calculated by the following relations:

$$P_d(w \equiv 0) = P_e(w \equiv 0) P_t(w \equiv 0),$$
$$P_d(w \equiv 1) = P_e(w \equiv 1) P_t(w \equiv 0),$$

To comprehend the fault detection probability concept, consider the digital circuit implemented as an n-input AND gate.

As for the general case, the procedure testing for the AND gate consists of applying sequences of random bits with probabilities $p(x_1=1)$, $p(x_2=1)$, ... $p(x_n=1)$ to its inputs and analysing its output response. Without loss of generality, suppose that $p(x_1=1)=p(x_2=1)=...=p(x_n=1)=p$.

For an n-input AND gate, the probability of fault $\equiv 0$ to occur at the primary input will be the same as the probability of a 1 to appear at the same input, i.e. $P_e(x_1 \equiv 0) = p(x_1=1) = p$ and, hence, $P_e(x_1 \equiv 1)$ are the probabilities of a 0 to appear at the specified input, where $P_e(x_1 \equiv 1) = p(x_1=0) = 1-p(x_1=1) = 1 - p$. A propagation condition for any fault at the primary input to an AND gate is path sensitization from the specified input to the gate output. The above condition is determined by the event $x_2 = 1$, $x_3 = 1$, ..., $x_n = 1$ whose probability is calculated as $p(x_2=1, x_3=1,...,x_n=1)$. Allowing for the independence of input random sequences, we finally obtain $P_t(x_1 \equiv w) = p(x_2=1)p(x_3=1)...p(x_n=1) = p^{n-1}$.

The probability $P_e(y \equiv 0)$ of a fault w to appear at the AND gate output is $p(y=1)$ for $w \equiv 0$ and $p(y=0)$ for $w \equiv 1$, which the probability $P_t(y \equiv w)$ being rigorously 1 since the AND gate output is the output node of the circuit.

For the straightforward structure of the digital circuit just discussed, fault manifestation and propagation are independent events and therefore the probability $P_d(f \equiv w)$ may be calculated by the following relation

$$P_d(f \equiv w) = P_e(f \equiv w) P_t(f \equiv w).$$

Table 6.1 gives the probabilities $P_e(f \equiv w)$, $P_t(f \equiv w)$ and $P_d(f \equiv w)$ for stuck-at faults at the i th input ($i=\overline{1,n}$) and output of AND and OR gates. Thus for $p=1/2$, by the relations of Table 6.1, the probability $P_d(x_1 \equiv 0)$ for a four-input AND gate is $p=1/2^4$ and $P_d(y \equiv 1) = 1-1/2^4$.

The concepts of fault detection, manifestation and propagation probabilities are to a great extent similar to those of controllability, observability and testability of a digital circuit. The only difference is that the newly introduced attributes of detection, manifestation and propagation probabilities are calculated both for a uniform distribution of test patterns and for a random distribution of input probabilities.

Table 6.1

$f \equiv w$	AND			OR		
	P_e	P_t	P_d	P_e	P_t	P_d
$x_i \equiv 1$	$1-p$	p^{n-1}	$p^{n-1}-p^n$	$1-p$	$(1-p)^{n-1}$	$(1-p)^n$
$x_i \equiv 0$	p	p^{n-1}	p^n	p	$(1-p)^{n-1}$	$p(1-p)^{n-1}$
$y \equiv 1$	$1-p^n$	1	$1-p^n$	$(1-p)^n$	1	$(1-p)^n$
$y \equiv 0$	p^n	1	p^n	$1-(1-p)^n$	1	$1-(1-p)^n$

Analytical estimation of probabilities P_d, P_e and P_t for a typical digital circuit becomes more complicated due to the dependence of the fault manifestation and propagation events on one of the circuit outputs. Therefore, in the majority of cases, only a fault detection probability is estimated.

Thus, the problem of the specified probability estimation reduces to calculation of a signal probability. The simple example of such approach is shown in Fig.6.3, where outputs of the original VLSI G and it faulty modifications G* are connected to EXCLUSIVE-OR gates. All changes at the G* outputs will be manifested at the output F_g of OR gate. Thus, the probability $P(F_g=1)$ will be the detection probability of the fault w in the G*.

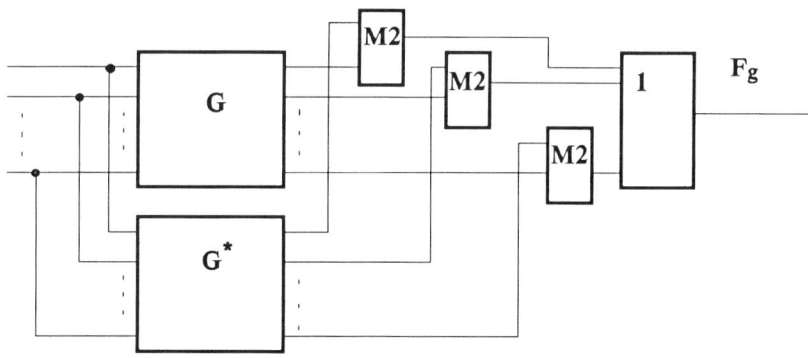

Fig. 6.3. Calculation of a signal probability

For the case of stuck-at faults this approach may be illustrated by Fig.6.4.

The digital circuit shown in Fig.6.4. consists of series-connected gates NAND(1), AND(2), OR(3) and NOR(4). Let the fault be a stuck-at fault $f_1 \equiv 0$, which propagates to the output along the path through gates 2-4. Fault manifestation at its origin is conditioned by a *1* signal at the output of the first gate; fault propagation to the output is conditioned by a *1* appearing at the primary input to gate 2 and *0* s at primary inputs to gates 3 and 4. The given set of conditions that determine fault observability at the circuit output may be united by inserting a dummy AND gate 5 whose output value will be *1* if and only if all the previous conditions have been satisfied. Hence, the probability $P(F_g=1)$ of the output value to be *1* for the dummy AND gate will be equal to that of the specified fault detection

$$P_d(f_1 \equiv 0) = P(F_g = 1),$$

which implies that the $f_1 \equiv 0$ detection probability corresponds to the signal probability of node F_g. Thus fault detection probability estimation reduces to finding a signal probability. Any of the techniques discussed earlier may be used for the purpose.

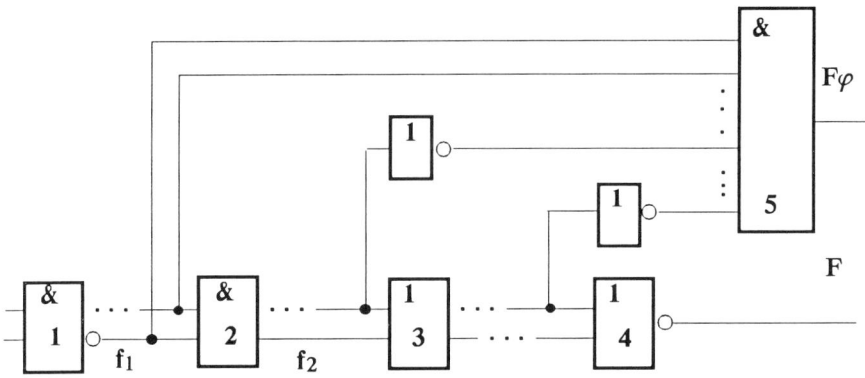

Fig. 6.4. Estimation of detection probability

When there are many fault propagation paths to the digital circuit outputs, the signal probability for the output node of a dummy AND gate will be used as a lower bound to the desired fault detection probability. Also, this eliminates the need to estimate the detection probability for all possible faults in the circuit. It will suffice to calculate the detection probability for representative faults, i.e. faults that lie on the input lines to the circuit and those associated with fanout points. Such faults are illustrated in Fig.6.5 and summarized in Table 6.2.

Table 6.2

Fault	P_d	P_d^*	Fault	P_d	P_d^*
$x_1 \equiv 0$	3/16	5/32	$f_1 \equiv 0$	3/16	5/32
$x_1 \equiv 1$	3/16	5/32	$f_1 \equiv 1$	1/8	1/8
$x_2 \equiv 0$	11/16	9/32	$f_2 \equiv 0$	3/16	5/64
$x_2 \equiv 1$	11/16	9/32	$f_2 \equiv 1$	3/16	1/16
$x_3 \equiv 0$	9/32	1/16	$f_3 \equiv 0$	11/32	3/16
$x_3 \equiv 1$	9/32	1/16	$f_3 \equiv 1$	1/8	5/64
$x_4 \equiv 0$	3/16	5/64	$f_4 \equiv 0$	3/16	3/16
$x_4 \equiv 1$	3/16	1/16	$f_4 \equiv 1$	1/8	5/64
$x_5 \equiv 0$	3/16	3/16	$f_5 \equiv 0$	7/16	3/8
$x_5 \equiv 1$	3/16	3/16	$f_5 \equiv 1$	5/16	9/32
			$f_6 \equiv 0$	7/16	25/64
			$f_6 \equiv 1$	3/16	3/16

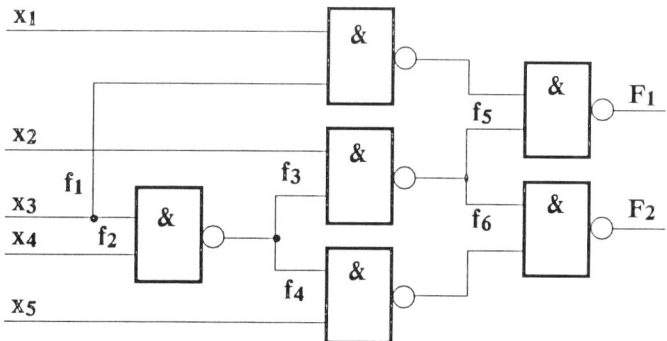

Fig. 6.5. A combinational circuit

The table gives an accurate estimate of fault detection probability P_d and its lower bound P_d^* obtained as a signal probability for the dummy AND gate for each case.

The analysis of fault detection probabilities makes it possible to estimate the effectiveness of random testing for a digital circuit as well as to determaine the minimal length of the test sequence.

6.4. Test Sequence Length Calculation

A major parameter of any testing and diagnosting system for digital circuits is the time consumed by the testing procedure, which is defined uniquely by a test sequence length for deterministic test design methods, the test length is determined by the required coverage for all possible faults in the circuit and does not exceed *1000* patterns on average. Note that fault coverage increases with number of test patterns. A similar relationship is evidently true for random testing as well.

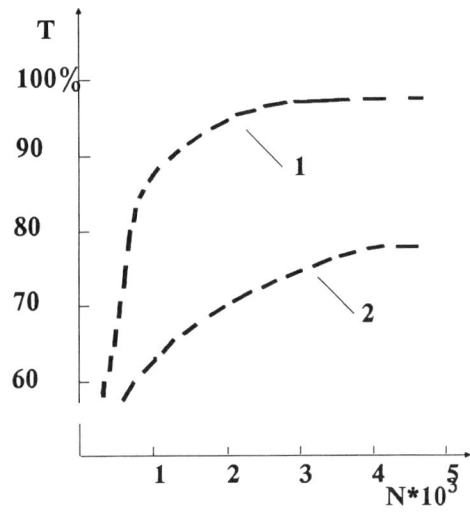

Fig. 6.6. Fault coverage for PLA-1 (1) and PLA-2 (2)

The relationship between fault coverage and number of test patterns represented by random number is shown in Fig.6.5. The figure implies that the fault coverage T achieved by generation of N random test patterns can be approximated as a function of an exponent

$$T = [1 - \exp(-\lambda \log_{10} N)]100\%$$

where λ is a constant that assumes a specific value for each circuit type. A similar relationship may be obtained by arguing on the basis of the fault whose detection probability is P_d, with P_d in this case characterizing the probability of detection of the specified fault by applying only one test pattern. With N distinct random patterns applied, the indicated probability will increase and in the general case will be defined by

$$P_d(N) = 1 - P_n(N), \qquad (6.8)$$

where $P_d(N)$ is the probability of fault detection by applying N test patterns and $P_n(N)$ is the non-detection probability of the fault specified. Considering that the generated test patterns are independent, we may calculate probability as the product of N terms $(1-P_d)$ representing the non-detection probability for a fault by applying a single test pattern. Subject to the latter, the relation (6.8) reduces to

$$P_d(N) = 1 - (1 - P_d)^N,$$

whence we can find the number of random test patterns for the specified test validity defined by $P_d(N)$ as

$$N = \frac{\ln[1 - P_d(N)]}{\ln(1 - P_d)}. \qquad (6.9)$$

For an n-input AND gate, the number of test patterns required to detect its fault $y \equiv 0$ (see Table 6.1) with probability $P_d(N)=0{,}99$ may be calculated by equation (6.10) subject to the equality $P_d = p^n$:

$$N \approx \frac{\ln(0.01)}{\ln(1 - p^n)},$$

where p is the probability of a 1 to appear in random patterns applied to this inputs to the AND gate. The analysis of the latter relation shows that the test sequence length N depends heavily on the form of the circuit under test (here on the nember n of AND gate inputs) and on the input probability distribution p.

The relationship (6.9) allows to determine the random test length. For this purpose we may be use the method of worst-case fault, i. e. the fault which has the minimal value of its detection probability P_d.

It should be mentioned that P_d from the equality (6.9) is the lower bound of the random test length. To detect all faults of the VLSI the random test length exceeding N defined by (6.9) must be used.

To find out what is the dependence N from detection probabilities of the all faults let us consider a very simple case i.e. digital circuit with two faults w_1 and w_2.

Suppose that random test pattern allows to define the fault w_1 with the probability P_1 and at the same time fails to detect the fault w_2. The probability P_2 is the detection probability of the only fault w_2, and P_3 is detection probability for faults w_1 and w_2. Thus the detection probability of the fault w_1 equals $P_1 + P_3$ and $w_2 - P_2 + P_3$.

The fault detection procedure may be treated as Markov process and described by Markov chain (Fig. 6.7). The states S_1 and S_2 represent the cases of detecting fault w_1 and w_2 respectively, and S_{12} both of them at the same time. The case when all faults go undetected is represented by the state S_0.

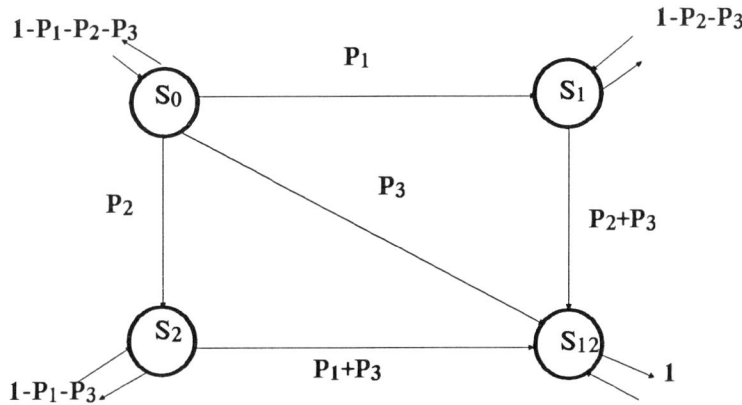

Fig. 6.7. The Markov chain

On the basis of Markov process theory it is known how to find the probability $P_{12}(N)$ of the state S_{12} as the result of applying N random test patterns. The initial states S_0, S_1, S_2 and S_{12} are described by $P_0(0)=1$, $P_1(0)=P_2(0)=P_{12}(0)=0$.

The Markov chain (Fig. 6.7) is described by the following equations

$$P_0(N) = [1-P_1 - P_2 - P_3]P_0(N-1),$$
$$P_1(N) = [1 - P_2 - P_3]P_1(N-1) + P_1P_0(N-1),$$
$$P_2(N) = [1 - P_1 - P_3]P_2(N-1) + P_2P_0(N-1),$$
$$P_{12}(N) = P_{12}(N-1) + P_3 P_0(N-1) + (P_2 + P_3)P_1(N-1) +$$
$$+ (P_1 + P_3)P_2(N-1),$$

The solution of the system is

$$P_{12}(N) = 1-(1 - P_1 - P_2)^N-(1 - P_2 - P_3)^N+(1 - P_1 - P_2 - P_3)^N \quad (6.10)$$

where $P_{12}(N)$ is the probability of detecting two faults w_1 and w_2. Expression $(1-P_1-P_3)^N$ is the probability of undetecting the fault w_1 $(1-P_2-P_3)^N$ fault w_2, and $(1-P_1-P_2-P_3)^N$ takes in account the case of undetecting both faults w_1 and w_2.

For determination of the minimal length of the random test we need to calculate the N, which satisfies the following inequality

$$1 - P_{12}(N) \le P_n(N), \quad (6.11)$$

$P_n(N)$ is the given value.

As an example let us calculate random test length for $P_n(N)=0,001$ $(P_d(N)= 1 - P_n(N) = 0,999)$.

We may consider three cases. The first one uses the method of the worst-case fault. So we get N_1 as the length of the random test from (6.11) for detection probability with the minimal value. In the second case N_2 is the length of the random test, when $P_3=0$ and N_3 - for $P_3=0,0005$. For all three cases detection probability of the fault w_1 equals $P_1+P_3=0,001$, and for fault w_2 P_2+P_3 takes the value from $0,001$ to $0,002$ in $0,0002$.

The results of calculation is shown in Table 6.3.

As we can see from Table 6.3, the random test has the maximal length for independent faults w_1 and w_2. The length of the test does not significantly depend on higher detection probabilities Than the minimal one.

Using the summary from the previous example let us present the theorem for estimating lower and upper bounds of the test length without proof.

Theorem 6.1. The length N of the random test, which ensures the undetection probability less or equal to $P_n(N)$, for detecting K independent faults each of which is described by the detection probability P_d, satisfies the following inequality

$$\frac{\ln\frac{P_n(N)}{K} - \ln[1 - \frac{P_n(N)}{2}]}{\ln(1-P_d)} \leq N \leq \frac{\ln\frac{P_n(N)}{K}}{\ln(1-P_d)}.$$

As we can see the difference between lower and upper bounds insignificant. Thus for calculation N, we may used the expression

$$N \approx \frac{\ln\frac{P_n(N)}{K}}{\ln(1-P_d)}.$$

which for real values of the $P_d << 1$ may be approximated by the following expression

$$N \approx \frac{\ln K - \ln P_n(N)}{P_d}.$$

Table 6.3

$P_1 + P_3$	$P_2 + P_3$	$P_n(N)$	N_1	N_2	N_3
0,001	0,001	0,001	6905	7597	7586
0,001	0,0012	0,001	6905	7120	7115
0,001	0,0014	0,001	6905	6964	6963
0,001	0,0016	0,001	6905	6920	6920
0,001	0,0018	0,001	6905	6909	6909
0,001	0,002	0,001	6905	6906	6906

As an example let us determine the random test length for the digital circuit with two faults whose detection probability is *0,001*. Then

$$N \approx \frac{\ln 2 - \ln 0{,}001}{10^{-3}} = 8 \cdot 10^3.$$

For $P_d = 10^{-2}$ length $N = 8 \cdot 10^2$

The last example shows us that the random test length depends on the fault detection probabilities.

6.5. Methods for Optimal Selection of Input Variable Probabilities

The optimal probability of input variables is that allowing one to obtain the minimal length N of random test sequence. There are some methods for finding optimal values for input probabilities.

First of all, let us consider the method for finding the optimal probability of obtaining the maximal length N of test sequence to ensure the specified detection probability for the worst-case fault.

The worst-case fault is that with the minimal value of its detection probability P_d. Thus the major problem to be solved in finding the optimal probabilities for input variables is to ensure the maximal possible value of the minimal fault detection probability in the circuit.

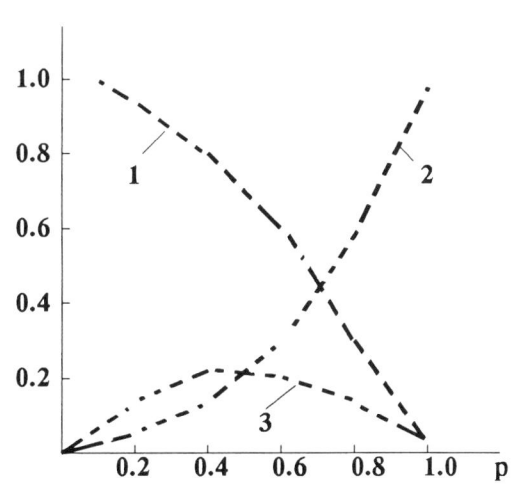

Fig.6.8. Detection probabilities for n-input AND gate

As an example demonstrating the estimation of optimal values for input probabilities, consider a n-input AND gate whose fault detection probabilities are given in Table 6.1. Probabilities plotted for

$P_d(x_i \equiv 1) = p^{n-1} - p^n$, $P_d(x_i \equiv 0) = P_d(y \equiv 0) = p^n$, and $P_d(y \equiv 1) = 1 - p^n$ ($n=2$) are shown in Fig. 6.8.

The analysis of dependences between P_d and p shows that

$$\max_p \min (p^{n-1} - p^n, 1 - p^n, p^n)$$

for $n=2$ is obtained with $p=0.5$ and for $n=3$ with $p=2/3=0.666$.

For a random n, we may define the optimal probability p of a 1 appear at the inputs to the AND gate by the following theorem.

Theorem 6.2. The optimal probability P_0 of a 1 to appear at the n inputs of the AND gate equals $(n-1)/n$.

Proof. The optimal probability P_0 must ensure the maximal possible value of the minimal fault detection probability of the circuit. There are three fault detection probabilities $p^{n-1} - p^n$, $1-p^n$ and p^n for the AND gate faults. So, for this case

$$\max_p \min (p^{n-1} - p^n, 1 - p^n, p^n)$$

is obtained with $p=P_0$.
Obviously,

$$\max_p \min (p^{n-1} - p^n, 1 - p^n, p^n) \leq \max_p (p^{n-1} - p^n).$$

The $\max_p (p^{n-1} - p^n)$ we may get for p which satisfies the equation

$$\frac{\partial (p^{n-1} - p^n)}{\partial p} = 0.$$

It value equals $(n-1)/n$.

For $p=(n-1)/n$ $p^{n-1} - p^n \leq p^n$ and $p^{n-1} - p^n \leq 1 - p^n$ so $P_0=(n-1)/n$ which ensure

$$\max_p \min (p^{n-1} - p^n, 1 - p^n, p^n)$$

For a digital circuit employing n-input OR gates, we may find P_0 in a similar manner. Here, P_0 equals $1/n$. Hence $P_0=0,333$ with $n=3$.

It has been shown that the optimal probability P_0 may have a new value to get minimal length N of the random test if we change the condition of its determination.

As an example let us consider determination of the optimal probability P_0 to get all patterns of AND gate test

$$\begin{matrix} 0 & 1 & 1 & \ldots & 1 & 1 & 1 \\ 1 & 0 & 1 & \ldots & 1 & 1 & 1 \\ 1 & 1 & 0 & \ldots & 1 & 1 & 1 \\ \ldots & \ldots & \ldots & \ldots & \ldots & \ldots & \ldots \\ 1 & 1 & 1 & \ldots & 0 & 1 & 1 \\ 1 & 1 & 1 & \ldots & 1 & 0 & 1 \\ 1 & 1 & 1 & \ldots & 1 & 1 & 0 \\ 1 & 1 & 1 & \ldots & 1 & 1 & 1 \end{matrix}$$

The very simple analysis shows us that the optimal probability P_0 is the ratio of the number of 1s in the particular column of patterns to number of patterns. As the result we get $P_0=n/(n+1)$ where n is the number of AND gate inputs.

For a n-input OR gates, we may find P_0 in a similar manner. Here P_0 equals $1/(n+1)$.

The optimal probabilities P_0 for AND and OR gates are used by the algorithm for determination of the optimal probabilities for any random circuit. The idea of the algorithm is to determine the input probabilities, which ensure the optimal value probabilities at the inputs of each elements of the circuit. The optimal (pseudooptimal) solution is the average value for the optimal probabilities of each elements.

The algorithm for determination of the optimal probabilities consists of the following steps.

1. At the element inputs the optimal probabilities are specified to the minimal test length N.

2. On the basis of the back-track method we calculates the probabilities at the circuit's inputs. To do the calculation we have to use the relation between the output and input probabilities of the elements. Thus, for an n-input AND gate the input probability p_i equals

$$p_i = p^{1/n}, \qquad (6.12)$$

where p_i is the probability of a *1* at the *i* th input of the AND gate. For OR gate

$$p_i = 1 - (1-p)^{1/n}, \qquad (6.13)$$

For inverter

$$p_i = 1 - p. \qquad (6.14)$$

For non-reconvergent fanout with the signal probabilities p_i, $i=\overline{1,k}$ at any fanout the probability p at common point equals

$$p = \frac{1}{k} \sum_{i=1}^{k} p_i.$$

3. The values of the input probabilities to each element of the circuit are put to the special table.

4. For each input of the circuit, calculate the average value of probabilities. It will equal to the optimal one.

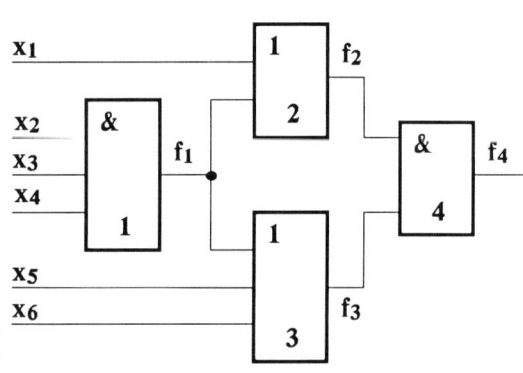

Fig. 6.9. A combinational circuit

Let's take as an example, a simple circuit with four elements (Fig.6.9). First of all define the optimal input probabilities for the 4th element. On the basis of theorem 6.2 $p(f_2=1) = p(f_3=1) = 0,5$. According to the equation (6.13) for elements 2 and 3, we get $p(x_1=1)=0,293$ $p(x_5=1)=p(x_6=1)=0,206$. and $p(f_1=1)$ = $(0,293 + 0,206)/2 = 0,25$. For the first element we use the relation (6.12). Results are given in the first row of the Table 6.4. Calculate for the rest of elements of the circuit are in Fig.6.9 in the same manner.

Table 6.4

N	p(x₁=1)	p(x₂=1)	p(x₃=1)	p(x₄=1)	p(x₅=1)	p(x₆=1)
4	0,293	0,63	0,63	0,63	0,206	0,206
2	0,5	0,794	0,794	0,794		
3		0,693	0,693	0,693	0,333	0,333
1		0,666	0,666	0,666		

As the result we get the pattern of probabilities for each input. Average value of the probabilities will be the optimal on for each input of the circuit. The result of the calculation is $p(x_1=1) = 0,397$, $p(x_2=1) = p(x_3=1) = p(x_4=1) = 0,696$, $p(x_5=1) = p(x_6=1) = 0,27$.

For the fault $x_2 \equiv 0$ (Fig.6.9), its detection probability, in the case of the eqiprobable random test, equals to *0,078* and the test length N according to (6.9) for $P_d(N)=0,001$ will be *85*. Under the same conditions, by using optimal distribution of probabilities, we get $N=22$.

The second example consists of two elements AND and OR (Fig.6.10). Using the last algorithm we'll get the optimal probabilities as

$$p(x_1 = 1) = p(x_2 = 1) = \ldots = p(x_n = 1) = \frac{1}{2}\left(\frac{n-1}{n} + \frac{1}{n}\right) = 0,5.$$

For each individual element the probability *0,5* is not optimal. It increases the length of the random test. We may achieve the significant reduction of length N by using two random tests with input probabilities $(n-1)/n$ and $1/n$. It should be noted, however, that the problem of selecting the optimal input probabilities for the general case is a linear optimization problem, which requires complex calculations.

A method for selecting the optimal values of input probabilities for random testing is suggested in some books. The idea behind the method is in maximizing the minimal detection probability for those circuit faults that are most difficult to detect (faults with minimal detection probabilities). It has been shown that the input lines to the circuit as well as lines at the origin of

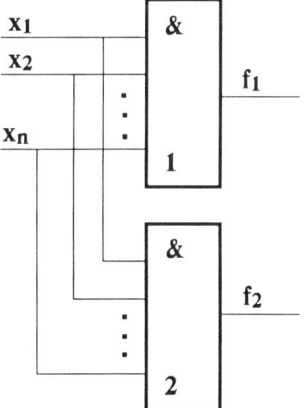

Fig. 6.10. A combinational circuit

fan-out points are the locations of faults with minimal detection probability. The optimal assigment of input vector probability is based on maximizing the objective, which represents the minimum fault detection probability. For this purpose, the so-called relaxation method has been used, but its assigment is not always optimal. However, the assigment obtained by the said method is always better than in the trivial case when input probabilities are *0,5*.

A similar problem of optimal selection of input probability distributions has been presented for the combinational circuit. It has been shown that for any digital circuit comprising memory elements the problem of calculating optimal input probabilities is more complex than that of generating deterministic test sequences. For this reason, the application of random techniques proves to be inefficient.

6.6. Automatic Search for Optimum Probability

<u>Stochastic Servointegrator.</u> As it has been noted earlier, the problem of finding optimal probabilities for input variables is a nonlinear optimization problem which requires rather complex calculations. However, the process of finding optimum probabilities may be significantly simplified by automatic search. The use of stochastic servointegrators, which serve as controllers in automated digital systems, is rather efficient for the purpose.

Fig.6.11 shows a block-diagram of stochastic servointegrator. It consists of a uniformly distributed pseudorandom number generator (PRNG), a comparison circuit (Comp.C), a reversible counter, and a combinational circuit (CC) used to produce input values of the counter in accordance with the truth table 6.5.

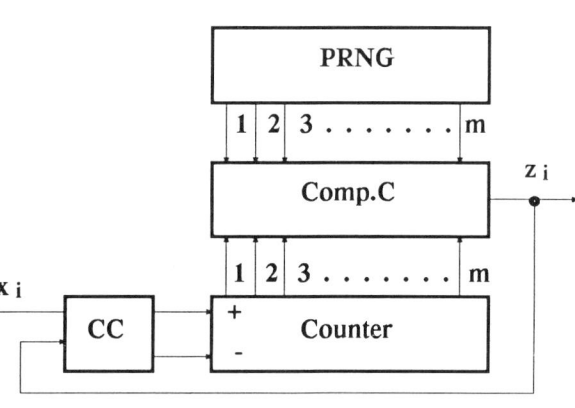

Fig. 6.11. Stochastic Servointegrator

The sequence of input values x_i is characterized by probability $p(x_i=1) = p_1$ which is constant in value. Bit z_i takes the

Table 6.5

x_i	z_i	+1	-1
0	0	0	0
0	1	0	1
1	0	1	0
1	1	0	0

value *1* if the counter is less than or equal to the code supplied from the PRNG output; otherwise $z_i=0$.

With the zero initial settings (counter content is zero) and the number of integrator steps being large, the following equality holds for $p(z_i=1)$

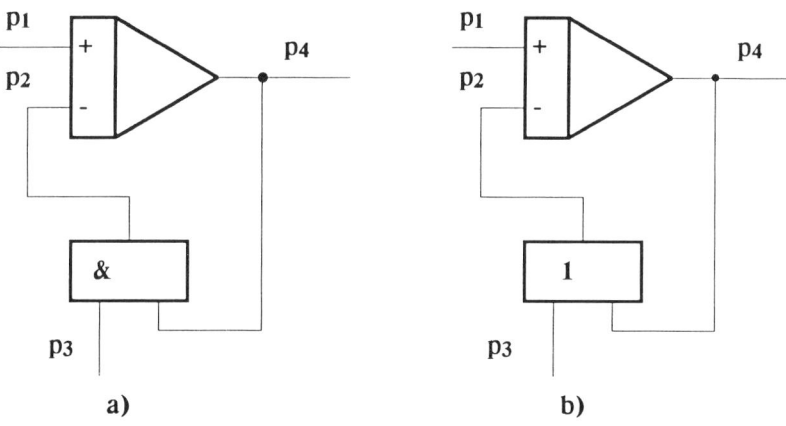

Fig. 6.12. The Stochastic Integrators for AND and OR gates

$$p(z_i = 1) = p_1(1 - \exp^{-k/2^m}). \quad (6.15)$$

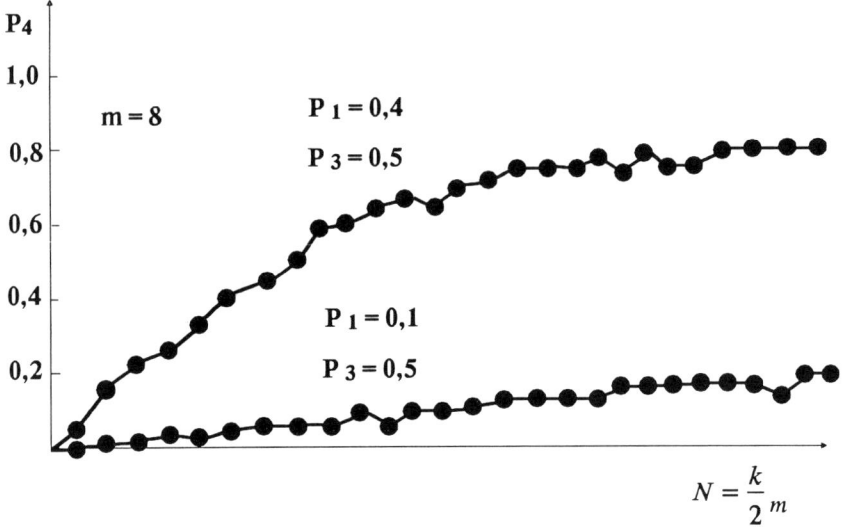

Fig. 6.13. The plots of p4 for AND gate

where *m* is the width of the PRNG and counter.

The limiting relation

$$\lim_{k \to \infty} p(z_i = 1) = p_1$$

is executed in the steady - state mode.

Thus for reasonably large values of k, probabilities $p(x_i=1)=p_1$ and $p(z_i=1)=p_2$ for sequences x_i and z_i applied to the inputs of the stochastic integrator assume identical values. This statement holds true for the negative feedback of integrator, with the gain assuming different values.

<u>Automatic search by means of a stochastic integrator.</u> Stochastic integrators can be used to search for optimum probabilities automatically. For example, for a two-input AND gate (Fig.6.12a) included in the negative feedback of the integrator in the steady-state mode, the equality $p_1=p_2$ holds true. Subject to equation (6.3), we can write $p_2=p_3\,p_4$ or eventually $p_1=p_3\,p_4$. the latter relation indicates that for constant probabilities p_1 and p_3, the stochastic integrator

searches for the probability p_4 for a one to appear at the second input of 2- input AND gate. The procedure of finding optimum values of p_4 is illustrated adequ-

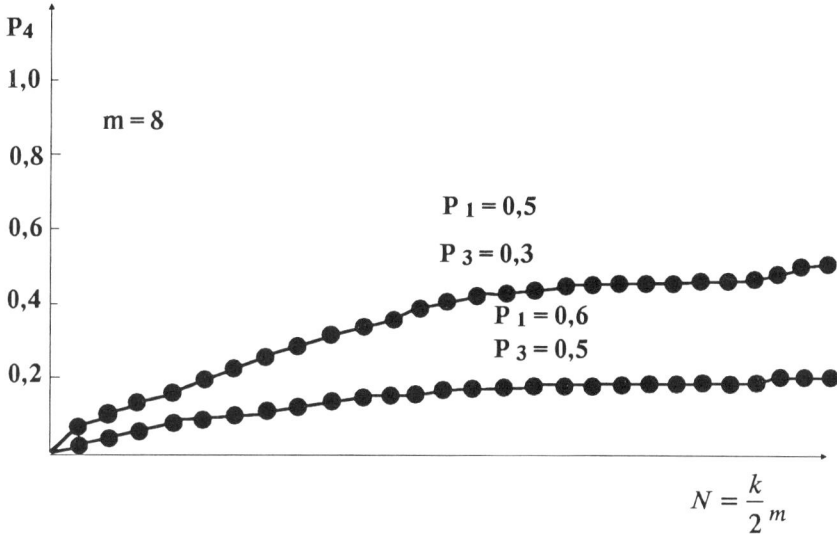

Fig. 6.14. The plots of p4 for OR gate

ately by plots for different values of p_1 and p_2 (Fig.6.13). In this case we have considered an example for $m=8$.

Similar results are shown in Fig.6.14 for an OR gate included in the negative feedback of the integrator.

The use of several stochastic integrators and the search procedure organized by the automatic control theory algorithms is a further outgrowth of the concept of automatic search for optimum probabilities.

Chapter 7

PSEUDORANDOM TESTING

7.1. Pseudorandom Sequences as the Test Sequences of the Circuit

Concept. The key characteristic of any test sequence is the fault coverage which defines as a ratio of detected faults for the full amount of the circuit faults. When applying random number sequences as the test patterns, of the VLSI test coverage can be expressed by value of $P_q(N)$. The probability $P_q(N)$ is the function of test length N, which does not decrease with N. It may be expressed as

$$P_q(N) = 1 - P_n(N),$$

where $P_n(N)$ - a fault escape probability function, which does not increase with increasing N for VLSI faults.

It should be noted, that the equality $P_q(\infty)=1$ satisfies for random test only and non-redundant VLSI. At the same time for pseudorandom test patterns (PRTPs) the following relation holds

$$\lim_{N \to \infty} P_q(N) = 1 - p^* \qquad (7.1)$$

where $0 \leq p^* \leq 1$ is a constant determined by the form of pseudorandom sequences used as well as the structure of the VLSI under test. Thus for a CC with the number of inputs n less then the number of outputs of the PRTP generator, the value of p^* may differ from 0 and is determined as the ratio of undetected faults fo their total number. Condition $p^* \neq 0$ is met due for the deterministic nature of the PRTP. Let us consider how it manifests itself in a specific case. Fig.7.1. shows a portion of the VLSI. The latter is tested by a PRTP generator, which generates an M-sequence by polynomial $\phi(x) = 1 \oplus x^3 \oplus x^4$. As evident from the figure, a $f_1 \equiv 0$ fault on the inputs of the EOR gate, a $f_3 \equiv 0$ on the OR gate and a briging fault on the inputs of the AND gate are not detected, since both inputs of the gates in question are applied with identical logic signals. The appearance of untestable portions in a sequential VLSI has been detailed in previous books. It has been proved that for any PRTP generator there can be found a sequential circuit that is untestable on inputs of some of its gates. Condition $p^*=0$ can be met by designing a PRTP generator for the specified VLSI or a class of such VLSIs. We shall consider the requirements for hardware pseudorandom test pattern generators (PTPGs) that satisfy the condition $p^*=0$ when testing a circuit that is a sequential synchronous one.

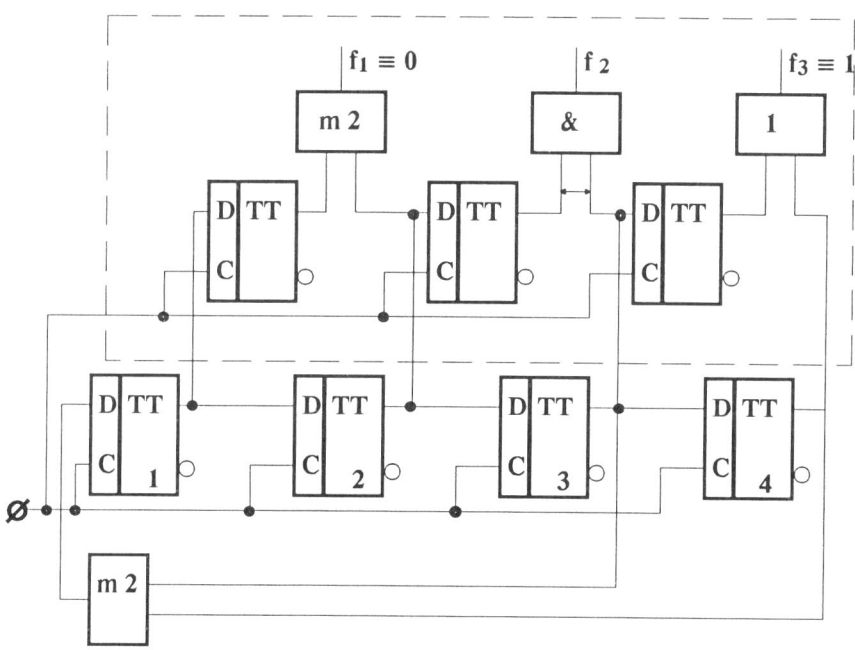

Fig.7.1. A portion of the VLSI

Any sequential circuit may be represented by the so-called Huffman model, for which $A(k)=a_1(k),...,a_c(k),...a_C(k)$ and $Y(k)=y_1(k),...,y_r(k),...,y_R(k)$ are, respectively, its input and output variables in the k-th cycle of operation, and $Z(k)=z_1(k),...,z_q(k),...,z_Q(k)$ are internal variables that define the circuit state. The circuit behaviour is described by the set of equations

$$y_r(k) = f_r\ [a_1(k),...,a_C(k),z_1(k),...,z_Q(k)] = f_r\ [A(k), Z(k)],\quad r = \overline{1,R},$$
$$z(k+1) = \varphi_q\ [a_1(k),...,a_C(k),z_1(k),...,z_Q(k)] = \varphi_q\ [A(k), Z(k)], \quad (7.2)$$
$$q = \overline{1,Q},\quad k = 0,1,2,...\ .$$

Assuming that the circuit has no feedback loops and makes use of only cynchronous memory elements performing the delay function, we may subdivide it into several levels of memory elements that have been functionally connected in series. Then the mathematical model of such a sequential circuit derived from the set of equations (7.2), when written in vector form, is

$$z_1(k+1) = \varphi_1\ [A(k)],$$
$$z_2(k+1) = \varphi_2\ [A(k), Z_1(k)],$$
$$z_3(k+1) = \varphi_3\ [A(k), Z_1(k), Z_2(k)], \quad (7.3)$$
$$\ldots\ldots\ldots\ldots\ldots\ldots\ldots\ldots\ldots\ldots\ldots\ldots$$
$$z_T(k+1) = \varphi_T\ [A(k), Z_1(k), Z_2(k),..., Z_{T-1}(k)],$$
$$Y(k) = f\ [A(k), Z_1(k), Z_2(k),..., Z_T(k)],$$

where $Z_1(k) \cup Z_2(k) \cup Z_3(k) \cup ... \cup Z_T(k) = Z(k)$. Thus the sequential circuit is represented as $T \leq Q$ levels of functionally dependent memory elements, with the state of a definite jth-level memory element $(j=\overline{1,T})$ in the $(k+1)$th cycle of circuit operation being dependent only on the states of preceding-level memory elements in the kth cycle and the values of input variables $A(k)$.

Rearrangement of the set of equation (7.3) results in

$$z_1(k+1) = \varphi_1\ [A(k)],$$
$$z_2(k+1) = \varphi_2\ [A(k), A(k-1)],$$
$$z_3(k+1) = \varphi_3\ [A(k), A(k-1), A(k-2)], \quad (7.4)$$
$$\ldots\ldots\ldots\ldots\ldots\ldots\ldots\ldots\ldots\ldots\ldots\ldots$$
$$z_T(k+1) = \varphi_T\ [A(k), A(k-1), A(k-2),..., A(k-T+1)],$$
$$Y(k) = f\ [A(k), A(k-1), A(k-2), ..., A(k-T)].$$

The set of equations (7.2) may now be represented in vector form as

$$Z(k) = \varphi\ [A(k), A(k-1), A(k-2), ..., A(k-T+1)],$$
$$Y(k) = f\ [A(k), A(k-1), A(k-2), ..., A(k-T)].$$

By using the above expressions, we may demonstrate that the value of switching function F_l on the l-th node of a sequential circuit is defined as

$$F_l(k) = F_l\,[A\,(k), A\,(k-1), A\,(k-2), ..., A\,(k-T)\,].$$

The occurrence of an untestable fault on the l-th node may be attributed to the fact that $F_l(k)$ has not been tested on all possible sets of vectors $A\,(k)$, $A\,(k-1)$, $A\,(k-2)$,..., $A\,(k-T)$ at generation of input test patterns $A\,(k)$. Although for random test patterns $A(k)$ all possible combinations of binary vectors $A(k)$, $A(k-1)$, $A(k-2)$,..., $A(k-T)$ can be basically derived, for PRTPs it is not the case. Thus for the circuit of Fig.7.1 the value of f_1 is defined as $f_1(k)=f_1\,[A(k), A(k-1)]$ where $A(k)=a_1(k)a_2(k)a_3(k)a_4(k)$ and $A(k-1)=a_1(k-1)a_2(k-1)a_3(k-1)a_4(k-1)$. By using the relations

$$\begin{aligned}
a_1(k) &= a_3(k-1) \oplus a_4(k-1), \\
a_2(k) &= a_1(k-1), \quad a_3(k) = a_2(k-1), \quad (7.5)\\
a_4(k) &= a_3(k-1), \quad k = 0,1,2,...
\end{aligned}$$

which describe the behaviour of a PRTPgenerator(Fig.7.1), we obtain

$$f_1(k) = f_1\,[a_1(k), a_2(k), a_3(k), a_4(k), a_2(k), a_3(k), a_4(k), a_1(k) \oplus a_4(k)\,].$$

Thus the values of f_1, f_2 and f_3 (Fig.7.1), which in the general case depend on eight binary variables, are not tested on all their possible combinations due to the functional dependence between $A(k)$ and $A(k-1)$ expressed by (7.5). Therefore, with the dependence between bits in patterns $A(k)$ and $A(k-1)$ causing certain combinations in the bit pattern sequence $a_1(k)a_2(k)a_3(k)a_4(k)a_1(k-1)a_2(k-1)a_3(k-1)a_4(k-1)$ to be eliminated, the given example (Fig.7.1) display untestable fault conditions. Exhaustive testing for the portion of a digital circuit in question is ensured by generating all possible binary combinations in two successive test patterns $A(k)$ and $A(k-1)$. In general, it is necessary to generate all possible binary combinations in $(T+1)$ succesive test patterns for a circuit containing T levels of functionally dependent sequential memory elements. This condition can be easily met by using test pattern generators based on physical concepts. Such devices are impracticable in testing systems because of the need for maximum speed, rather high length and reproducibility of a test signal.

The fact that very long hardware-generated PRTPs have found extensive applications may be attributed to the above conditions. They are used in testing systems based on probability aproaches, signature analysis as well as in self-testing of LSI/VLSI circuits. A PRTP used in many cases is an M-sequence for which the condition for generating all possible $(T+1)$ n-bit combinations in the $(T+1)$ th successive test pattern will be satisfied, provided the n-bit test patterns are uniformly distributed by the $(T+1)$-dimensional distribution. Then $p^*=0$ holds for any VLSI comprising not more than T functionally dependent memory levels and no feedbacks.

7.2. Estimating Fault Coverage for Random and Pseudorandom Tests

PRTP based on the M-sequences may be used for random and pseudorandom tests. For the case of random testing test patterns are uniformly distributed by the desired dimensional distribution. This allows us to use the probabilistic feature of the PRTP generators. In the second case the deferministic nature of the PRTP is used.

Let us consider the difference between the random and pseudorandom testing. For the purpose we'll take a n-input CC with the total amount of faults equals to Q.

A classical example of random and pseudorandom testing is the use PRTP generators based on M-sequences. Thus for random testing a M-sequence generator, which characteristic polynomial has the degree $m>>n$ is usually used. For pseudorandom testing, the degree of characteristic polynomial satisfies the equality $m=n$. Let us estimate the efficiency of the random and pseudorandom testing by comparing their fault coverage - the main characteristic of any testing method.

At the beginning consider the main peculiarity of random and pseudorandom testing realization. For the purpose remind that for the n-input CC $L=2^n$ test patern which cover all n-bit combination are used. In the case of random testing, the next test pattern chosen from the entire set of patterns equals L.

In the case of the j-th pseudorandom testing, the next test pattern chosen from the rest of the test patterns equals $L-j$ patterns. The full set of patterns for pseudorandom testing equals $L=2^m-1$ where $m=n$ and the code $000...0$ is forbidden. This difference in the generating the next test pattern has a significant impact on the efficiency of random and pseudorandom testing.

Let us define the probability P_j of the first fault detection as the result of applying the j-th test pattern. The first $j-1$ patterns faild detect this fault. Suppose l out of L patterns detect the fault, than for random testing we get

$$P_j = \left(\frac{L-l}{L}\right)^{j-1} \frac{l}{L} = \left(1 - \frac{l}{L}\right)^{j-1} \frac{l}{L} \qquad (7.6)$$

where $\frac{L-l}{L}$ - is the probability fault escape after applying one test pattern; l/L - is the detection probability.

Corresponding probabilities for pseudorandom testing depend on the number j of the patterns. The fault escape probability as the result of applying t pattern is defined as a ratio

$$\frac{L-l-t+1}{L-t+1}, \qquad 1 \leq t \leq j-1,$$

and the detection pobability as $l/(L-j+1)$. As the result

$$P_j = \left(\frac{L-l}{L}\right)\left(\frac{L-l-1}{L-1}\right)\left(\frac{L-l-2}{L-2}\right) \cdots \left(\frac{L-l-(j-2)}{L-(j-2)}\right)$$
$$\left(\frac{l}{L-(j-1)}\right) = \frac{C_{L-j}^{l-1}}{C_L^l}, \qquad j \geq 1. \qquad (7.7)$$

Here we asume that the set of pseudorandom test patterns includes the code $000...0$, so $L=2^n$.

For the purpose estimating the fault coverage of the circuit, let us introduce the distribution of the detect abilities as vector $H = h_1, h_2, ..., h_l$, where the element h_l of the vector is the number of faults detected by l out of L patterns. Moreover, introduce the random variable d_{ji} which equals 1 if fault w_i is detected first by the j-th pattern only and 0 if for all j patterns w_i remains undetected. Thus the total number G_j of "new" faults (the faults detected firstly by the j-th pattern) will be defined by the relation

$$G_j = \sum_{i=1}^{Q} d_{ji}, \qquad (7.8)$$

where Q is the total number of the faults.

On the basis of the value of P_j and the vector H define the probabilistic estimation of G_j. It may be present as mean

$$M[G_j] = M\left\{\sum_{i=1}^{Q} d_{ji}\right\} = \sum_{i=1}^{Q} M(d_{ji}) = \sum_{i=1}^{Q} p_{ji} = \sum_{l=1}^{L} h_l P_j, \quad (7.9)$$

where p_{ji} is the probability of equality $d_{ji}=1$.
Substituting the value P_j from (7.6) and (7.7) into (7.9) we get, that

$$M[G_j] = \sum_{l=1}^{L} h_l \left(1 - \frac{l}{L}\right)^{j-1} \left(\frac{l}{L}\right) \quad (7.10)$$

for random testing and

$$M[G_j] = \sum_{l=1}^{L} h_l \frac{C_{L-j}^{l-1}}{C_L^l} \quad (7.11)$$

for pseudorandom testing.

Then the mean of fault coverage C_N will be defined as the ratio of the number of faults to be detected to the total amount of faults detected by the test of length N. Corresponding expression for this estimation has the form

$$M[C_N] = \frac{1}{Q}\left(\sum_{j=1}^{N} M(G_j)\right). \quad (7.12)$$

Substituting (7.10) into (7.12) we get

$$M[C_N] = 1 - \frac{1}{Q}\sum_{l=1}^{L}\left(1 - \frac{l}{L}\right)^N h_l. \quad (7.13)$$

for random testing and the same estimation for the pseudorandom testing

$$M[C_N] = 1 - \frac{1}{Q}\sum_{l=1}^{L-N} h_l \frac{C_{L-N}^l}{C_N^l}. \qquad (7.14)$$

Differences between the last two relations show that the random and pseu-

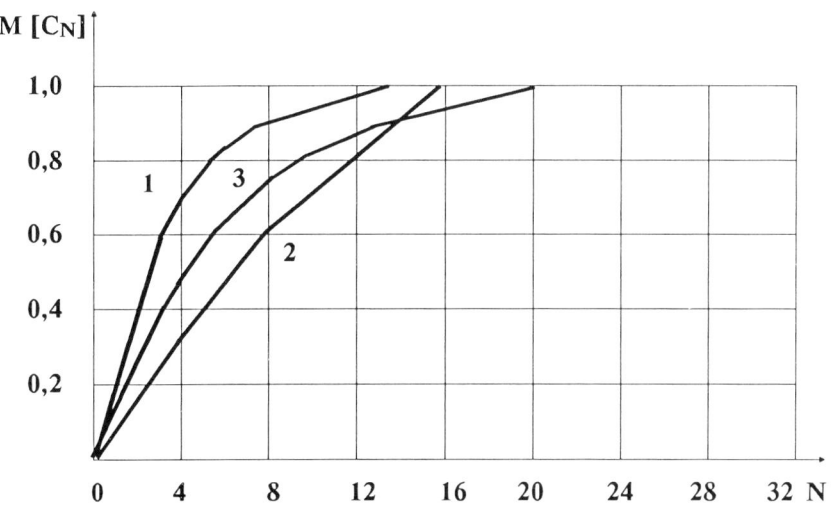

Fig. 7.2. The plots

dorandom testing differ in quality.

As an example consider the circuit of the full four-input adder for which there exists the set of $L=2^4=16$ test patterns. The total number of faults equals Q ($Q=90$), the vector $H=h_1h_2h_3h_4h_5h_6h_8=1,11,2,43,21,4,8$. The plot of $M[C_N]$ for the case of the pseudorandom testing are shown at the Fig.7.2-1. Fig.7.2.-2 and Fig.7.2.-3 show the plots for $M[C_N]$ for circuit of Schneider counterexample and adder-2. The values for the two circuits are: $L=16$ and 32; $Q=44$ and 142; vector $H=h_1h_2h_3h_{14}=29,19,1,1$ and $H=h_4h_8h_{16}=86,24,32$.

More detailed investigation of the numerical characteristics of the random and pseudorandom testing was carried out by E.McCluskey.

7.3. Pseudorandom Test Length Calculation

As is shown the main arguments for random test length calculation are the detection probability of the fault P_d as the result of applying the first test pattern and the desired probability of its detection during N patterns $P_d(N)$. For the pseudorandom testing, the probability of fault detection by the first pattern is determined by the same value P_d according to relation

$$P_d^1 = P_d = \frac{l}{L}, \qquad (7.15)$$

where l - is the number of test patterns out of L detected the fault. The probability of fault detection by the second pattern will be defined as the product of the probability to escape detection of the fault by the first pattern by the ratio $l/(L-1)$. The last ratio is the probability of fault detection by the second pattern when the first pattern failed for defect the fault. The second pattern will be generated out of $L-1$ patterns. There are the differences from random testing. At the result

$$P_d^2 = (1 - P_d^1)\frac{l}{L-1} = (1 - P_d^1)\left[\frac{L}{L-1}\right] P_d. \qquad (7.16)$$

For the N-th pattern we get

$$P_d^N = (1 - P_d^1 - P_d^2 - \ldots - P_d^{N-1}) P_d\left[\frac{l}{L-N+1}\right] \qquad (7.17)$$

All probabilities in expression (7.17) are the probabilities of independent events. Taking in to account the last comments, define the probability of fault detection by the N successive pseudorandom test patterns. As the result

$$P_d(N) = \sum_{j=1}^{N} P_d^j = \sum_{j=1}^{N}\left[(1 - \sum_{k=1}^{j-1} P_d^k) P_d \frac{L}{L-j+1}\right]. \qquad (7.18)$$

The last expression may be rearranged as

$$P_d(N) = \frac{1 - [(L-N)!(L-P_dL)!]}{L!(L-P_dL-N)!} = \frac{1-[(L-N)!(L-1)!]}{L!(L-1-N)!} =$$

$$1 - \frac{\prod_{j=0}^{N-1}(L-l-j)}{\prod_{j=1}^{N-1}(L-j)}. \tag{7.19}$$

For desired values of $P_d(N)$ and P_d let us calculate pseudorandom test lengh N using the expression (7.19). Here the value of l equals N. Experimental results using the expression (7.19) are shown in Tables 7.1. and 7.2. where the test length is calculated on the basis of the worst-case fault. Tables 7.1 and 7.2 show the desired length N for random (R) and pseudorandom (P) testing respectively for a five-input CC and eight-input CC.

Table 7.1.

$P_d(N)$	P_d											
	1/64		2/64		4/64		8/64		16/64		32/64	
	R	P	R	P	R	P	R	P	R	P	R	P
0.90	147	58	73	44	36	28	18	16	9	8	4	4
0.95	191	61	95	50	47	33	23	19	11	10	5	5
0.99	293	64	146	58	72	43	35	27	17	15	7	7
0.999	439	64	218	62	108	52	52	35	25	20	10	10

Table 7.2.

$P_d(N)$	P_d											
	1/64		2/64		4/64		8/64		16/64		32/64	
	R	P	R	P	R	P	R	P	R	P	R	P
0.90	589	231	294	175	147	112	73	64	36	34	18	17
0.95	766	244	382	199	191	135	95	79	47	43	23	22
0.99	1177	254	588	23o	193	175	146	111	72	63	35	33
0.999	1765	256	881	248	439	210	218	147	106	88	52	147

The main result which follows from the tables is that the length of the pseudorandom test in any case less then the length of random one.

The length of random test may be the upper bound of the length of the pseudorandom test. To get a more precise value of pseudorandom test length, take into account the relation (7.19).

7.4. Structured VLSI Design with Built-In Random and Pseudorandom Tests

As it has been noted earlier, for the specified efficiency of VLSI test defined by $P_d(N)$, the length N both for random and pseudorandom test depends to a large degree on the value of P_d, i.e. the probability of worst-case fault detection techniques for random tests with different probabilities of a zero or one to appear in the test pattern bits to increase P_d. Similar results may be obtained by using the random test principles at the VLSI design stage. The increase of P_d for the worst-case fault is mainly influenced by observability, controllability and testability of the entire VLSI and its individual nodes. The same as for deterministic tests, one must maximize observability and controllability to improve the efficiency of random and pseudorandom tests used. For the purpose, we can use different techniques of structured VLSI design (refer to Chapter 2) thereby improving its characteristics under study to a large extent. When the structured solutions fail to produce the required results, VLSI-implemented logic modification procedure can be used. Circuit engineering solutions applied make it possible to increase the probability of a worst-case fault to be detected by improving observability and controllability.

The VLSI logic modification procedure consists of the following stages:

1. Calculate the probability of fault detection P_d for the entire set of faults. Then determine the values of signal probabilities on all VLSI nodes. Find the accurate or lower estimates both for P_d and signal probabilities.

2. Calculate the limiting value P_d^* for the VLSI nodes such that for P_d less than P_d^* the VLSI design is to be modified.

3. For the faults with $P_d < P_d^*$ such that $P_d \neq 0$ (meaning that the node is not logically redundant), find the type of VLSI modification in the following order of priority:

a) if controllability is high enough and $P_d < P_d^*$ modify the VLSI design to improve observability.

b) if controllability and P_d are equally low, modify the VLSI design to improve controllability.

c) if controllability and P_d do not meet the requirement, and P_d differs from the threshold value P_d^* for more than the controllability from its specification, increase both controllability and observability.

4. Calculate the values of P_d and signal probabilities for the VLSI. If they meet the specification, terminate the procedure. Otherwise go to item 3.

The inclusion of random testing is crucial in designing the VLSIs which represent microprocessor packages. In parallel with the use of classic solutions for structured VLSI design there were originally formulated specific requirements governed by the nature of test sequences.

First requirement. The entire set of possible codes used to specify an operation to be performed by the microprocessor must be legal. By this is meant that there must be no illegal operation codes which might make the microprocessor behavior unpredictable at implementation of random or pseudorandom tests. For the purpose, certain operations may be coded ambiguously, i.e. a subset of operation codes rather than a unique code is assigned to one or more operations.

Second Requirement. Data exchanged between blocks must be uniformly distributed and uncorrelated in the VLSI test mode.

To meet the above requirements, we propose to randomize test sequences by using modulo-2 adders. By way of example let us consider the ALU of Fig.7.3, which is applied with a test pattern x from data bus in the test mode. In the general case, when random testing is not provided, the ALU has no modulo-2 adders between register P_1 and the combinational portion of ALU. The absence of adders is equivalent to applying a variable $a=0$ to their secondary inputs. Let us consider the case. Assume that x is a binary

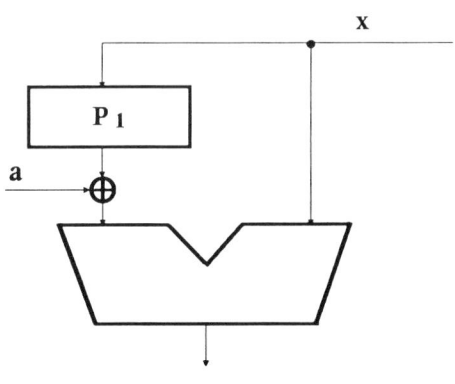

Fig. 7.3 ALU

variable, which takes the value of 0 or 1. One can readily see that the sequence of test patterns x, x will be followed by input values x, x produced at the ALU inputs. Both input codes are terms. After the pair of terms x,x has been generated, the next pair of terms corresponding to $\bar{x}$, $\bar{x}$ or $\bar{x}$, x cannot be obtained. This may be attributed to correlation between data applied to both adder inputs due to a single-step delay in the buffer register P_1.

Inability of generating different pairs of test patterns involves degradation in the random testing efficiency caused by the failure to detect faults that transform the combinational circuit into the sequential. Considering the above fact, it is necessary to eliminate the dependency between the ALU terms in random test mode. As may be seen from Fig.7.3, the ALU structure is completed with two-input modulo-2 adders which secondary inputs are applied with a pseudorandom variable $a \in \{0, 1\}$ thus upsetting the severe dependence between data applied to the ALU inputs.

Third Requirement. In the test mode, microoperations of microprocessor must be generated randomly. To implement this requirement as well as the preceding two, hardware overhead are to be imposed on microprocessor structure. Hardware overhead to implement a self-testing microprocessor design by the above requirements is only 14%.

Chapter 8

EXHAUSTIVE TESTING

8.1. Exhaustive Testing Principles

Exhaustive testing is an outgrowth of the pseudorandom testing. Similar to other techniques discussed in the preceding sections, it is based on the use of pseudorandom test sequences. A distinguishing feature of exhaustive testing consists in the use of only deterministic pseudorandom sequences during the test experiment. In this case probability properties of a pseudorandom test pattern sequence are practically fully ignored.

Exhaustive testing is widely used to design self-testing VLSI chips that implement one of the design for testability techniques. This kind of testing is most frequently used for the LSSD implementation. In such a case, a self-testing VLSI is a structure represented by a single shift register and a combinational portion partitioned into multiple combinational subcircuits in the test mode.

Exhaustive testing consists in producing all possible input patterns for each input to a n-input combinational circuit or for each combinational subcircuit in the circuit.

Consider in more detail exhaustive testing for the VLSI structure consisting in the test mode of shift register R1 of length w and combinational circuit consisting of r subcircuits G_i, $i=\overline{1,r}$, with the i th subcircuit having n_i inputs and v_i outputs (Fig.8.1). Let G have n inputs and v outputs. Normally $v_i=1$, hence $v=r$. In the general case, the self-testing VLSI structure consists of a more compex combination of shift registers and combinational circuits. However, a structure of Fig.8.1, for which the exhaustive test is to be produced, can always be singled out.

For the VLSI circuit of Fig.8.1, the problem of exhaustive testing consists in generating a sequence of bits $y(k) \in \{0,1\}$, $k=\overline{1,l}$, applied to an input of shift register R1 of length w to ensure generation of all possible binary patterns for any subcircuit G_i, $i=\overline{1,r}$, in G. This problem can be solved by different methods. However, the major measure of their effectiveness is the complexity of constructing an exhaustive test generator and the length of test produced by the generator.

One of the most efficient solutions available is the use of a M-sequence as $\{y(k)\}$. Then hardware overheads will be m bits of R1 and some modulo-2 adders. The value m, which is the degree of generating polynomial $\phi(x)$, is most frequently calculated from the inequality

$$deg\,\phi(x) > w. \qquad (8.1)$$

Then the length of test sequence N is defined by

$$N = 2^{deg\,\phi(x)} - 1 > 2^w.$$

It is evident that when N test patterns are generated by the primitive polynomial whose high degree satisfies the equation (8.1), all possible binary patterns, i.e. exhaustive tests, will be produced for all subcircuits in the circuit of Fig.8.1.

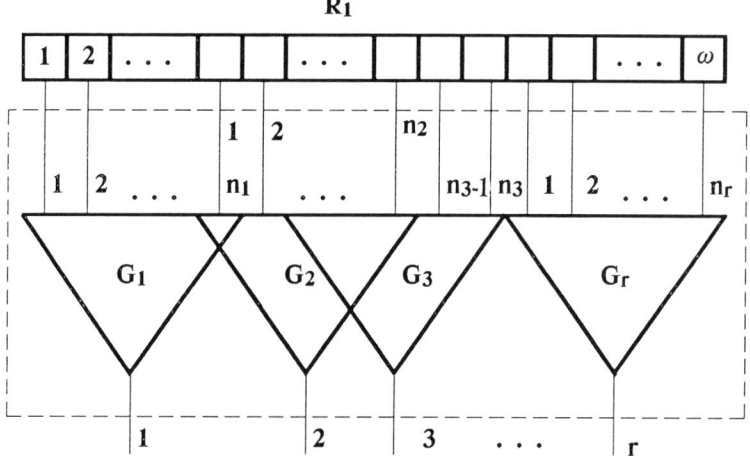

Fig. 8.1. The VLSI circuit

Actual values w may be from *100* to *1000* thereby making impossible exhaustive test generation. Therefore expression (8.1) is used most often as an upper bound on the degree m of generating polynomial $\phi(x)$ that allows one to implement exhaustive testing. The lower bound can be calculated from the inequality

$$deg\phi(x) > \max_i n_i. \qquad (8.2)$$

Note that the inequality (8.2) gives a necessary but not sufficient condition to be satisfied by the generating polynomial $\phi(x)$. In fact, for the combinational portion of circuit G in Fig.8.2, which consists of a single subcircuit G_1 with $n=3$ inputs, it is impossible to ensure all possible input patterns by using $\phi(x)=1\oplus x^3\oplus x^4$ with $deg\phi(x)=4>3$. Then there will be no input patterns *001*, *010*, *100* and *111* on inputs 1,2 and 3 of G_1, which is defined by the form of the selected generating polynomial $\phi(x)$.

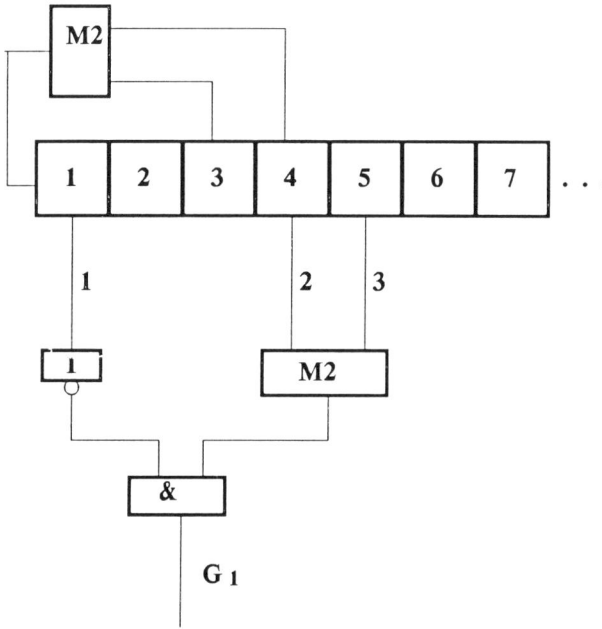

Fig. 8.2. The Combinational Portion of Circuit G

Based on the shift-and-add property of the M-sequence generated by $\phi(x)=1\oplus x^3\oplus x^4$, we can demonstrate that the relation $a_1(k)=a_4(k)\oplus a_5(k)$, $k=\overline{1,L}$, $L=2^4-1$ is met for values $a_1(k)$, $a_4(k)$ and $a_5(k)$ of shift-register positions *1,4* and *5*. As a consequence, some input patterns are missing and, hence, certain faults such as $\equiv 0$ are undetectable at the output of a two-input AND gate.

Thus it is necessary to select a suitable polynomial $\phi(x)$ to generate the sequence $\{a(k)\}$, $k=1$ to 2^m-1, which meets the conditions below.

Chapter 8

1. On generating a M-sequence $\{a(k)\}$, described by $\phi(x)$, all possible n_i-bit patterns are produced in the positions of shift register R1 (see Fig.7.1) whose numbers Q_j^i, $j=\overline{1,n_i}$, are determined by the topology of connections between the inputs of the i th subcircuit G_i.

2. The high degree m of generating polynomial $\phi(x)$ must assume the minimal possible value defined by the inequality

$$\max_i n_i < m \leq w + 1,$$

or, subject to generation of a zero n_i-bit code, where $i=\overline{1,r}$, as a result of initial setting of R1, by the inequality

$$\max_i n_i \leq m \leq w$$

To ensure the above conditions for the generating polynomial $\phi(x)$, the circuit under test should be subjected to a rigorous analysis.

This analysis can be eliminated from the procedure of synthesizing the required sequence $\{a(k)\}$ through the definition of its generating polynomial only by increasing the period and respectively the length N of the test sequence. Here we can confine ourselves to the relation (8.1) or expression

$$deg\,\phi(x) = \max_i (max_j Q_j^i - \min_j Q_j^i) + 1, \qquad (8.3)$$

where Q_j^i is the number of position of R1 whose output is connected to the j-th input, $j=\overline{1,n_i}$, of the i th subcircuit G_i.

For the example shown in Fig.8.2, the set of combinational subcircuits consists of a single subcircuit G_1 for which $j=\overline{1,3}$ and $Q_1^1 = 1$, $Q_2^1 = 4$, and $Q_3^1 = 5$. From equation (8.3), we obtain deg $\phi(x) = Q_3^1 - Q_1^1 + 1 = 5$; hence, the application of $\phi(x)=1 \oplus x^2 \oplus x^5$, for example, ensures exhaustive testing of the circuit G_1 under test (Fig.8.2). The M-sequence period $L=31$ is the basic value that determines the test length N calculated for $m<w$ by the formula:

$$N = L + w - m = 2^m + w - m - 1. \qquad (8.4)$$

This case takes account of the fact of exhaustive test generation for the subcircuits of G, which are connected randomly to R1 of length w. As a result, for $w=7$, $N=33$.

Thus by equation (8.3), we can calculate the high degree of polynomial $\phi(x)$ that ensures exhaustive testing of the circuit under test without its analysis. Then any primitive polynomial $\phi(x)$ whose degree is defined by (8.3) will meet the requirements imposed on such polynomials.

We may easily demonstrate that the combinational circuit G_1 (Fig.8.2) may also be exhaustively tested by application of polynomial $\phi(x)=1 \oplus x^2 \oplus x^3$. In fact, all possible 3-bit combinations are applied to the three input lines to combinational circuit G_1 in accordance with the time chart for the states of memory elements in the PRPG shift register (Fig.8.3). This time the length of test sequence $N=11$ will be significantly shorter than that for the case of polynomial $\phi(x)=1 \oplus x^2 \oplus x^5$.

As is seen, an attempt to eliminate the procedure of synthesizing an optimal M-sequence in terms of its minimally possible period causes the test sequence length N to be significantly increased. At the same time, by artificially synthesizing the M-sequence, we can find the polynomial $\phi(x)$ whose high degree is defined by $m=max_i\ n_i$. Below we shall estimate the complexity of finding a polynomial with the minimally possible high degree.

Table 8.1

Cycle N	Position number in shift register					Circuit input number		
	1	2	3	4	5	1	2	3
1	1	0	0	0	0	1	0	0
2	0	1	0	0	0	0	0	0
3	1	0	1	0	0	1	0	0
4	1	1	0	1	0	1	1	0
5	1	1	1	0	1	1	0	1
6	0	1	1	1	0	0	1	0
7	0	0	1	1	1	0	1	1
8	1	0	0	1	1	1	1	1
9	0	1	0	0	1	0	0	1

8.2. Complexity of Exhaustive Test Sequence Generator Design

A major task in designing exhaustive test sequence generators is to find a generating polynomial $\phi(x)$ with the minimal high degree m for an M-sequence which provides for generation of all possible binary combinations on all sets $\{Q_1^j, Q_2^j, \ldots, Q_{n_i}^j\}$ of positions in the shift register of the VLSI chip under test. Note that the set $\{Q_1^j, Q_2^j, \ldots, Q_{n_i}^j\}$ for specific i determines the positions of shift register R1 that are applied with the inputs of subcircuit G_i of the combinational part of the VLSI chip in off-line testing mode. Consider the solution of the problem for the case when the combinational part of the VLSI chip consists of a single subcircuit G_1 whose layout of input connections is described by the set $\{Q_1, Q_2, \ldots, Q_{n_i}\}$.

In the first place, let us associate the set of shift register position numbers $\{Q_1, Q_2, \ldots, Q_n\}$ with the values of symbols $a(k), a(k+\tau_1), a(k+\tau_2), \ldots, a(k+\tau_{n-1})$ of the desired M-sequence, where $\tau_j = Q_{j+1} - Q_1$, $j=\overline{1,n-1}$, $Q_{j+1} > Q_j$. Then the original problem of exhaustively covering the combinational VLSI part G_1 reduces to making the symbols $a(k), a(k+\tau_1), a(k+\tau_2), \ldots, a(k+\tau_{n-1})$ of the desired M-sequence linearly independent. Its generating polynomial $\phi(x)$ has the high degree m defined by (8.2) whose minimal value is n.

Let us estimate the probability of a situation that a randomly selected polynomial $\phi(x)$ will allow to implement exhaustive testing. As a measure of testing efficiency, let us introduce the probability $P_r(m,v)$ of producing all possible binary combinations in $v \leq m$ random

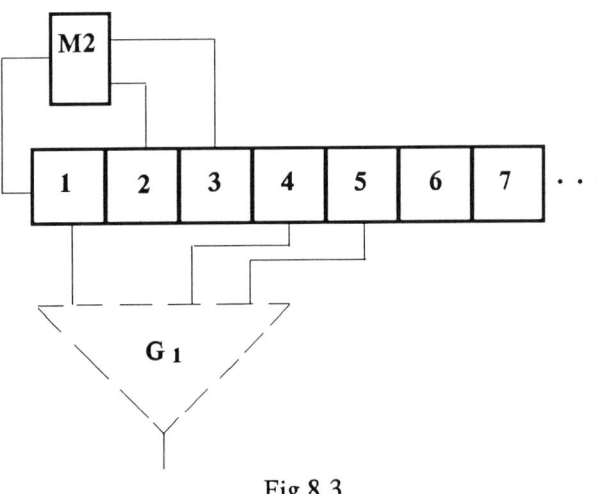

Fig.8.3

positions of the shift register of length $w \geq m$ for which the following theorem holds true.

Theorem 8.1. When an M-sequence described by $\phi(x)$ for which $deg\phi(x)=m$, where $m<w\leq 2^m-1$, is generated, the probability $P_r(m,v)$ of producing all possible binary combinations in $\leq m$ positions of the w-position shift register is defined by

$$P_r(m,v) = \prod_{t=3}^{v} \frac{2^m - 2^{t-1}}{2^m - t} \qquad (8.5)$$

<u>Proof.</u> Suppose that all possible binary combinations may be produced in the $(t-1)$th of v positions of the shift register. This is an indication of linear independence of M-sequence symbols associated with the $(t-1)$th register position. Estimate the probability that a randomly chosen tth position will form a linearly independent group of M-sequence symbols together with any subset of $(t-1)$th positions.

For the prespecified $(t-1)$th positions of the shift register, there are $2^{t-1}-1$ nonempty subsets of positions and appropriate subsets of M-sequence symbols. Owing to the fact that the $(t-1)$th symbol of the M-sequence forms a linearly independent set, any pair of their subsets will also be linearly independent. Each of the 2^{t-1} nonempty subsets will be associated with only one distinct tth position such that a M-sequence symbol produced in it is linearly dependent on the original subset. An exception is the $(t-1)$ subsets consisting of a single shift-register position each, due to the fact that the condition $w\leq 2^m-1$ is satisfied. Hence the total number of occurrences when t symbols of the M-sequence are linearly dependent is

$$2^{t-1} - 1 - (t-1) = 2^{t-1} - t$$

and the number of positions where the tth symbol might be formed is $2^m-1-(t-1)=2^m-t$. With the equally likely choice of the tth position in the shift register of length $w=2^m-1$, the probability $P(m,t)$ that the tth symbol of the M-sequence will be linearly independent of any subset from the $(t-1)$th preselected pattern may be obtained by

$$P(m,t) = \frac{2^m - t - 2^{t-1} + t}{2^m - t} = \frac{2^m - 2^{t-1}}{2^m - t}, \quad t \leq m. \qquad (8.6)$$

For $w<2^m-1$, the relation (8.6) may be used as the expected value of probability $P(m,t)$ and extended to the general case $w \leq 2^m-1$.

The derivation of (8.6) is based on the assumption that the $(t-1)$th symbol of an M-sequence is as linearly independent as any subset of its symbols. Since the probability $P(m,t-1)$ of such an event is calculated by expression (8.6) and based on the linear independence of $(t-2)$ symbols and any of their subsets, the probability $P_r(m,v)$ of producing all possible binary combinations in v out of m positions of the shift register will be determined by the probability that all subsets of symbols in sets consisting of one, two, three, etc., and v symbols of the M-sequence are linearly independent at a time. Following from the relation $w \leq 2^m-1$, we finally obtain

$$P_r(m,v) = P(m,1) P(m,2) P(m,3) \ldots P(m,v).$$

Since $P(m,1)=P(m,2)=1$,

$$P_r(m,v) = \prod_{t=3}^{v} P(m,t) = \prod_{t=3}^{v} \frac{2^m - 2^{t-1}}{2^m - t}.$$

The value of $P_r(m,v)$ is evidently a function of variables m and v, where $v \leq m$, and it depends uniquely on their specific ratio. We examine this dependence of $P_r(m,v)$ for $m=10$, 20 and 30. The estimates for $P_r(m,v)$ are summarized in Table 8.2. From the table it follows that, the value of $P_r(m,v)$ depends only on the difference $m-v$. This is supported by the estimates of Table 8.3 for the values of $m=10$ to 20 and $m-v=0$ to 5.

From the table it follows that for values $m>14$, $P_r(m,m)=0,29$. This implies that the use of primitive polynomial $\phi(x)$ whose $deg\phi(x)=m$ allows one to produce all possible binary combinations in $v=m$ specified positions of the VLSI shift register (see Fig.8.1) with probability 0.29. To improve the probability of exhaustive coverage of the VLSI combinational part connected to v positions of the register, it is advisable to use the polynomials whose high degree $m>v$. Thus, for example, for $m=14$ the probability of producing all possible binary combinations in the specified $v=10$ positions of the VLSI shift register is 0.94 whereas it is 0.29 for $m=10$.

The outcome of Theorem 8.1 may be interpreted as follows. A random primitive polynomial $\phi(x)$ allows one to implement VLSI exhaustive testing with probability $P_r(m,v)$, where v is the number of inputs to the VLSI

Table 8.2

v	m=10	m=20	m=30
2	1,000000	1,000000	1,000000
4	0,984361	0,999995	0,999999
6	0,959216	0,999959	0,999999
8	0,798514	0,999791	0,999999
10	0,305077	0,999077	0,999999
12		0,996173	0,999996
14		0,984555	0,999984
16		0,938913	0,999939
18		0,770227	0,999756
20		0,288846	0,999023
22			0,996099
24			0,984456
26			0,938790
28			0,770101
30			0,288788

Table 8.3

| m | m-v | | | | | |
	0	1	2	3	4	5
10	0,31	0,60	0,80	0,91	0,96	0,98
12	0,29	0,59	0,78	0,89	0,95	0,98
14	0,29	0,58	0,77	0,88	0,94	0,97
...	...	...	...	...	...	...
20	0,29	0,58	0,77	0,88	0,94	0,97

combinational part. This probability $P_r(m,v)$ implies that random polynomial $\phi(x)$ provides for exhaustive coverage for the r-input combinational VLSI part. The search for an appropriate polynomial $\phi(x)$ among the set of polynomials whose high degree is m will be characterized by rather high probability $P_r(m,m)=0.29$ for its finding as a result of the first polynomial selected at random. In this case, $v=m$ and is the number of inputs to a single subcircuit G_1 in the combinational VLSI part.

8.3. Exhaustive Testing by Preweighted Vectors

The problem of designing exhaustive tests that has been discussed in the preceding sections consists in finding a M-sequence generator such that exhaustive search is implemented on the specified bit positions of w long register R1. More generally we can formulate the problem as follows: for the specified w and $n<m$ an exhaustive test sequence generator is to be designed for any n of w register positions. In other words, the generated test sequence must ensure exhaustive search for binary combinations on any subset of n bits in w- long register.

An ingenious method used for solving the problem is based on the use of constant weight vectors where weight is taken to mean the number of unity components in a vector. The principle of the method is made evident by the following theorem.

Theorem 8.2. Sequence T of binary test patterns of length w provides for obtaining all possible binary combinations on any $n<w$ of w random bits if it contains all w-dimensional vectors of weight v such that $v = c \bmod (w-n+1)$ for constant c which satisfies the inequality $0 \leq c \leq w-n$.

An important corollary to the above theorem is the fact that there exist $w-n+1$ solutions for T. Let us consider some of them for the special case of w and n. Initially consider the case where $w=5$, $n=2$ and hence $w-n+1=4$. According to the theorem 8.2, c varies from 0 to $w-n=3$ and, for example, for $c=0$, the test pattern T consists of vectors whose weight v satisfies the identity $v=0 \bmod 4$. Having resolved the latter identity, we obtain v equal to 0 and 4. Thus the test pattern T consists of vectors T_0 weighting 0 and vectors T_4 weighting 4. As a result we obtain

$$T^0 = T_0 \cup T_4 = \begin{vmatrix} 0 & 0 & 0 & 0 & 0 \end{vmatrix} \cup \begin{vmatrix} 0 & 1 & 1 & 1 & 1 \\ 1 & 0 & 1 & 1 & 1 \\ 1 & 1 & 0 & 1 & 1 \\ 1 & 1 & 1 & 0 & 1 \\ 1 & 1 & 1 & 1 & 0 \end{vmatrix} = \begin{vmatrix} 0 & 0 & 0 & 0 & 0 \\ 0 & 1 & 1 & 1 & 1 \\ 1 & 0 & 1 & 1 & 1 \\ 1 & 1 & 0 & 1 & 1 \\ 1 & 1 & 1 & 0 & 1 \\ 1 & 1 & 1 & 1 & 0 \end{vmatrix}.$$

For another value of c, let it be 2, we obtain

$$T^0 = T_2 = \begin{vmatrix} 1 & 1 & 0 & 0 & 0 \\ 0 & 1 & 1 & 0 & 0 \\ 0 & 0 & 1 & 1 & 0 \\ 0 & 0 & 0 & 1 & 1 \\ 1 & 0 & 1 & 0 & 0 \\ 0 & 1 & 0 & 1 & 0 \\ 0 & 0 & 1 & 0 & 1 \\ 1 & 0 & 0 & 1 & 0 \\ 0 & 1 & 0 & 0 & 1 \\ 1 & 0 & 0 & 0 & 1 \end{vmatrix}.$$

Exploring this example further we can construct test patterns T^0 and T^1 for $c=1$ and $c=3$. As a result we obtain that the minimum number of vectors contains patterns T^0 and T^1 and the whole collection T^0, T^1, T^2 and T^3 comprises the complete set of 5-bit vectors. Subsets T^0, T^1, T^2 and T^3 which are test patterns, represent disjoint subsets of a complete set of 5-dimensional binary vectors. The latter statement which holds true for the general case, makes it possible to estimate the minimal size of test T^c. Considering that the minimal cardinality of test T^c must not exceed the average number of test patterns for all solutions, we may demonstrate that

$$N^*_{min} \leq \frac{2^w}{w-n+1}. \quad (8.7)$$

The above expression gives the upper estimate of the minimal number of test patterns.

This estimate can be refined only the specific cases of relations between w and n. Thus by theorem 8.2 the minimum test for $n \leq w/2$ will be obtained with $v=int(n/2)$ and $int(n/2)+w-n+1$. Then

$$N^*_{min} \approx C_w^{(n/2)} + C_w^{n-(n/2)-1}.$$

Table 8.4 gives the minimum test length N_{min} and its estimate N^*_{min} against w and n. The analysis of results presented in the table reveals that the present technique is inapplicable for large values of w and n. At the same time, however, the test patterns for small values of w and n are relatively short and can be readily generated by shift registers thereby making them feasible for self-test VLSI design.

Table 8.4

n	v	N_{min}	N^*_{min}
10	3	20	128
10	6	165	205
10	9	512	512
20	3	40	58254
20	6	1330	69905
20	12	54264	116508
20	18	349525	349525
40	3	80	$2,89 \times 10^{10}$
40	6	10060	$3,14 \times 10^{10}$
40	10	749398	$3,55 \times 10^{10}$
40	20	$1,121 \times 10^9$	$5,24 \times 10^{10}$
80	3	160	$1,55 \times 10^{22}$
80	6	85320	$1,61 \times 10^{22}$
80	10	$25,62 \times 10^6$	$1,70 \times 10^{22}$
80	20	$1,88 \times 10^{12}$	$1,98 \times 10^{22}$

Then, for mutually prime values of w and v, it will take

$$\frac{(2w-1)N}{w} \approx 2N$$

cycles to produce all test patterns rather than wN shifts required for generation of all N tests. For the purpose, a cyclic shift register with a modulo-2 gate at its input (Fig.8.4) is used. The input sequence $a(k)$ to the shift register determines the rules of generating the full set of test patterns for test T.

By way of example let us consider the timing chart of exhaustive test generation.

$$T^3 = T_3 = \begin{vmatrix} 1 & 1 & 1 & 0 & 0 \\ 0 & 1 & 1 & 1 & 0 \\ 0 & 0 & 1 & 1 & 1 \\ 1 & 1 & 0 & 1 & 0 \\ 0 & 1 & 1 & 0 & 1 \\ 1 & 0 & 1 & 1 & 0 \\ 0 & 1 & 0 & 1 & 1 \\ 1 & 1 & 0 & 0 & 1 \\ 1 & 0 & 0 & 1 & 1 \\ 1 & 0 & 1 & 0 & 1 \end{vmatrix}.$$

for the scheme of Fig.8.2, when $w=5$ and $n=2$. Table 8.5 gives one of the versions of $a(k)$ and the timing chart of shift register states for the shift register producing test patterns in T^3.

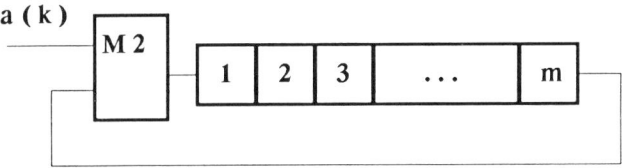

Fig. 8.4. The Cyclic Shift Register

It is assumed that the shift register (Fig.8.4) is initially set to sero. The first test pattern is produced within the first three cycles. Their furthe order is determined by the values of $a(k)$. Generation of *10* test patterns of T^3 results in only *3* excessive patterns.

Table 8.5

k	a(k)	Register state	Excessive pattern
0		0 0 0 0 0	
1	1	1 0 0 0 0	x
2	1	1 1 0 0 0	x
3	1	1 1 1 0 0	
4	0	0 1 1 1 0	
5	0	0 0 1 1 1	
6	0	1 0 0 1 1	
7	0	1 1 0 0 1	
8	1	0 1 1 0 0	x
9	1	1 0 1 1 0	
10	0	0 1 0 1 1	
11	0	1 0 1 0 1	
12	0	1 1 0 1 0	
13	0	0 1 1 0 1	

A major advantage of the technique in question is the possibility of its implementation together with one of the VLSI design for testability techniques e.g. LSSD. This eventually makes possible the self-test VLSI design.

8.4. Iterative Algorithm for Exhaustive Test Generation

An iterative procedure provides for formalizing exhaustive test generation for large w. It uses the test for small w at the first iteration to produce all possible combinations of n bits out of w bits. The resulting test will also exhaustively test any n out of w inputs to the VLSI chip under test.

Let us introduce some concepts required to describe the iterative test generation procedure.

Exhaustive Testing

Assume that $P(i)$ is a cyclic version of vector consisting of m elements such that

$$P(i) = (i, i+1, i+2, \ldots, m-1, 0, 1, \ldots, i-2, i-1) \quad (8.8)$$

According to equation (8.8), for $i=0,1,2$ and $m-1$, we obtain

$$P(0) = (0, 1, 2, \ldots, m-2, m-1), \quad P(1) = (1, 2, 3, \ldots, m-1, 0),$$
$$P(1) = (2, 3, 4, \ldots, 0, 1), \quad P(m-1) = (m-1, 0, 1, \ldots, m-3, m-2),$$

Also, let us introduce vector $C(i)$, $C(i)=(i,i,\ldots,i)$, $i \in \{0,1,2,\ldots,m-1\}$

For $m=q^j$, where q is a primitive number and $j>1$, we can construct the $(q+1) \times m^2$ dimensional matrix $M(m^2)$ of the form

$$M(m^2) = \begin{vmatrix} C(0), C(1), & C(2), & \ldots, C(m-1), \\ P(0), P(0), & P(0), & \ldots, P(0), \\ P(0), P(1), & P(2), & \ldots, P(m-1), \\ P(0), P(2), & P(4), & \ldots, P[2(m-1)], \\ \ldots, & \ldots, & \ldots, \ldots, \\ P(0), P(q-1), & P[2(q-1)], & \ldots, P[(m-1)(q-1),] \end{vmatrix} \mod m, (8.9)$$

which contains no (2x2)-dimensional submatrix with identical columns. For example, for $m=3$, the matrix $M(3^2)$ will appear by equation (8.9) as

$$M(9) = \begin{vmatrix} C(0) & C(1) & C(2) \\ P(0) & P(0) & P(0) \\ P(0) & P(1) & P(2) \\ P(0) & P(2) & P(4) \end{vmatrix} \mod 3 = \begin{vmatrix} C(0) & C(1) & C(2) \\ P(0) & P(0) & P(0) \\ P(0) & P(1) & P(2) \\ P(0) & P(2) & P(1) \end{vmatrix} =$$

$$= \begin{vmatrix} (000) & (111) & (222) \\ (012) & (012) & (012) \\ (012) & (120) & (201) \\ (012) & (201) & (120) \end{vmatrix} \quad (8.10)$$

The matrix $M(m^2)$ is used to construct an iterative procedure for the test $T(m^2)$ to be built on the basis of exhaustive test $T(m)$ by replacing each

$M(m^2)$ element by the column of matrix $T(m)$ whose number corresponds to the element of matrix $M(m^2)$. For the above example we can construct a matrix $T(9)$ by using $M(9)$ and the original matrix $T(3)$ which has the following form

$$T(3) = \begin{vmatrix} 0 & 0 & 0 \\ 0 & 1 & 1 \\ 1 & 0 & 1 \\ 1 & 1 & 0 \end{vmatrix}. \quad (8.11)$$

Then

$$T(9) = \begin{vmatrix} 0 & 0 & 0 & 0 & 0 & 0 & 0 & 0 & 0 \\ 0 & 0 & 0 & 1 & 1 & 1 & 1 & 1 & 1 \\ 1 & 1 & 1 & 0 & 0 & 0 & 1 & 1 & 1 \\ 1 & 1 & 1 & 1 & 1 & 1 & 0 & 0 & 0 \\ 0 & 0 & 0 & 0 & 0 & 0 & 0 & 0 & 0 \\ 0 & 1 & 1 & 0 & 1 & 1 & 0 & 1 & 1 \\ 1 & 0 & 1 & 1 & 0 & 1 & 1 & 0 & 1 \\ 1 & 1 & 0 & 1 & 1 & 0 & 1 & 1 & 0 \\ 0 & 0 & 0 & 0 & 0 & 0 & 0 & 0 & 0 \\ 0 & 1 & 1 & 1 & 1 & 0 & 1 & 0 & 1 \\ 1 & 0 & 1 & 0 & 1 & 1 & 1 & 1 & 0 \\ 1 & 1 & 0 & 1 & 0 & 1 & 0 & 1 & 1 \\ 0 & 0 & 0 & 0 & 0 & 0 & 0 & 0 & 0 \\ 0 & 1 & 1 & 1 & 0 & 1 & 1 & 1 & 0 \\ 1 & 0 & 1 & 1 & 1 & 0 & 0 & 1 & 1 \\ 1 & 1 & 0 & 0 & 1 & 1 & 1 & 0 & 1 \end{vmatrix}.$$

The iterative procedure for exhaustive test generation is based on the following theorem.

Theorem 8.3. If the original exhaustive test $T(m)$ provides for all possible binary combinations of any n out of m bits of register of length $w \le m$, where $m=q^j$, such that q is a prime number, and $j \ge 1$ and $m \ge int(n^2/4)$, test $T(m^2)$ constructed in accordance with any combination $int(n^2/4)+1$ of rows of $M(m^2)$ will exhaustively cover any n out of m^2 register positions.

An example of such test is test $T(9)$ that has been produced in accordance with the first two rows of matrix $M(9)$ and test $T(3)$, which is exhaustive for $n=2$. Here $m=q=3$.

As a result we obtain the new test

$$T(9) = \begin{vmatrix} 0\,0\,0 & 0\,0\,0 & 0\,0\,0 \\ 0\,0\,0 & 1\,1\,1 & 1\,1\,1 \\ 1\,1\,1 & 0\,0\,0 & 1\,1\,1 \\ 1\,1\,1 & 1\,1\,1 & 0\,0\,0 \\ 0\,0\,0 & 0\,0\,0 & 0\,0\,0 \\ 0\,1\,1 & 0\,1\,1 & 0\,1\,1 \\ 1\,0\,1 & 1\,0\,1 & 1\,0\,1 \\ 1\,1\,0 & 1\,1\,0 & 1\,1\,0 \end{vmatrix},$$

which provides for all possible binary combinations of any two bits out of nine. Note that $T(3)$ is an exhaustive test for $(3=q)$- bit register. The solution of the same problem by test $T(9)$ for a higher-dimensional register can be explained by the increase in dimensionality, i.e. the number of test patterns N. Let us find the length N of test $T(m^{2^\gamma})$ where γ is the number of iterations if $N=B$ for $T(m)$. Then for $T(m^2)$ we obtain $N=B\,[int\,(n^2/4)+1]$, and for the general case, i.e. $T(m^{2^\gamma})$ we obtain

$$N = B\,[\,int\,(n^2/4) + 1\,]^\gamma$$

Rearrangement of latter expression will eventually produce

$$N = B\,[2^\gamma\,]^{\log_2\,[int(\frac{n^2}{4})\,+\,1]}. \qquad (8.12)$$

For the example being considered, $B=4$ and $N=4\,2^\gamma$.

A major problem in the usage of the iterative algorithm is selection of an original exhaustive test whose dimentionality affects the resulting test length to a large extent.

Only particular cases of exhaustive tests whose dimensionality is low are known for small w's. Among them is the test:

$$T(4) = \begin{vmatrix} 0 & 0 & 0 & 0 \\ 0 & 0 & 1 & 1 \\ 0 & 1 & 0 & 1 \\ 0 & 1 & 1 & 0 \\ 1 & 0 & 0 & 1 \\ 1 & 0 & 1 & 0 \\ 1 & 1 & 0 & 0 \\ 1 & 1 & 1 & 1 \end{vmatrix}$$

providing for the exhaustive search for any *n=3* out of *4* bits, and the test:

$$T(8) = \begin{vmatrix} 0 & 0 & 0 & 0 & 0 & 0 & 0 & 0 \\ 0 & 0 & 1 & 1 & 1 & 0 & 0 & 1 \\ 0 & 1 & 0 & 1 & 0 & 1 & 1 & 1 \\ 1 & 0 & 0 & 0 & 1 & 1 & 1 & 0 \\ 1 & 1 & 1 & 0 & 0 & 0 & 1 & 1 \\ 1 & 1 & 1 & 1 & 1 & 1 & 0 & 0 \end{vmatrix}.$$

A significant drawback of the technique in question is the complexity of test pattern generation, especially for the self-test VLSI chips that implement one of the LSSD modifications.

Chapter 9

SIGNATURE ANALYSIS

9.1. Signature Analysis as a Binary Polynomial Division Algorithm

A dominating data compression technique used at compact testing is signature analysis. It is carried out by the use of special devices, signature analyzers, which are designed for detecting errors caused by the VLSI faults in data streams being analyzed.

Various data compression models and algorithms have been used to describe the procedure of data compression in a signature analyzer. The model that is most commonly used implements the idea of data compression as binary polynomial division. In this case the binary sequence

$$[y(k)] = y(l-1)\, y(l-2)\, y(l-3) \ldots y(0),$$

where $y(k) \in \{0,1\}$, $k = \overline{0, l-1}$ may have the form of binary polynomial

$$\chi(x) = y(l-1)\, x^{l-1} \oplus y(l-2)\, x^{l-2} \oplus y(l-3)\, x^{l-3} \oplus \ldots \oplus y(1)\, x^1 \oplus y(0). \quad (9.1)$$

Thus, for example, sequence *10011* may be represented as a binary polynomial $\chi(x) = 1x^4 \oplus 0x^3 \oplus 0x^2 \oplus 1x^1 \oplus 1x^0 = x^4 \oplus x^1 \oplus 1$. The maximum degree of the polynomial $\chi(x)$ is called its degree. In the above example the polynomial degree is *4*.

For a binary polynomial of (9.1), there are defined modulo-2 arithmetic operations. By adding two polynomials $x^3 \oplus x^1 \oplus 1$ and $x^2 \oplus x^1 \oplus 1$ we obtain the resulting polynomial $x^3 \oplus x^2$.

Addition is equivalent to subtraction. For these operations the following equality is met:

$$x^r \oplus x^r = x^r - x^r = 0.$$

Multiplication of binary polynomials is similar to arithmetic multiplication except for modulo-2 addition performed to obtain the resulting product. For example, by multiplying two polynomials $x^4 \oplus x^2 \oplus x^1 \oplus 1$ and $x^3 \oplus x^2 \oplus 1$, we obtain

$$
\begin{array}{r}
x^4 \oplus x^2 \oplus x^1 \oplus 1 \\
x^3 \oplus x^2 \oplus 1 \\
\hline
x^4 \oplus x^2 \oplus x^1 \oplus 1 \\
x^6 \oplus x^4 \oplus x^3 \oplus x^2 \\
x^7 \oplus x^5 \oplus x^4 \oplus x^3 \\
\hline
x^7 \oplus x^6 \oplus x^5 \oplus x^4 \oplus x^1 \oplus 1
\end{array}
$$

Division of polynomials results in a quotient and a remainder that are related by the classical equation

$$\chi(x) = q(x)\psi(x) \oplus S(x), \qquad (9.2)$$

where $\chi(x)$ is the dividend; $\varphi(x)$ is the divisor, $q(x)$ is the quotient and $S(x)$ is the remainder. Dividing polynomial $\chi(x) = x^7 \oplus x^6 \oplus x^5 \oplus x^4 \oplus x^2 \oplus 1$ by polynomial $\psi(x) = x^3 \oplus x^2 \oplus 1$ we obtain

$$
\begin{array}{r|l}
x^7 \oplus x^6 \oplus x^5 \oplus x^4 \oplus x^2 \oplus 1 & \; x^3 \oplus x^2 \oplus 1 \\
x^7 \oplus x^6 \oplus x^4 & \\
\cline{1-1}\cline{2-2}
x^5 \oplus x^2 & x^4 \oplus x^2 \oplus x^1 \oplus 1 \\
x^5 \oplus x^4 \oplus x^2 & \\
\cline{1-1}
x^4 & \\
x^4 \oplus x^3 \oplus x^1 & \\
\cline{1-1}
x^3 \oplus x^1 \oplus 1 & \\
x^3 \oplus x^2 \oplus 1 & \\
\cline{1-1}
x^2 \oplus x^1 &
\end{array}
$$

where $q(x) = x^4 \oplus x^2 \oplus x \oplus 1$; $S(x) = x^2 \oplus x^1$.

Data stream representation in the form of polynomial is only used for convenience of investigating the properties of coding/decoding networks that are widely used in the theory of cyclic codes. The polynomial representation may be replaced with an original binary representation of data streams upon which the above described operations can be performed. For the example considered earlier, similar conversions during division will take the form

```
1 1 1 1 0 1 0 1        | 1 1 0 1
1 1 0 1                ---------
---------               1 0 1 1 1
    1 0 0 1
    1 1 0 1
    -------
      1 0 0 0
      1 1 0 1
      -------
        1 0 1 1
        1 1 0 1
        -------
          1 1 0
```

Here the quotient is *110* which corresponds to the polynomial representation $S(x)=x^2 \oplus x$.

Division can be formalized as data compression on a shift register with internal modulo-2 adders.

Fig.9.1 shows an example of dividing data stream *11110101* described by polynomial $\chi(x)=x^7 \oplus x^6 \oplus x^5 \oplus x^4 \oplus x^2 \oplus 1$ whose compression is performed by the shift register described by polynomial $\psi(x)=x^3 \oplus x^2 \oplus 1$. The initial state of storage elements in the shift register is assumed to be *000*. The data stream *11110101* to be compressed is successively applied to the shift register input, causing storage elements to change their state as per the timing chart of Fig.9.1. The remainder $S(x)$ of polynomial $\chi(x)=x^7 \oplus x^6 \oplus x^5 \oplus x^4 \oplus x^2 \oplus 1$ divided by polynomial $\psi(x)=x^3 \oplus x^2 \oplus 1$ is set on the shift register's storage elements and assumes the value $S(x)=x^2 \oplus x$ as a polynomial or $S(x)=110$ as a binary code of polynomial coefficients. The correspondence between $S(x)$ and the remainder of $\chi(x)$ divided by $\psi(x)$ may be supported by the earlier considered examples of division.

The scheme of a shift register described by polynomial $\psi(x)$ and employed for compressing binary data by the division algorithm is called a signature analyzer. Compression results in a signature, which is the remainder $S(x)$.

144   Chapter 9

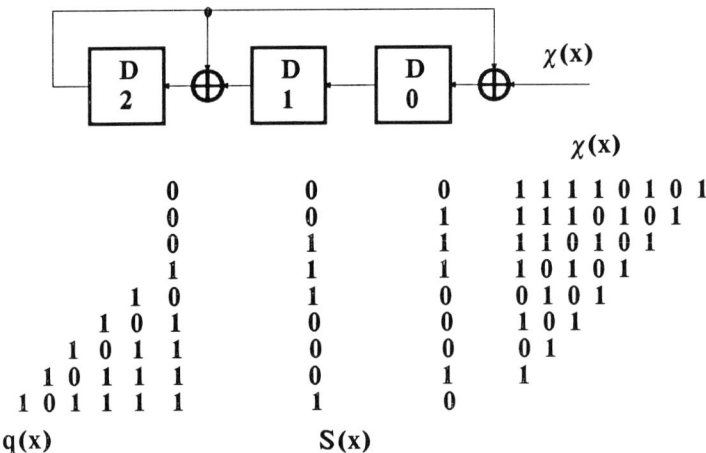

Fig. 9.1. Shift register with internal modulo-2 adders

For practical implementation of the signature analyzer described by polynomial $\psi(x)=x^3 \oplus x^2 \oplus 1$, as well as for any other polynomial, there exists an alternative structure (see Fig.9.2), which is preferable from the hardware implementation standpoint. However, $S(x)$ resulting from data stream convolution by the signature analyzer with external modulo-2 adders does not coincide with the remainder, i.e. $S'(x) \neq S(x)$. At the same time, there is a one-to-one correspondence between $S'(x)$ and $S(x)$ that may be expressed in the general case as

$$S(x) = \begin{vmatrix} \alpha_m & 0 & 0 & \ldots & 0 & 0 \\ \alpha_{m-1} & \alpha_m & 0 & \ldots & 0 & 0 \\ \alpha_{m-2} & \alpha_{m-1} & \alpha_m & \ldots & 0 & 0 \\ \ldots & \ldots & \ldots & \ldots & \ldots \\ \alpha_2 & \alpha_3 & \alpha_4 & \ldots & \alpha_m & 0 \\ \alpha_1 & \alpha_2 & \alpha_3 & \ldots & \alpha_{m-1} & \alpha_m \end{vmatrix} S'(x) \quad (9.3)$$

here $S'(x)$ results from convolution by the signature analyzer described by polynomial $\phi(x)$; $S(x)$ is the remainder of the polynomial $\chi(x)$ divided by polynomial $\psi(x)$, which is the inverse of $\phi(x)$; and $\alpha_i \in \{0,1\}$, $i=\overline{1,m}$. are coefficients of polynomial $\psi(x)$.

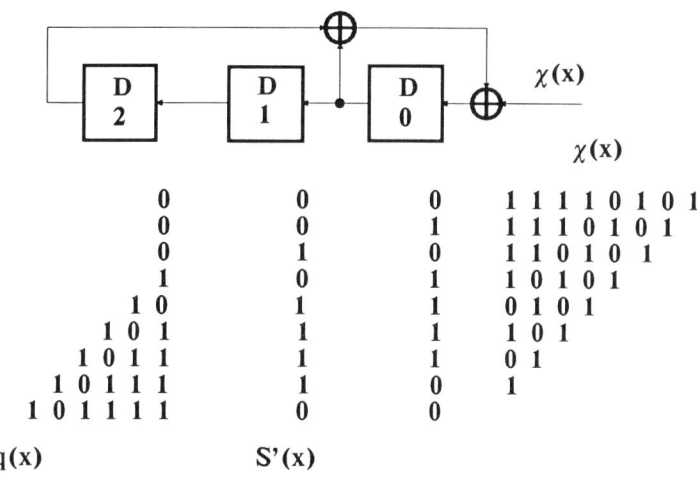

Fig. 9.2. The alternative structure of the shift register

For the specific case of Fig.9.1 and 9.2, signature $S(x)$ results from dividing $\chi(x)$ by $\psi(x)=x^3 \oplus x^2 \oplus 1$ and $S'(x)$ results from compressing the input data stream on the analyzer with external modulo-2 adders described by $\varphi(x)=x^3 \oplus x \oplus 1$ where $\psi(x)=\varphi^{-1}(x)$. The relation (9.3) for $S(x)=110$ and $S'(x)$ will have the form

$$S(x) = \begin{vmatrix} \alpha_3 & 0 & 0 \\ \alpha_2 & \alpha_3 & 0 \\ \alpha_1 & \alpha_2 & \alpha_3 \end{vmatrix} \quad S'(x) = \begin{vmatrix} 1 & 0 & 0 \\ 1 & 1 & 0 \\ 0 & 1 & 1 \end{vmatrix} \begin{vmatrix} 1 \\ 0 \\ 0 \end{vmatrix} = \begin{vmatrix} 1 \\ 1 \\ 0 \end{vmatrix}$$

The conventional signature analyzer scheme includes a shift register (1) and a modulo-2 adder (2) whose inputs are connected to the outputs of register positions in accordance with the generating polynomial $\phi(x)$ (Fig.9.3). The signature analyzer control signals are START, CLOCK and STOP. The START and STOP signals establish the time window within which data compression is performed. The START signal resets all storage elements in the shift register (1) to zero and the register starts bitwise shifting bit by bit to the right by the CLOCR signal. There at the analyzer performs binary polynomial division. To describe the procedure, we can use other mathematic expressions describing signature generation. For the perpose, we can use the same mathematic approach as for describing the behavior of M-sequence generators. In this case the behavior of signature analyzer described by polynomial $\varphi(x) = 1 \oplus x^1 \oplus x^2 \oplus \ldots \oplus x^{m-1} \oplus x^m$ is expressed as

$$\begin{vmatrix} a_1(k) \\ a_2(k) \\ a_3(k) \\ \ldots \\ a_m(k) \end{vmatrix} = \begin{vmatrix} \alpha_1 & \alpha_2 & \alpha_3 & \ldots & \alpha_{m-1} & \alpha_m \\ 1 & 0 & 0 & \ldots & 0 & 0 \\ 0 & 1 & 0 & \ldots & 0 & 0 \\ \ldots & & & & & \\ 0 & 0 & 0 & \ldots & 1 & 0 \end{vmatrix} \begin{vmatrix} a_1(k-1) \\ a_2(k-1) \\ a_3(k-1) \\ \ldots \\ a_m(k-1) \end{vmatrix} \oplus \begin{vmatrix} y(k) \\ 0 \\ 0 \\ \ldots \\ 0 \end{vmatrix},$$

where $y(k) \in \{0,1\}$ is the k th bit of compressed sequence $\{y(k)\}$, $k=\overline{1,l}$; $\alpha_i \in \{0,1\}$ are the coefficients of generating polynomial and $a_i(k-1) \in \{0,1\}$ is the content of the ith storage element in the shift register (1) in the $(k-1)$ th cycle. The initial state of the register is $a_1(0)a_2(0)a_3(0)\ldots a_m(0)=000\ldots0$. The latter expression can be represented in a more compact form as a set of equations

$$a_i(0) = 0, \quad i = \overline{1,m};$$
$$a_1(k) = y(k) \oplus \sum_{i=1}^{m} {}^{\oplus} \alpha_i a_i(k-1); \qquad (9.4)$$
$$a_j(k) = a_{j-1}(k-1), \quad j = \overline{2,m}, \quad k = \overline{1,l},$$

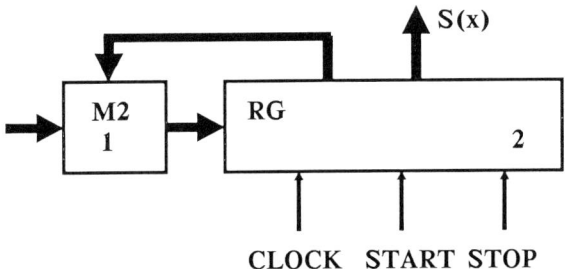

Fig. 9.3. Signature analyzer

where l is normally taken to be equal to or less than 2^m-1, hence it determines the length of sequence to be compressed. When l cycles of signature analyzer operation elapse, the binary code, which is a signature $S(x)=a_1(l)a_2(l)a_3(l)...a_m(l)$ mapped into m-bit code, will be fixed on its storage elements.

Relations (9.3) and (9.4) that describe the signature generating algorithm prove the simplicity of signature analyzer implementation. As an example, consider the HP5004A signature analyzer described by polynomial $\varphi(x) = 1 \oplus x^7 \oplus x^9 \oplus x^{12} \oplus x^{16}$. This analyzer has been considered an unofficial standard to similar devices for a long time. Its scheme comprises a five-input modulo-2 adder and a *16*-position shift register as well as several gates for generation of START and STOP signals. The shift-register positions are grouped into four tetrads, whose contents define the value of hexadecimal signature position.

9.2. Structured Design of Signature Analyzers

For signature analysis application it is necessary to estimate its detectability in relation to specific fault types in a circuit under test. The outcome of this idea is the number and configuration of faults that may be detected by specific analyzer structure. Such studies are normally based on an assumption that specific fault type usually involves distortion of circuit output responses.

Thus choosing the most efficient signature analyzer involves the use of trial-and-error method. This results in an analyzer which detects the maximum number of faults in a circuit. It is currently topical to develop an analytical method of signature analyzer design. The starting point for solving the problem is to estimate the length of a signature capable to detect faults in a specific digital circuit with the required probability. Let us consider the process of solving the problem.

We shall initially try to find out the relationship between signature analysis efficiency and the number of faults Q that may be detected by the method in a digital circuit and the length $m=deg\ \phi(x)$ of a signature analyzer where $\phi(x)$ is a generating polynomial of the analyzer. As a measure of the signature analysis efficiency we shall choose the probability P_d of detecting a circuit fault by the signature analyzer. The probability P_d is expressed as

$$P_d = (1 - \frac{1}{2^m}). \qquad (9.5)$$

On the assumption that there may be Q distinct faults in the circuit under test their detection will characterize the probability $P_d(Q)$ estimated by

$$P_d(Q) = (1 - \frac{1}{2^m})^Q. \qquad (9.6)$$

Fig 9.4 depicts the relationship between $P_d(Q)$ and Q for different m where the number of faults Q is replaced with their relative value $q=Q/2^m$. By analyzing the plots for $m=2,4,5$, we may deduce that when m increases for all q's, $P_d(q)$ tends to limits (with $m=\infty$). For this purpose we shall find the limit

$$\lim_{m \to \infty} (1 - \frac{1}{2^m})^{q 2^m}. \qquad (9.7)$$

Subject to

$$\lim_{m \to \infty} (1 - \frac{1}{2^m})^{q 2^m} = \lim_{2^m \to \infty} (1 - \frac{1}{2^m})^{q 2^m}$$

and substituting $2^m = x$ we obtain

$$\lim_{x \to \infty} (1 - \frac{1}{x})^{qx} = e^{-q}.$$

Thus we shall finally find for (9.7) that

$$\lim_{m \to \infty} P_d(Q) = \lim_{m \to \infty} (1 - \frac{1}{2^m})^{q 2^m} = e^{-q}. \qquad (9.8)$$

Applying the estimate obtained for $P_d(q)$, we can write the general expression for any m. For the purpose we substitute the right side of expression (9.6) for e^{-q}, where $q=Q/2^m$, and obtain

Signature Analysis

$$P_d(Q) = e^{-Q/2^m}. \quad (9.9)$$

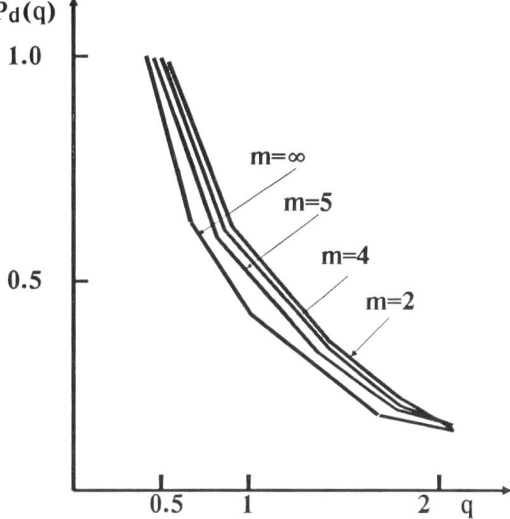

Fig. 9.4. Relationship between Q and Pd(Q)

The relation (9.9) determines the efficiency of applying signature analysis $P_d(Q)$ for detection Q faults in a digital circuit by a m-bit signature analyzer. With the number of circuit faults Q and their detection probability $P_d(Q)$ used as initial data, we can determine signature analyzer length m, which ensures the required efficiency (9.9).

As a result, we obtain the expression

$$\ln P_d(Q) = -\frac{Q}{2^m},$$

which, when rearranged, takes the form

$$2^m = -\frac{Q}{\ln P_d(Q)}.$$

By taking the *log*s and rearranging we shall finally obtain

$$m = \log_2 Q - \log_2 \ln \left[\frac{1}{P_d(Q)}\right]. \quad (9.10)$$

Taking in account of the fact that m must be only integer, we finally write

$$m \geq int\left[\log_2 Q - \log_2 \ln \left(\frac{1}{P_d(Q)}\right)\right]. \quad (9.11)$$

The latter inequality is a relation allowing to calculate the key signature analyzer parameter, i.e. its length m which defines the high degree of its primitive generating polynomials.

There is another rough method of estimating the probability $P_d(Q)$. Based on equation (9.6), we can obtain the relations for finding the probability of detecting the specified number of faults Q. Considering the relation for $Q=1,2,3,...$, we obtain

$$P_d(1) = 1 - \frac{1}{2^m};$$
$$P_d(2) = (1 - \frac{1}{2^m})^2 = 1 - \frac{2}{2^m} + \frac{1}{2^{2m}}.$$

In the latter relation we can neglect the term $1/2^{2m}$ for comparatively large m. Then

$$P_d(2) = 1 - \frac{2}{2^m}.$$

Reasoning along similar lines, we can write

$$P_d(3) = 1 - \frac{3}{2^m}.$$

for $Q=3$.

Thus the general relationship for the probability of detecting Q faults by a m-bit signature analyzer assumes the form

$$P_d(Q) = 1 - \frac{Q}{2^m}. \qquad (9.12)$$

Whence we obtain

$$m > int[\log_2 Q - \log_2(1 - P_d(Q))]. \qquad (9.13)$$

By comparing the relations (9.11) and (9.13) we may note that they differ only by arguments in the second term. With the required probability values $P_d(Q)$ approximating one, calculations by formulas (9.11) and (9.13) produce the same result. The accuracy of such calculations depends on the accuracy of an involved parameter Q, i.e. the number of distinct faults in a digital circuit. This parameter definition is a rather complicated problem which for the single stuck-at fault case can be expressed as

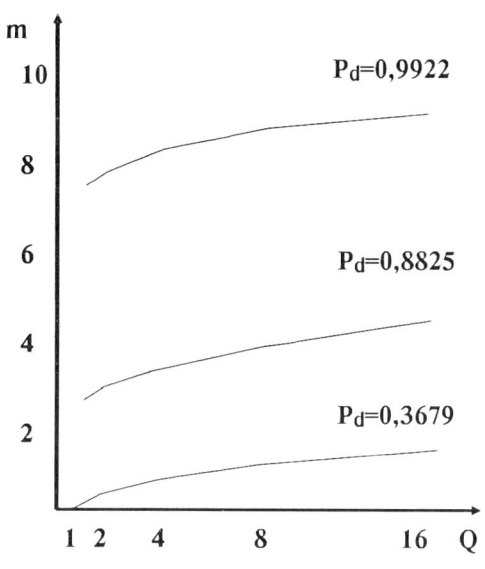

Fig. 9.5. Relationship between m and number of faults Q

$$Q = 2n, \quad (9.14)$$

where n is the number of nodes in a digital circuit.

Let us consider an example of a digital circuit with single stuck-at faults. Their total number $Q=2^{10}$. For the circuit, we must construct a signature analyzer which detects faulty circuit state with probability $P_d=0.9922$.

By the formulas (9.11) and (9.13) we obtain $m=10+7=17$. Therefore, applying the primitive polynomial $\phi(x)$ of degree $m=17$ for signature analyzer construction, we may detect $Q=2^{10}$ faults with $P_d=0.9922$.

Fig. 9.5 shows the relationships between signature analyzer lengths depending on the number of faults Q for the specified probability $P_d(Q)$. Such plots can be used as nomograms to determine the minimum length of signature analyzer as a function of fault detection probability $P_d(Q)$ and the total number of distinct faults Q in a digital circuit.

9.3. Quadratic Signature Analyzer Property

The distinctive property of a signature analyzer is that its efficiency is independent of the form of reference sequence to be compressed. Actually, whether a fault is detected or undetected is determined by its shape. Therefore the linear signature analyzer of fixed structure will not detect a well-defined set of errors whose shape is described by a primitive polynomial associated with the analyzer. For example, the signature analyzer described by a primitive polynomial with degree $m=3$ will not detect $V_n=15$ errors in the sequence of length $l=2-1=7$. And for $\phi(x) = 1 \oplus x^2 \oplus x^3$ the errors of Table 9.1 described by polynomial $e(x)$, where $q(x)$ is the quotient of $e(x)$ divided by $\phi(x)$, will be undetectable. The fact that the error escaped detection is $S(x)=0$, which is the remainder from dividing $\phi(x)$ by $e(x)$.

Table 9.1.

N	e(x)	q(x)	N	e(x)	q(x)
1	0 0 0 1 1 0 1	0 0 0 1	9	0 1 0 1 1 1 0	0 1 1 0
2	0 0 1 1 0 1 0	0 0 1 0	10	1 0 1 1 1 0 0	1 1 0 0
3	0 1 1 0 1 0 0	0 1 0 0	11	0 1 1 1 0 0 1	0 1 0 1
4	1 1 0 1 0 0 0	1 0 0 0	12	1 1 1 0 0 1 0	1 0 1 0
5	0 1 0 0 0 1 1	0 1 1 1	13	1 1 0 0 1 0 1	1 0 0 1
6	1 0 0 0 1 1 0	1 1 1 0	14	1 0 0 1 0 1 1	1 1 1 1
7	1 0 1 0 0 0 1	1 1 0 1	15	1 1 1 1 1 1 1	1 0 1 1
8	0 0 1 0 1 1 1	0 0 1 1			

For any error of Table 9.1 the equality

$$S\,[\chi(x)] = S\,[\chi(x) \oplus e(x)] \qquad (9.15)$$

holds true. Taking in account of the linearity property of modulo addition,

$$S\,[\chi(x) \oplus e(x)] = S\,[\chi(x)] \oplus S\,[e(x)] = S\,[\chi(x)].$$

Thus the signature $S[\chi_e(x)]$ of the sequence with error $e(x)$ matches the reference signature $S[\chi(x)]$, thereby pointing to the fact that the error in question is undetectable.

The idea behind quadratic signature analysis is to relate detection of specific error $e(x)$ with the shape of the reference compressed sequence $y(k) \in \{0,1\}$, $k=\overline{0,l-1}$, associated with polynomial $\chi(x)$. It is implemented by the nonlinear binary vector multiplication. The procedure of signature generation in the quadratic signature analyzer consists in finding its value by the expression

$$S(x) = z(0)\,z(1) \oplus z(2)\,z(3) \oplus \ldots \oplus z(2t-2)\,z(2t-1), \quad (9.16)$$

where $z(v)=y(vm)y(vm+1)y(vm+2)\ldots y(vm+m-1)$, $v=\overline{0,2t-1}$, and $l=2mt$ is the length of sequence $\{y(k)\}$. In the expression (9.16) addition corresponds to bit-to-bit modulo-2 addition, and multiplication corresponds to modulo-2 multiplication of primitive polynomial $\varphi(x)$ with degree m. Operation $z(v)z(v+1)$ results in the remainder from dividing the product obtained at multiplication by polynomial $\varphi(x)$. Let us consider such a multiplication for the case when $z(v)=0101$ and $z(v+1)=1001$. Each of cofactors is associated with a binary polynomial: polynomial $x^2 \oplus 1$ for $z(v)$; polynomial $x^3 \oplus 1$ for $z(v+1)$. By multiplying them we obtain

$$
\begin{array}{rrrrr}
 & & & x^2 \oplus & 1 \\
 & & x^3 \oplus & & 1 \\
\hline
 & & & x^2 \oplus & 1 \\
x^5 \oplus & & x^3 & & \\
\hline
x^5 \oplus & & x^3 \oplus & x^2 \oplus & 1 \\
\end{array}
$$

Considering that polynomial $\varphi(x)=1 \oplus x \oplus x^4$, we obtain $x^4=1 \oplus x$ and $x^5=x \oplus x^2$. Once $x^5=x \oplus x^2$ has been substituted, we obtain

$$z(v)z(v-1)\,mod\,\varphi(x) = (x^2 \oplus 1)(x^3 \oplus 1)\,mod\,\varphi(x) =$$
$$= x^5 \oplus x^3 \oplus x^2 \oplus 1\,mod\,\varphi(x) = x^3 \oplus x \oplus 1 \; = 1\,0\,1\,1.$$

Now we can find the value of signature for the sequence $y(k)=10100 10$ 101010011 in accordance with expression (9.16) and primitive polynomial

$\phi(x) = 1 \oplus x \oplus x^4$ (with $m=4$). Since $l=16=2\;4\;2$, where $m=4$, $t=2$, then $z(0)=1010$, $z(1)=0101$, $z(2)=0101$, $z(3)=0011$.

Hence

$$S(x) = [z(0)z(1)]\,mod\,(1 \oplus x \oplus x^4) \oplus [z(2)z(3)]\,mod\,(1 \oplus x \oplus x^4) =$$
$$[(x^3 \oplus x)(x^2 \oplus 1)]\,mod\,(1 \oplus x \oplus x^4) \oplus$$
$$[(x^2 \oplus 1)(x \oplus 1)]\,mod\,(1 \oplus x \oplus x^4) =$$
$$(x^5 \oplus x)\,mod\,(1 \oplus x \oplus x^4) \oplus (x^3 \oplus x^2 \oplus x \oplus 1)\,mod\,(1 \oplus x \oplus x^4) =$$
$$(x \oplus x^2 \oplus x) \oplus (x^3 \oplus x^2 \oplus x \oplus 1) = (x^3 \oplus x \oplus 1) = 1011.$$

Modulo multiplication of a primitive generating polynomial is most time-consuming operation used to find signature $S(x)$ by quadratic signature analyzer. Technically, it can be performed by a shift register described by polynomial $\phi(x)$ as seen from the scheme of Fig.9.6. The first operand is written in the m-bit static register R1, and the second operand is successively applied to the primary inputs of m two-input AND gates. The secondary inputs of the same gates are connected to the outputs from R1 positions. In m cycles the product $z(v)\,z(v-1)\,mod\,\phi(x)$ will be generated on the linear shift register. High efficiency of the quadratic signature analyzer can be substantiated by the following theorem.

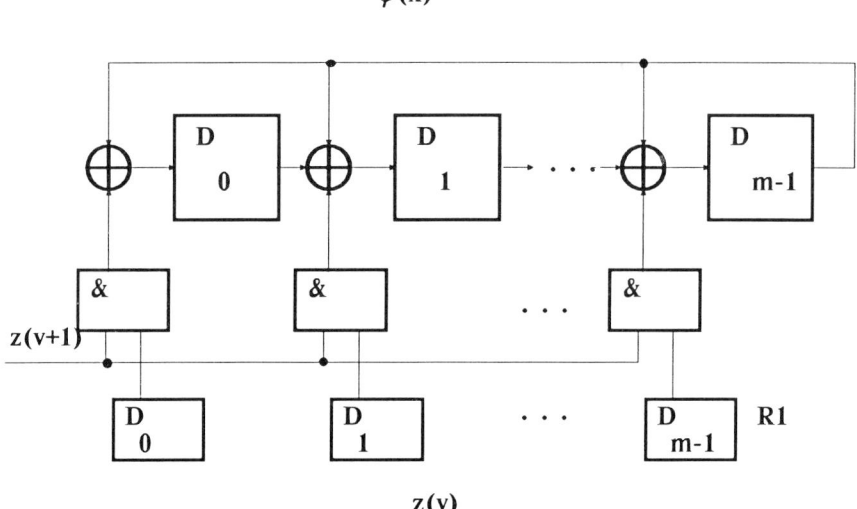

Fig. 9.6. Quadratic signature analyzer

Theorem 9.1. The probability $P_n(e)$ of failing to detect an error described by polynomial $e(x) \neq 0$ in binary data stream $y(k)$, $k=\overline{0, l-1}$ by analyzing the signature expressed as

$$S(x) = \sum_{v=0}^{t-1} {}^{\oplus} [z(2v)z(2v+1)] \bmod \varphi(x)$$

where $z(r) = y(rm) y(rm+1) y(rm+2) ... y(rm+m-1)$, $r=\overline{0, 2t-1}$, $l=2mt$ and m is the high degree of primitive generating polynomial $\varphi(x)$, satisfies the following equation

$$P_n(e) = \frac{1}{2^m}.$$

For the case of classical linear signature analysis, the probability $P_n(e)$ assumes the value 0 or 1 for specific error e. The theorem holds under the assumption that a reference data sequence to be compressed is equiprobable.

For comparative evaluation of the effectiveness of quadratic signature analyzer we shall consider the experimental data obtained by M.G.Karpovsky for a 8-and 12-bit arithmetic-logical unit. They were constructed from typical 4-bit adder circuits. For each adder, all possible single struck-at faults were modeled and tested for detectability both by signature analyzer (SA) and quadratic signature analyzer (QSA) with $m=4$ and 8. Twenty distinct pseudorandom test sequences of equal length were used as test stimuli. The averaged percentages of undetectable errors are given in Table 9.2.

Table 9.2

ALU	Pseudorandom test length	Averaged test coverage, %,				
		without comp- ression	with m of			
			4	4	8	8
			SA	QSA	SA	QSA
8-bit	300	5.7	11.8	11.4	6.1	5.9
12-bit	500	0.7	7.2	6.6	1.2	1.1

As seen from Table 9.2, the quadratic signature analyzer that has been used in the discussed example is more valid and conforms to the theorem.

At the same time if the observed signature obtained by quadratic signature analyzer differs from the reference one the only practically available information is that the sequence is in error. The signature produced by a multi-functional signature analyzer based on stochastic integrators is more informative.

9.4. Multifunctional Signature Analyzer

For more detail on the behavior of a faulty digital circuit, it is reasonable to use a multifunctional signature analyzer (Fig.9.7). This analyzer allows one to determine some properties of the sequences produced on the nodes of the circuit under test and thus to estimate the serviceability of each gate individually.

Two modes of multifunctional analysis are possible depending on the value of control signal $V \in \{0,1\}$. With $V=1$, the analyzer consists of two functionally independent devices. The first device is formed of a shift register (1) and a modulo-2 adder (4) representing a conventional serial signature analyzer structure. The value of signature $S(x)=a_1(l)a_2(l)...a_m(l)$ is formed on its storage elements by the following linear relation (9.4):

$$a_i(0) = 0, \quad i = \overline{1,m};$$
$$a_1(k) = y(k) \oplus \sum_{i=1}^{m} {}^{\oplus} \alpha_i \, a_i(k-1);$$
$$a_j(k) = a_{j-1}(k-1), \quad j=\overline{2,m} \quad k=\overline{1,l},$$

where $y(k) \in \{0,1\}$ is the k th character of the sequence to be compressed, $\alpha_i \in \{0,1\}$ is the i th coefficient of generating polynomial $\phi(x)$, l is the length of tested sequence, which normally satisfies the condition $l \leq 2^m - 1$, and is specified by the control unit (10) as the number of clock pulses C_1.

The second device is formed of a reversible block counter (3), which is a totalizing meter counting the number $C(x)$ of 1s in the sequence $\{y(k)\}$, $k=\overline{1,l}$, by the expression

$$C(x) = \sum_{k=1}^{l} y(k) \qquad (9.17)$$

with $l \leq 2^m - 1$ where $m = \deg(x)$ and determines the width of counter 3. Units 5 to 9 (Fig.9.7) are standard computing components.

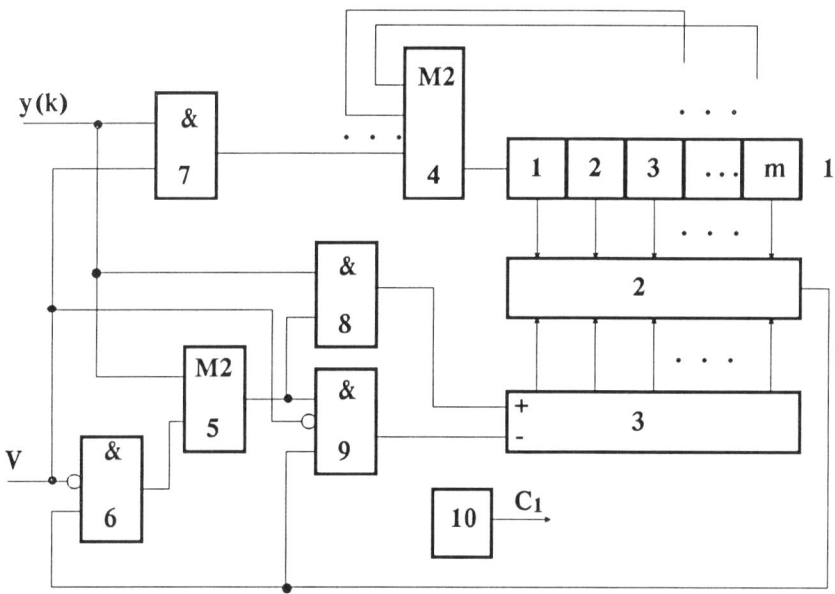

Fig. 9.7. Multifunctional signature analyzer

When the control signal $V=0$, the multifunctional signature analyzer structure is transformed into that of a stochastic servo-integrator consisting of a pseudorandom number generator (PRNG) employing units 1 and 4, a comparison circuit (CC) (2) and a reversible counter (RC) The inputs of the latter are applied with the tested sequence $\{y(k)\}$ and the sequence from the CC output. The latter causes a *1* signal to be produced at the comparison circuit output if the binary code for a successive random number is less than the current code stored by RC. Otherwise, a *0* signal is produced.

With $l < 3 \cdot 2^m$, the RC contents will indicate the probability $P(y) = P[y(k)=1]$ for a 1 signal to appear in the sequence $\{y(k)\}$. Thus by using the multifunctional signature analyzer (Fig. 9.7) for testing the sequence $\{y(k)\}$ we may obtain three major characteristics of the sequence: signature $S(x)$, number of 1s $C(x)$ in the sequence, and the probability $P(y)$ of their appearance. Let us evaluate these characteristics for some typical sequences $\{y(k)\}$, and determine their interrelation for NOT, EOR, OR and AND gates. For a zero sequence $y(k)=0$, $k=\overline{1,l}$, the values of the associated characteristics take the following form:

$$S(x) = a_1(l)a_2(l)...a_m(l) = 000...0,$$
$$C(x) = 000...0, \qquad (9.18)$$
$$P(y) = 000...0.$$

At the same time, for the sequence $y(k)=1$, $k=\overline{1,l}$, we shall have

$$S(x) = a_1(l)a_2(l)...a_m(l) = 000...0 \quad \text{for} \quad l = 2^m-1,$$
$$S(x) = a_1(l)a_2(l)...a_m(l) = 000...1 \quad \text{for} \quad l = 2^m-2, \qquad (9.19)$$
$$C(x) = l, \quad l \le 2^m-1,$$
$$P(y) = 1, \quad l > 3\,2^m.$$

To define the value $S(x)$ for the sequence $y(k) = 2\lceil k/2 \rceil - k$, $k = \overline{1,l}$ where $\lceil k/2 \rceil$ is the nearest integer that is greater than or equal to $k/2$, let us examine the following theorem.

Theorem 9.2. The value of signature $S_l(x)$ for the sequence of symbols $y_1(k) = 2\lceil k/2 \rceil - k$, $k = \overline{1,2^m-1}$ corresponds to that for its reciprocal and is determined by the relation

$$S_1(x) = a_1^1(2^m-1)\,a_2^1(2^m-1)\,...\,a_m^1(2^m-1) = 1\,1\,1\,...\,1.$$

<u>Proof.</u> Suppose that signature $S_1(x) = a_1^1(2^m-1)\,a_2^1(2^m-1)\,...\,a_m^1(2^m-1)$ corresponds to the sequence of symbols $y_1(k) = 2\lceil k/2 \rceil - k$, $k = \overline{1,2^m-1}$ and signature $S_2(x) = a_1^2(2^m-1)\,a_2^2(2^m-1)\,...\,a_m^2(2^m-1)$ corresponds to its reciprocal. The values of these signatures are identical since the summed sequence $[y(k)]=[y_1(k)] \oplus [y_2(k)]$ is a unit sequence which has the zero signature $S(x)=000...0$. Hence $S_1(x) \oplus S_2(x)=000...0$ and

Signature Analysis

$$a_i^1(2^m-1) = a_i^2(2^m-1), \quad i = \overline{1,m} \qquad (9.20)$$

Considering that the first zero symbol of sequence $\{y_2(k)\}$ does not affect the final value of signature $S_2(x)$, we may deduce that $\{y_2(k)\}$ compressed by the signature analyzer produces a result that is equivalent to compressing the first 2^m-2 symbols of $\{y_1(k)\}$. Therefore, the values of symbols $a_i^2(2^m-1) \in [0,1]$, $i = \overline{1,m}$, in signature $S_2(x)$ may be used for determining $S_1(x)$ by (9.4). Then, subject to

$$a_i^1(2^m-2) = a_i^2(2^m-1),$$

we obtain

$$\begin{aligned} a_1^1(2^m-1) &= y_1(2^m-1) \oplus \sum_{i-1}^{m} {}^{\oplus} \alpha_i a_i^2(2^m-1), \\ a_j^1(2^m-1) &= a_{j-1}^2(2^m-1), \quad j = \overline{2,m}. \end{aligned} \qquad (9.21)$$

Substituting the obtained values $a_i^1(2^m-1)$ into the set of equations (9.20) and subject $y_1(2^m-1) = 2\lceil 2^{m-1}-2^{-1} \rceil - 2^m+1 = 2^m-2^m+1 = 1$, we shall finally have

$$\begin{aligned} a_1^2(2^m-1) &= 1 \oplus \sum_{i-1}^{m} {}^{\oplus} \alpha_i a_i^2(2^m-1), \\ a_1^2(2^m-1) &= a_2^2(2^m-1) = a_3^2(2^m-1) = \ldots a_m^2(2^m-1). \end{aligned} \qquad (9.22)$$

An evident solution of the system of equation (9.22) is that all its unknowns $a_i^2(2^m-1)$, $i=\overline{1,m}$, be one. Thus, $S_1(x)=111...1$.

Based on Theorem 9.2, the sequence of the form $\{y(k)\}=101010...01$ will have the following integral characteristics:

$S(x) = a_1(l)a_2(l)a_3(l)...a_m(l) = 111...1$ for $l = 2^m-1$ and $l = 2^m-2$,
$C(x) = 2^{m-1}$ for $l = 2^m-1$, (9.23)
$P(y) = \frac{1}{2}$ for $l > 3 \cdot 2^m$.

The above characteristics can be similarly defined for more complex forms of sequence $\{y(k)\}$.

We may establish the interrelation of characteristics $S(x)$, $C(x)$ and $P(y)$ for some logic elements. We shall represent the signatures in the unitary code as before, whereas the values $C(x)$ and $P(y)$ will be represented in the positional code. Let us consider first the NOT gate, for which the relations

$$S_f(x) = S_1(x) \oplus S(1),$$
$$C_f(x) = l - C_1(x), \qquad (9.24)$$
$$P(x) = 1 - P_1(y), \quad l > 3 \cdot 2^m,$$

are satisfied, where $S_f(x)$, $C_f(x)$ and $P_f(x)$ are the characteristics of the sequence of length l formed at the output of the NOT gate that realizes the function $f = \bar{x}$; $S_1(x)$, $C_1(x)$ and $P_1(y)$ are the respective characteristics of its input sequence; and $S(1)$ represents the signature of a sequence that consists of l 1s

For a multi-input EOR gate, subject to the linearity, we may obtain

$$S_f(x) = \sum_{i=1}^{n} \oplus S_i(x). \qquad (9.25)$$

The expressions for $C_f(x)$ and $P_f(y)$ bear very complex functional dependence on the values of $C_i(x)$ and $P_i(y)$, $i = \overline{1,n}$, of the n-input EOR gate and therefore the calculations of $C_f(x)$ and $P_f(y)$ are rather cumbersome. Thus, for $n=2$, $P_f(y)$ has the form:

$$P_f(y) = P_1(y) + P_2(y) - 2 P_1(y)P_2(y).$$

Therefore, to check whether the EOR gate behaves normally, its signatures are usually analyzed by the expression (9.25).

Signature Analysis

At the same time, the behavior of AND and OR gates cannot be tested by their signatures $S(x)$, although the relations for checking the specified gate behavior may be obtained for $C(x)$ and $P(y)$. Thus for an OR gate, we may show that the value $C_f(x)$ satisfies the relation

$$\max[C_1(x), C_2(x), \ldots C_n(x)] \le C_f(x) \le Q_1, \qquad (9.26)$$

where

$$Q_1 = \begin{cases} C_1(x) + C_2(x) + \ldots + C_n(x), & C_1(x) + C_2(x) + \ldots + C_n(x) < l, \\ l, & C_1(x) + C_2(x) + \ldots + C_n(x) \ge l, \end{cases}$$

and n is the number of inputs to the OR gate. The relation (9.26) holds true for a random functional dependence of input sequences in the gate of interest. Under the assumptions made, the following inequality is satisfied for the same gate similar to (9.26):

$$\max[P_1(y), P_2(y), \ldots, P_n(y)] \le P_f(y) \le Q_2, \qquad (9.27)$$

where

$$Q_2 = \begin{cases} P_1(y) + P_2(y) + \ldots + P_n(y), & P_1(y) + P_2(y) + \ldots + P_n(y) < 1, \\ 1, & P_1(y) + P_2(y) + \ldots + P_n(y) \ge 1. \end{cases}$$

At the same time, the relation

$$P_f(y) = 1 - \prod_{i=1}^{n}[1 - P_i(y)],$$

which assumes the form

$$P_f(y) = P_1(y) + P_2(y) - P_1(y)P_2(y).$$

with $n=2$, holds true for $P_f(y)$ when the symbols of the input sequences of an OR gate are independent.

Moreover, for input sequences whose unit symbols are nonconcurrent events, the equality $P_f(y) = Q_2$ is satisfied.

Relation (9.26), the same as (9.27), holds true for random input sequences although in this case the value of $C_f(x)$ can be estimated to a higher accuracy, provided that there is additional information on test stimuli, for example. This may be illustrated by the implementation of syndrome testing.

For an AND gate that realizes the function $f = x_1 x_2 ... x_n$ the following interrelation holds true for integral characteristics $C(x)$ of its input and output sequences:

$$Q_3 \leq C_f(x) \leq \min [C_1(x), C_2(x),...,C_n(x)] \quad (9.28)$$

where

$$Q_3 = \begin{cases} 0, & C_1(x) + C_2(x) + ... + C_n(x) \leq l, \\ C_1(x) + C_2(x) + ... + C_n(x) - l, & C_1(x) + C_2(x) + ... + C_n(x) > l. \end{cases}$$

Similarly to relation (9.28), for the gate in question the inequality

$$Q_4 \leq P_f(y) \leq \min [P_1(y), P_2(y), ..., P_n(y)] \quad (9.29)$$

where

$$Q_4 = \begin{cases} 0, & P_1(y) + P_2(y) + ... + P_n(y) \leq 1, \\ P_1(y) + P_2(y) + ... + P_n(y) - 1, & P_1(y) + P_2(y) + ... + P_n(y) > 1. \end{cases}$$

holds true.

Thus, the expressions obtained for the integral characteristics of sequences produced on the nodes of NOT, EOR, AND and OR gates allow one to check their behavior.

When the above gates behave normally, the experimental characteristics must satisfy the relations (9.24) to (9.29). At the above relations are not satisfied for any of the characteristics being tested, we may infer that the gate under test is faulty.

Further improvement of signature analysis efficiency and resolution expressed as the fault localization capability may be attained by modifying the basic design of signature analyzer as well as by applying the new techniques for compressing the sequences under test.

Chapter 10

SIGNATURE ANALYSIS EFFICIENCY

10.1. Estimation of Signature Analysis Efficiency

The fault coverage in a digital circuit primarily depends on the quality of test stimuli. When the designated fault does not manifest itself in the output responses of the circuit as a distortion of its characters, it is undetectable by signature analysis, which is nothing more than an efficient data stream compression technique. Therefore if the data stream carries no information on the fault, neither will the compressed stream.

Thus by signature analysis efficiency we shall basically mean its capability of error detection in the data stream being compressed. This property may be estimated both by deterministic and probabilistic approaches. The most commonly used is the probabilistic approach, which consists in determination of probability P_n of failing to detect errors in the data sequence being analyzed. For the first time the value of P_n has been estimated for the models of equiprobable reference output sequence and equiprobable error sequence. For binary polynomials describing the mentioned sequences as well as for their prototypes the following equality is met

$$\chi_e(x) = \chi(x) \oplus e(x), \quad (10.1)$$

where $\chi(x)$ is the polynomial describing the reference data stream; $e(x)$ is the error polynomial, and $\chi_e(x)$ is the polynomial describing the error sequence. It is evident that the error escapes detection only if $e(x)$ is associated with the zero signature $S(x)$ or in other words, if $e(x)$ is divisible by $\phi(x)$, which describes the signature analyzer structure. By using the polynomial division algorithm as a body of mathematics for signature generation, we may demonstrate that for a l-bit dividend a $(l-m)$-bit quotient and a m-bit remainder (signature) are obtained. Then the correspondence between polynomial $e(x)$ and polynomial zero which characterizes an error-free case in the output sequence, is estimated by the equality of the m-bit signature to zero. However, the zero signature may be associated with 2^{l-m} distinct quotients. This implies that $2^{l-m}-1$ l-bit error sequences are associated with a single (zero) sequence described by $e(x)$. Considering the equiprobability of error data streams, we may infer that $2^{l-m}-1$ error sequences initiating the zero signature are undetectable. Thus, the probability P_n of failing to detect errors in the data stream being analyzed is calculated as

$$P_n = \frac{2^{l-m}-1}{2^l-1}, \quad (10.2)$$

where 2^l-1 is the total number of probable error polynomials $e(x)$.
The expression (10.2) for $l>>m$ may be simplified to give

$$P_n \approx \frac{1}{2^m}. \quad (10.3)$$

The value of P_n may come as a convincing reason for the high efficiency of signature analysis. In fact, for the Hewlett-Packard signature analyzer, $Pn = 0.0000152...$, which proves that the efficiency of the signature analyzer is sufficiently high.

The value $P_d=1-P_n=0.999984...$ may be found in practically every report on signature analysis, but only some of the reports realistically estimate the relation, which prevents one from assessing the advantages of signature analysis. A simple example, which shows that the integral estimate (10.3) is unsuitable as a criterion of signature analysis efficiency, consists of the fact that, with the use of a technique based on the analysis of only m of l bits by ignoring $l-m$ bits of the sequence under test and under the assumption adopted while deriving the relation (10.2), the probability P_n will be calculated also by the same formula.

As a more accurate measure of signature analysis effectiveness, we may use the dependence of error escape probability P_n^μ on error multiplicity μ. For the numerical values P_n^μ of the following theorem holds.

Theorem 10.1. The probability for an error of multiplicity μ, $\mu \in \{1,2,3...,2^m-1\}$, to escape detection by the signature analyzer described by a primitive polynomial $\phi(x)$, whose high degree is m, can be calculated for equally distributed μ-multiple error by the recurrence formula

$$P_n^1 = 0, \quad P_n^2 = 0,$$

$$P_n^\mu = \frac{V^{\mu-1} - V_n^{\mu-1} - V_n^{\mu-2}(2^m + 1 - \mu)}{\mu V^\mu}, \qquad \mu = \overline{3, 2^m - 1},$$

where $V_n^{\mu-1}$ is the number of undetectable errors of multiplicity $\mu-1$, $V^\mu = C_{2^m}^\mu$.

For high values of m the following asymptotic result presented as a theorem holds.

Theorem 10.2. For value of P_n found by theorem 10.1 the following limiting relation is met:

$$\lim_{m \to \infty} P_n^\mu = \frac{1}{2^m}. \qquad (10.4)$$

Practical results suggest that even for m=7 the value of P_n^μ differs from $1/2^m$ by 0.02%.

When deriving the values of P_n^μ, the use of primitive polynomials is of fundamental importance, but it is not mandatory for obtaining an integral estimate of P_n by formula (10.3). The two latter theorems basically prove the necessity of using primitive polynomials for generation of signature analyzers, otherwise, probability distribution P_n^μ will be nonuniform.

An important deduction from expression (10.4) is the fact that it is invariant as to the compressed sequence and the form of primitive polynomial $\phi(x)$. The only parameter that effects the values of P_n^μ is the high degree m of $\phi(x)$.

An ingenious technique of evaluating signature analysis efficiency is the one based on the use of model of single errors in the bits of compressed stream. By this technique, the probability p of an error in any single bit may assume different values within the range from p=0 to p=1. Let us estimate the probability P_n for an error to escape detection by this model. Similar to the earlier described approaches to P_n estimation, we take that an error sequence will escape detection only if the resulting signature $S(x)$ is zero (or any other prespecified value). Taking account of the latter remarks and assuming that an error occurrence in each bit of the l-long sequence to be compressed is an independent event, we can demonstrate that P_n satisfies the following inequalities.

$$p^m < p_n < (1-p)^m, \quad 0 > p > \frac{1}{2},$$
$$(1-p)^m < p_n < p^m, \quad \frac{1}{2} > p > 1. \quad (10.5)$$

where m is the length of signature analyzer $(m<<1)$.

The boudary values of P_n are the function of probability p. It may be noticed that for $p=1/2$ we obtain correct values of P_n corresponding to an equiprobable model for all possible errors in a sequence to be compressed. However, the estimates of P_n are incorrect for the values of p approaching zero and one. To examine the efficiency of signature analysis on small and high values of p, Markov chain theory or z-transformation may be used.

The simplest example of a Markov chain that describes the signature analyzer behavior is the circuit of Fig.10.1 for $m=2$ and primitive generating polynomial $\phi(x) = 1 \oplus x \oplus x^2$.

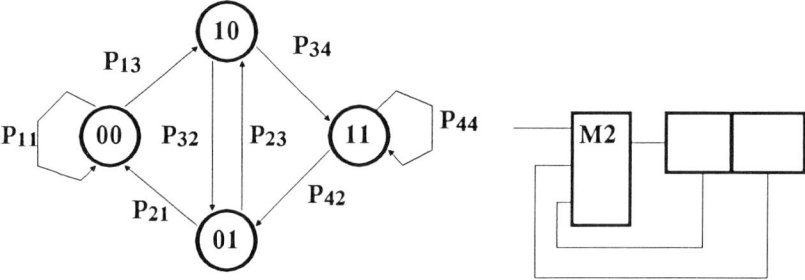

Fig. 10.1. Signature analyzer

Probabilities P_{ij}, $i,j=\overline{1,4}$, characterize analyzer transitions from state i to state j. Complete probabilistic description of a Markov chain consists in the use of transition probability matrix P. For the example of Fig. 10.1 the matrix has the form

$$P = \begin{vmatrix} P_{11} & P_{12} & P_{13} & P_{14} \\ P_{21} & P_{22} & P_{23} & P_{24} \\ P_{31} & P_{32} & P_{33} & P_{34} \\ P_{41} & P_{42} & P_{43} & P_{44} \end{vmatrix}.$$

Let us find numeric values of P_{ij}. Suppose that the analyzer's state is *00* and the probability for a one to appear in the compressed error sequence defined by $e(x)$ is p. Then the probability for a zero to appear in specific error bit is defined then as $1-p$. The analyzer can transit from state *00* (first state) to only state *00* (original first state) and *10* (third state). It can never change from *00* to *01* (second state) or *11* (fourth state). Hence $P_{12}=P_{14}=0$, and transition from *00* to *00* is only possible with the zero input, so $P_{11}=1-p$. Transition from *00* to *10* is only possible with one input, therefore $P_{13}=p$. In a similar way we can find the values for the remaining single-step transition probabilities. As a result we obtain the matrix

$$P = \begin{vmatrix} 1-p & 0 & p & 0 \\ p & 0 & 1-p & 0 \\ 0 & p & 0 & 1-p \\ 0 & 1-p & 0 & p \end{vmatrix},$$

which is the bistochastic transition probability matrix describing the behavior of the Markov chain of Fig.10.1.

Therefore the signature analysis behavior can generally be defined by a transition probability matrix that is a bistochastic matrix. By using matrix P, we can determine the probability for an analyzer to assume any of its 2^m states (m=analyzer length). In the general case, the probability depends on a probability vector $B(0) = b_1(0), b_2(0), \ldots, b_{2^m}(0)$ for the analyzer initial state and the number of operation cycles. Then

$$B(1) = b_1(1), b_2(1), \ldots, b_{2^m}(1) = [b_1(0), b_2(0), \ldots, b_{2^m}(0)]P = B(0)P,$$

$$B(2) = B(1)P = B(0)P^2,$$

where $b_i(1)$ is the probability for the analyzer to assume state i upon completion of one operation cycle.

In the general case

$$B(l) = B(0)P^l$$

$$\sum_{v=1}^{2^m} b_v(r) = 1, \quad r \in [0,1,2,\ldots,l].$$

For vector $B(r)$, the asymptotic result associated with the following theorem holds true.

Theorem 10.3. With the specified probability p for a one to appear in the error sequence where $p\neq 0$, $p\neq 1$ and error occurrence in its specific bit be an independent event, the final probability vector for a m-long analyzer to assume any of its 2^m states does not depend on the initial condition (vector $B(0)$) and appears as

$$\frac{1}{2^m}, \frac{1}{2^m}, \frac{1}{2^m}, \ldots \frac{1}{2^m}.$$

An important corollary of the theorem is the relation

$$P_n = \frac{1}{2^m},$$

which indicates that the probability P_n of failing to detect an error in quantity $1/2^m$ is equally likely for the model of distortion of each symbol in the compressed sequence with probability p, $p\neq 0$ and $p\neq 1$.

The above theorem and its corollary produce only the asymptotic result that holds true for sufficiently large l. The relation for probability P_n is no less important for small l. To estimate the rate of convergence for P_n to its asymptotic value, we may use the following functional

$$\lambda(P^l) = \max_c (\max_{i,j} |a_{ic} - a_{jc}|),$$

where P is a bistochastic circulant matrix that describes the signature analyzer behavior; a_{ij} is the value of matrix element P^l located in the i-th row and j-th column. For a $2^m \times 2^m$, matrix whose elements assume the values 0, p and $1-p$, the above functional appears as

$$\lambda(P^l) = \gamma \, |1-2p|^{int(l/m)}, \qquad (10.6)$$

Signature Analysis Efficiency

where $\gamma \leq 1$. It is evident that the rate of convergence for P_n to its asymptotic value depends to a greater degree on p. The closer is p to $1/2$ the lesser number of time steps l is required to obtain the asymptotic result.

The functional described by equation (10.6) corresponds to any polynomial of degree m. At the same time, the primitivity of generating polynomial may have a pronounced effect on signature analysis effectiveness. According to the theorem (10.3), the asymptotic value of P_n does not depend on the form of polynomial of degree m, whereas the values of P_n for small l largely depend on the polynomial used. Figs.10.2 and 10.3 show the values of P_n that is a function of the compressed sequence length for a non-primitive polynomial $\phi(x)=1 \oplus x^8$ and a primitive polynomial $\phi(x)=1 \oplus x^3 \oplus x^5 \oplus x^6 \oplus x^8$, respectively. Practical results suggest the necessity of primitive polynomials. In fact, comparing the plots of Fig 10.2 and 10.3 we may notice that the probability for an error to escape detection is closer to its asymptotic value for primitive polynomials.

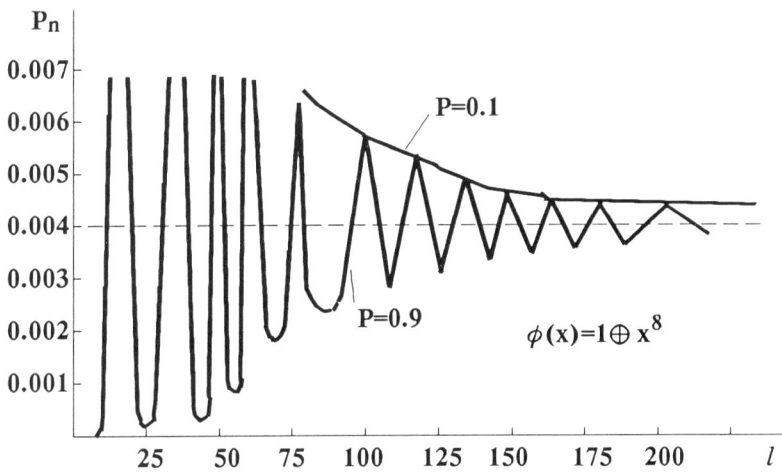

Fig 10.2. The plot of Pn for non-primitive polynomial

A severe functional dependence of probability P_n on l for the polynomial $\phi(x)=1 \oplus x^8$ appears as

$$P_n = 2^{-m} \frac{[1 + (1 - 2p)^{\lfloor l/m \rfloor}]^{m - l \bmod m}}{[1 + (1 - 2p)^{\lceil l/m \rceil}]^{l \bmod m} - (1-p)^l}. \qquad (10.7)$$

There is a technique that allows to calculate the probability P_n or estimate it for an arbitrary polynomial $\phi(x)$. Based on this technique, we can demonstrate that the inequality

$$P_n \leq 2^{-m}[1 + |1 - 2p|^{\lfloor l/m \rfloor}] - (1-p)^l. \qquad (10.8)$$

holds true for primitive polynomials.

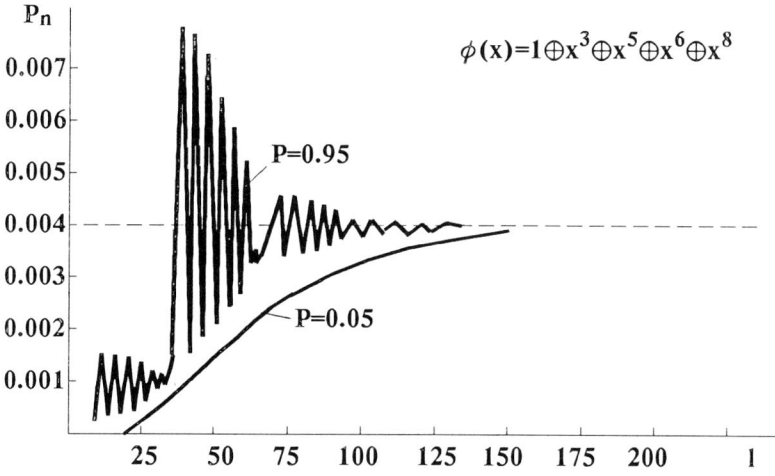

Fig. 10.3. The plot of Pn for primitive polynomial

It is evident that for large l $\lfloor l/m \rfloor \approx \lceil l/m \rceil$ and hence P_n determined by formula (10.7) equal to the top estimate of P_n obtained by formula (10.8) for primitive polynomials. The difference can be observed only for small ls (Fig.10.2 and 10.3).

Based on the Markov chain theory, we obtained an accurate value of P_n that comprises two components P_n^∞ and P_n^l, respectively. For an asymptotic and transient value of P_n. For the generating polynomial $\phi(x)$ of degree m, this expression takes the form

$$P_n = P_n^\infty + P_n^l = \frac{1}{2^m} + \frac{1}{2^m} \sum_{i=1}^{2^m-1} (1-2p)^{w(i,l)} - (1-p)_n^l \qquad (10.9)$$

where $w(i,l)$ is the number of ones in the m th bit of M-sequence generator described by the polynomial $\phi(x)$ that has been obtained for l time steps of its activity, starting with the i th state. Thus the expression (10.9) takes account of the used polynomial form as $w(i,l)$ for the probability of error escape. The relation (10.9) as well as the preceding results allow us to conclude that the use of primitive polynomials offers advantages over non-primitive polynomials.

It is evident that the output response of a faulty digital circuit does not necessarily correspond to the selected model. The nature of output distortion will strongly depend on the form of test stimuli and fault type in the circuit.

10.2. Examination of Error Occurrence in Output Responses of Digital Circuits

The validity of error analysis depends on probability distribution of error sequences to occur depending on error multiplicity. On the one hand, probability distribution determination lows a most efficient error analyzer to be generated for the specific digital circuit; on the other hand, it makes possible to obtain a realistic estimate of any compact testing technique validity.

According to the probability distribution of different multiplicity errors to occur, we may select the analyzer such that its maximum detection probability corresponds to the most probable error sequence. Therefore, we may select the error analyzer with the maximum detectability for the given digital circuit.

Let us consider fault manifestation as an error of specific multiplicity for the class of two-level digital circuits that implement the functions specified in the orthogonal disjunctive normal form (ODNF). As a test sequence, we shall take a trivial test consisting of 2^n all possible binary patterns, where n is the number of variables affecting the Boolean function implemented by the circuit. The multiplicity of errors may vary from 1 to 2^n.

The Boolean function $F(x_1, x_2,...,x_n)$ that has been represented in ODNF can be implemented as a widely used structure based on AND gates with the inverting inputs and on a multi-input OR gate (Fig.10.4). The function F is characterized by l conjunctions f_j of g_j variables each, $j=\overline{1,l}$. The whole set of single-bit stuck-at faults in a two-level combinational circuit (Fig.10.4) can be represented as a union of four subsets M_1, M_2, M_3 and M_4. The subset $M_1=\{F\equiv 1, F\equiv 0\}$ consists of two stuck-at faults on an output node in the circuit.

The subset $M_2=\{f_1\equiv 1, f_1\equiv 0, f_2\equiv 1, f_2\equiv 0, \ldots, f_l\equiv 1, f_l\equiv 0\}$ represents the faults on intermediate nodes of the circuit. The subset M_3 consists of faults on input nodes of multi-input AND gates, where the input node x_{ji} for specific values $j\in\{1, 2, \ldots, l\}$ and $i\in\{1, 2, \ldots, n\}$ may assume either the values $x_i, \overline{x}_i$ or none. In the general case, $M_3=\{x_{11}\equiv 1, x_{11}\equiv 0, x_{12}\equiv 1, x_{12}\equiv 0, \ldots, x_{ln}\equiv 1, x_{ln}\equiv 0\}$. Faults $\{x_1\equiv 1, x_1\equiv 0, x_2\equiv 1, x_2\equiv 0, \ldots, x_n\equiv 1, x_n\equiv 0\}$ form the subset M_4.

With no fault in the digital circuit $\sum_{i=1}^{l} 2^{n-g_i}$ ones, where each term of the

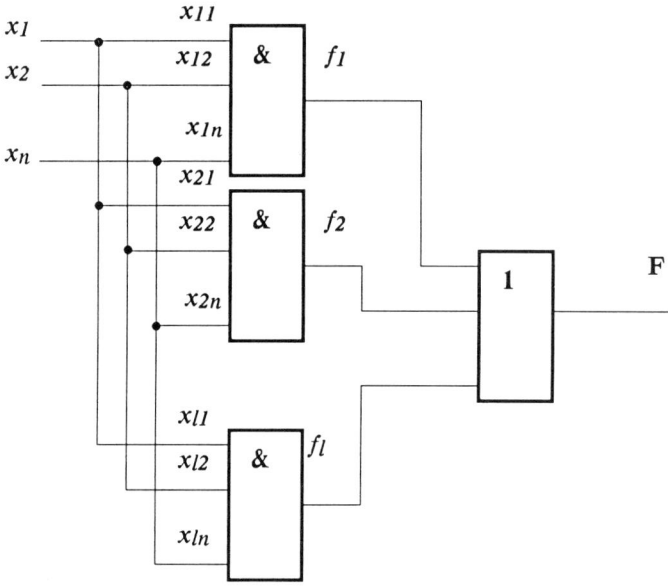

Fig.10.4. A two-level digital circuit

sum corresponds to the number of ones generated by the i-th AND gate, and $2^n - \sum_{i=1}^{l} 2^{n-g_i}$ zeros are produced on its output. Therefore, a reference output sequence for the circuit of Fig. 10.4 contains $\sum_{i=1}^{l} 2^{n-g_i}$ ones and $2^n - \sum_{i=1}^{l} 2^{n-g_i}$ zeros.

When a $F \equiv 0$ occurs, the output response of the circuit will consist of 2^n zero symbols thereby indicating to a $\sum_{i=1}^{l} 2^{n-g_i}$-multiple error. A $F \equiv 1$ initiates a $2^n - \sum_{i=1}^{l} 2^{n-g_i}$-multiple error to appear. Thus the subset of M_1 is mapped over the error subset N_1, which consists of a single $\sum_{i=1}^{l} 2^{n-g_i}$-multiple and a single $(2^n - \sum_{i=1}^{l} 2^{n-g_i})$-multiple error.

The appearance of fault $f_j \equiv 0$ from the subset M_2 distorts unit symbols produced by conjunction f_j. The fault causes an error of multiplicity 2^{n-g_j} to appear. Faults $f_j \equiv 1$, $j = \overline{1,l}$, that are equivalent to $F \equiv 1$. Thus the subset M_2 will be mapped in the error subset N_2 consisting of l errors of multiplicity $(2^n - \sum_{i=1}^{l} 2^{n-g_i})$ and l errors of multiplicity (2^{n-g_j}), where $j = \overline{1,l}$.

A fault $x_{ji} \equiv 0$ appeared on the AND gate input causes different errors to appear depending on whether there is a negation at the gate input in question. Thus, for $x_{ji} = x_i$, the ones associated with conjunction f_j will be inverted in the output sequence and for $x_{ji} = \overline{x}_i$, conjunction f_j will be expanded relative to x_i, i.e. an extra number of ones, which is the difference between that produced by f_j for a known good circuit and that initiated concurrently by other conjunctions will appear at the circuit output. Therefore, for inputs $x_{ji} = x_i$ of the j-th AND-gate, faults $x_{ji} \equiv 0$ will cause $g_j \cdot p_j$ errors of multiplicity 2^{n-g_j}, $j = \overline{1,l}$, where p_i is the number of variables with negation in conjunction f_j, to appear. For $x_{ji} \neq x_i$, the fault will cause an error of multiplicity

$$2^{n-g_j} - \sum_{k_{ij}} 2^{n-m_{ij}} \qquad i = \overline{1,n} \quad j = \overline{1,l} \quad (10.10),$$

where k_{ij} is the number of terms which sum is the number of nonzero products of conjunction f_j with $\overline{x}_i$ replaced for x_i by the remaining conjunctions of $F(x_1,x_2,...,x_n)$ m_{ij} is the number of factors in the resulting products. The total number of such errors is $\sum_{j=1}^{l} p_j$. Also, the sum includes the errors, that may be masked by other conjunctions of F.

We can similarly demonstrate that a $x_{ji}\equiv 1$ fault, with $x_{ji}=x_i$, will cause error of multiplicity

$$2^{n-g_j} - \sum_{k_{ij}'} 2^{n-m_{ij}'} \quad i=\overline{1,n}, \quad j=\overline{1,l}, \quad (10.11)$$

where coefficients k_{ji}' and m_{ji}' are defined the same as k_{ji} and m_{ji}, to appear. The total number of such errors will be $\sum_{j=1}^{l}(g_j-p_j)$. If there is no variable x_i in a conjunction, no error occurs. With $x_{ji}=\bar{x}_i$, p_j errors of multiplicity 2^{n-g_j}, $j=\overline{1,l}$, appear for the j-th AND gate. Thus the set of errors caused by faults of subset M_3 consists of the number of errors g_j, $j=\overline{1,l}$, of multiplicity 2^{n-g_j} plus $\sum_{j=1}^{l} p_j$ errors of different multiplicity described by (10.10), plus $\sum_{j=1}^{l}(g_j-p_j)$ errors of multiplicity described by (10.11).

A $x_i \equiv 0$ fault at the circuit input causes an error of multiplicity

$$\sum_{j \in t_i} 2^{n-g_{ij}} + \sum_{j \in t_i'} (2^{n-g_{ij}'} - \sum_{k_{ji}} 2^{n-m_{ji}}) \quad (10.12)$$

to occur.

The first term of (10.12) is the sum of terms whose number is the number of F conjunctions containing x_i without negation and forming the subset t_i, and g_{ij} is the number of variables in the such conjunction. The second term is the sum of terms whose number is the number of F conjunctions containing the inverted value of x_i and forming the subset t_i', and g_{ij}' is the number of variables in such conjunction. For each conjunction of subset t_i', it is necessary to subtract the ones to be aliased by the unity values of other F conjunctions (subtracted under the summation sign of the second term in (10.12), where coefficients k_{ji} and m_{ji} are described for (10.10), from the extra ones introduced by the fault in question. The number of errors equals the number of variables in F.

Signature Analysis Efficiency 175

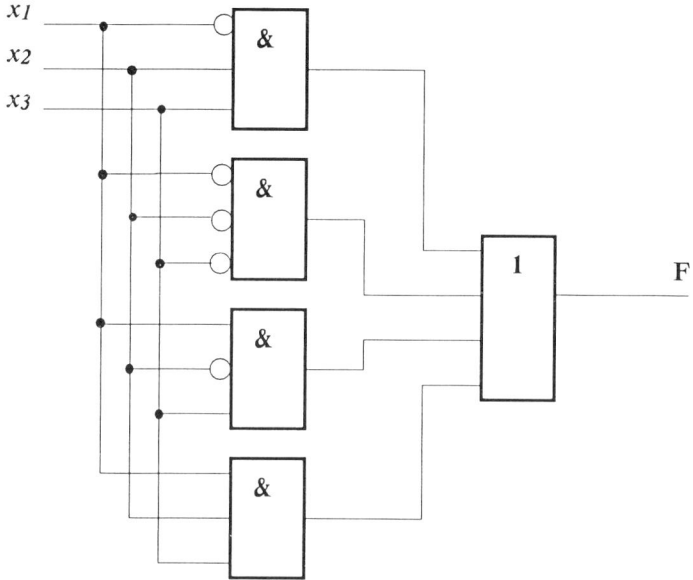

Fig.10.5 A two-level digital circuit

In the same way we can derive the expression for the multiplicity of errors caused by the fault $x_i \equiv 1$:

$$\sum_{j \in t_i'} 2^{n-g_{ij}'} + \sum_{j \in t_i} (2^{n-g_{ij}} - \sum_{k_{ji}} 2^{n-m_{ji}'}) \qquad (10.13)$$

For the fault class M_4, the number of errors is $2n$. Therefore the faults in subset M_4 are mapped onto the subset of errors N_4 whose number is $2n$ and the multiplicity is determined by (10.12) and (10.13).

Consider a special case of the problem of finding error multiplicities for the full disjunctive normal form (FDNF). When represented in that form, the Boolean function $F(x_1, x_2, ..., x_n)$ is characterized by l constituents of unity and v_i pairs of adjacent constituents relative to variable x_i, $i = \overline{1, n}$.

For the subset of faults M_1, the expression $\sum_{i=1}^{l} 2^{n-g_i}$ equals l since $g_i = n$, therefore the fault subset will be mapped onto the error subset N_1, consisting of a single l-multiple and a single $(2^n - l)$-multiple error.

The same as before we can obtain the error subset N_2 consisting of l single-bit and l errors with multiplicity (2^n-l) for the fault subset M_2.

For the subset M_3, 2^{n-g_i}-multiple errors whose number is $\sum_{j=1}^{l} g_j$, with $g_j=n$, are transformed into nl single-bit errors. Expressions (10.11) and (10.13) can assume l or 0 depending on whether the constituent of unity has a neighboring constituent relative to x_i. Therefore, the expression $\sum_{j=1}^{l} p_j$ is transformed into $\sum_{i=1}^{n} r_i$ or, subject to the masked errors, into $\sum_{i=1}^{n} r_i - \sum_{i=1}^{n} v_i$, where r_i is the number of negations for x_i in $F(x_1,x_2,...,x_n)$. The

Table 10.1.

N_k	Error multiplicity	Number of errors
N_1	l	1
	2^n-l	1
N_2	1	l
	2^n-1	l
N_3	1	$2nl-2\sum_{i=1}^{n} v_i$
N_4	$l-v_1$	2
	$l-v_2$	2
	...	...
	$l-v_n$	2

term $\sum_{j=1}^{l} (g_j-p_j)$ is transformed to $nl - \sum_{i=1}^{n} r_i - \sum_{i=1}^{n} v_i$, subject to the masked errors.

Thus the subset of errors N_3 consists of single-bit errors whose number is

$$nl + \sum_{i=1}^{n} r_i - \sum_{i=1}^{n} v_i + nl - \sum_{i=1}^{n} r_i - \sum_{i=1}^{n} v_i = 2nl - 2\sum_{i=1}^{n} v_i$$

We can similarly show that the expression (10.12) for $x_i \equiv 0$ takes the form $l-r_i+r_i-v_i=l-v_i$, and the expression (10.13) for $x_i \equiv 1$ takes the form $r_i+l-r_i-v_i=l-v_i$.

Summarizing the obtained results, we can form the general table of single-bit stuck-at faults mapped onto output response errors of a digital circuit described by the function specified in the FDNF (Refer to Table 10.1).

For a two-level digital circuit described by

$F(x_1,x_2,x_3)=\bar{x}_1 x_2 x_3 + \bar{x}_1 \bar{x}_2 \bar{x}_3 + x_1 \bar{x}_2 x_3 + x_1 x_2 x_3$ (Fig.10.5), the entire set of single-bit stuck-at faults of Table 10.1 is mapped onto the set of errors of Table 10.2.

For a digital circuit described by an arbitrary expression, errors and their multiplicity are determined in the same way as for FDNF.

Therefore, analysis of digital circuit behavior on fault occurrence testifies that the probability of an error versus multiplicity is arbitrary for various objects under test. The problem of finding it for specific circuit is computationally complex and mostly infeasible. This proves signature analyzer advantage over other compact testing methods as to test validity since signature analysis features stable error-escape probability depending on multiplicity, and hence is efficient for various objects under test.

Table 10.2

Error multiplicity	1	3	4
Number of errors	4	4	8

10.3. Signature Analysis Efficiency Estimation Techniques

The signature data compression technique was compared analytically against other compact testing techniques. The analysis is based on estimating the probability P_n of failing to detect errors in the data stream for the two compact testing techniques, i.e. signature analysis and transition counting. The estimation of P_n is carried out for both techniques under sufficiently general assumptions: 1) any reference data stream is an equally probable event; 2) any error bit configuration may occur with equal probability.

Under these assumptions, as has been shown in the preceding subsection, the value of P_n may be obtained by (10.3), and it is $1/2^m$, where m is the signature length, for long enough streams. For the transition counting, the value of P_n is also calculated as the ratio of all undetectable errors in the data stream of length l to all possible errors. However, the number of undetectable errors cannot be determined in a straightforward manner. This is due to the nonlinearity of the transition count operation, and hence the dependence of the number of undetectable errors on the form of the reference stream. In fact, the number of undetectable errors in the stream with r transitions is the function of r. This affects the final value of P_n

$$P_n \approx \frac{1}{(\pi l)^{1/2}}. \qquad (10.14)$$

For $l=2^m$, which corresponds to the length of m-bit checksums, we finally obtain

$$P_n \approx \frac{1}{(\pi\, 2^m)^{1/2}},$$

Hence the probability of failing to detect an error by transition counting is sufficiently higher than that for signature analysis. Apart from (10.14), a major reason for the use of signature analysis is the probability P_n^1 of failing to detect single errors, which is $1/2$ for transition counting and 0 for signature analysis. Thus by comparing the relations (10.3) and (10.14) as well as the estimates for P_n^1, we may conclude that signature analysis is more efficient over other compact testing techniques.

However, a slight variation in the rather general initial assumptions may produce the opposite result. Thus when it is assumed that generation of sequences with r transitions, where $r \in \{0, 1, 2, ..., l\}$, is an equally likely event, the probability P_n will be calculated as

$$P_n = \frac{1}{l+1} \sum_{r=0}^{l} \frac{C_l^r - 1}{2^l - 1}, \qquad (10.15)$$

where $C_l^r - 1$ is the number of error sequences consisting of r transitions.

At the same time, for a particular digital circuit the number r of transitions in its reference response is known in advance, and the same probability P_n will appear as

$$P_n = \frac{C_l^r - 1}{2^l - 1}, \qquad (10.16)$$

for transition counting. Therefore, apart from the estimate of P_n as per (10.14), we may use other estimates, for example (10.15) and (10.16), and compare them against the appropriate value for signature analysis. In so doing, consider that the value of P_n for signature analysis is invariant relative to the sequence compressed. Whence it follows, in particular, that the expression (10.13) also holds true for the assumptions adopted when the relations (10.15) and (10.16) were derived. Owing to this the given compact testing techniques may be compared under other similar conditions. Thus for the case of equiprobability of generating a sequence of r transitions we may demonstrate that the ratio of P_n for the two techniques will be 1 with $l=2^m-1$. This suggests that transition counting and signature analysis techniques are equivalent. This example is a good illustration of the fact that forced assumptions may lead to quite different conclusions on the effectiveness of the same techniques.

The comparison between signature analysis and transition counting for the actual case with known transition count r in the sequence under test makes it possible to estimate the probability ratio P_n as

$$\frac{2^{l-m}-1}{C_l^r - 1}. \qquad (10.17)$$

It is evident that for different r values the ratio (10.17) varies over a wide range. For the ratio values less than 1, signature analysis is more efficient; otherwise, transition counting is used.

Consider some examples demonstrating the efficiency of signature analysis and transition counting as a function of r. Let $l=2^m-2$ and the appropriate probability values P_n for both techniques be calculated by expressions (10.3) and (10.15). For $m=3$ and $m=4$, we obtain the plots of Figs.10.6 and 10.7, respectively. The full line shows the probability P_n plot for signature analysis, and the broken 'curve' shows the plot for transition counting.

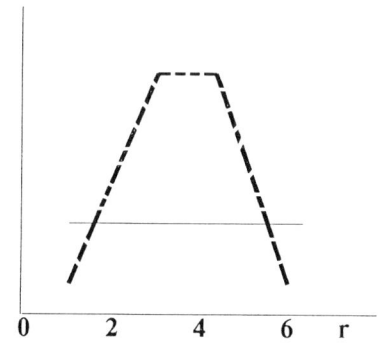

Fig. 10.6. The probability Pn plot for m=3

From the plots obtained we may deduce that with $m=3$ the signature analysis efficiency estimate is greater than that of transition counting only for $r=2,3,4$ and 5 transitions in the data stream being analyzed. For $r=1$ and 6, transition counting is more efficient. At the same time, with $m=4$, the number of r's, for which the latter outperforms signature analysis, increases substantially. In fact, signature analysis is only efficient for $r=5,6,7,8,9$ and 10, which is the lesser portion of all possible r's. The plot of

(10.17) against $r/(2^m-2)$ makes it evident that the range of r, for which signature analysis efficiency exceeds that of transition counting is, narrower. The range becomes narrower with the increase in the tested sequence length, which is 2^m-2 in this case. In Fig.10.8, plot 1 is given for $m=3$, plot 2 for $m=4$ and plot 3 for $m=5$.

When using the relation (10.17) for estimating signature analysis efficiency, the following two things should be considered. First, the expression (10.17) may only be used for the case when the number of transitions r is known in the reference sequence. Secondly, such comparative analysis is only true for the hypothesis that the error bit sequences caused by a circuit failure are equally likely.

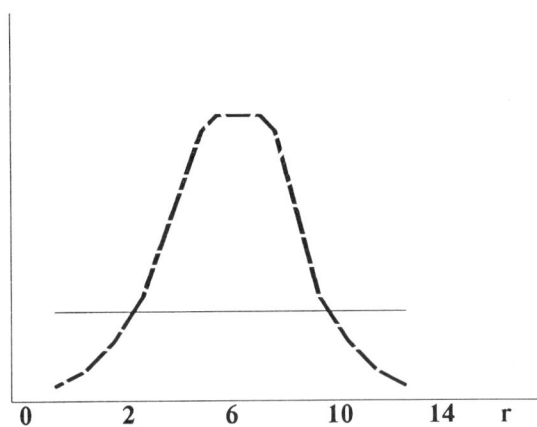

Fig. 10.7. The probability Pn plot for m=4

The relation (10.17), when used for deriving more general conclusions on signature analysis efficiency, may produce invalid results. Thus, we may deduce that transition counting is superior to signature analysis since for sufficiently long sequences l the range of efficient r is narrow for signature analysis. Note that for $m=\infty$ the range of efficient P_n is insignificant for signature analysis. Moreover, we may readily show that for $m=\infty$ the range of variable $r/(2^m-2)$, for which the relation (10.17) is less than 1, tends to the same value of 0.5. This may be explained by the fact that the

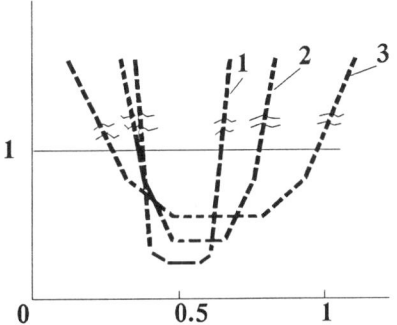

Fig. 10.8. Dependence of efficiency the signature analysis

number of r-transition sequences changes with r, which is not accounted for by (10.17). In fact, from the quick analysis of plots in Fig.10.7 we may conclude that, with $m=4$, transition counting is more suitable since it is more efficient for $r=1, 2, 3, 4, 11, 12, 13$ and 14, whereas signature analysis is efficient only for six values out of 14. However, the total number of sequences

$$C_{15}^1 + C_{15}^2 + C_{15}^3 + C_{15}^4 + C_{15}^{11} + C_{15}^{12} + C_{15}^{13} + C_{15}^{14},$$

for which the transition counting technique is more preferable turns out to be less than the number of sequences C_{15}^5 that consist of $r=5$ transitions only. This fact has been taken into account in deriving the relation (10.14).

Thus by estimating the signature analysis efficiency we came to the conclusion that any assumption of equiprobability of any error sequence, the same as any other general assumption, prevents us from obtaining even an approximate signature analysis estimation compared to other compact testing techniques. The principal consideration in comparing the techniques is the presence of more detailed characteristics of compressed sequences associated with their possible errors.

A more universal approach to comparing compact testing techniques is the one, which takes account of tested sequence properties.

The principle of the approach is that the efficiency of any compact testing technique may be estimated by using the error escape probability P^μ, which depends on the error multiplicity $\mu \in \{1,2,3,...,l\}$, where l is the length of sequence to be tested.

The indicated error escape probability for multiplicity μ may be obtained by

$$P^\mu = P_v^\mu P_n^\mu,$$

where P_v^μ is the probability of an possible μ-multiple error of multiplicity μ; P_n^μ is the probability of failing to detect an error of multiplicity μ. P_v^μ is defined as a ratio of practically possible μ-multiple errors to the number of all possible errors of any multiplicity in its output response

$$P_v^\mu = \frac{Q_v^\mu}{\sum_{\mu=1}^{l} Q_v^\mu}. \quad (10.18)$$

and the value P_n^μ is defined as a ratio of μ-multiple errors undetectable by specific compact testing technique to the number of all possible μ-multiple errors

$$P_n^\mu = \frac{Q_n^\mu}{C_l^\mu} \quad (10.19)$$

Then the error escape probability P_n will be

$$P_n = \sum_{\mu=1}^{l} P^\mu = \sum_{\mu=1}^{l} P_v^\mu P_n^\mu \quad (10.20)$$

In the following we shall use P_n as an efficiency measure for compact testing techniques. The probability P_n depends on an arbitrary probability distribution P_v^μ dependent on specific characteristics of a circuit under test, the set of all possible errors for the circuit and the test sequence as well as on probability distribution P_n^μ characterizing the applied compact testing technique.

Considering the fact that correct probability distribution P_v^μ is difficult if not impossible to obtain, we shall find the limit estimates for some compact testing techniques. For the purpose we estimate the probability P_n for each of the three techniques: signature analysis, ones counting and a trivial compression technique. In the following we shall discuss the test sequences of length $l=2^m-2$ where m is the signature length.

The minimum values of P_n obtained by (10.14) will be zero for the three techniques of interest since there may exist an error multiplicity μ for each of them such that $P_n^\mu = 0$. Then for $\mu = 1, 2$ the probability values will be $P_n^1 = P_n^2 = 0$ for signature analysis. For $P_v^\mu = 0$ where $\mu = 2i$, $i = \overline{1, \min(s, l-s)}$, and s is the number of ones in the output response, the probability $P_n = 0$ for ones counting since the technicue detects all odd errors in the sequences under test.

Signature Analysis Efficiency

For the trivial compression technique, P_n, if at least one distortion of an output sequence has fallen within m bits under test which holds for errors of greater multiplicity than l-m.

Let us find the maximum estimates of P_n for the three compression techniques in question. For the start we shall prove the following lemma.

Lemma 10.1. The maximum probability P_n of failing to detect an error sequence by a specific compact testing technique equals to the maximum value of probability from distribution P_n^μ of failing to detect μ-multiple errors, for $\mu = \overline{1,l}$, by this technique.

Proof. From the expression (10.18) it follows that for probabilities P_n^μ, $\mu = \overline{1,l}$, the equality

$$P_\nu^1 + P_\nu^2 + \ldots + P_\nu^l = 1. \quad (10.21)$$

hold true.

Then depending on specific values of P_ν^μ satisfying the equation (10.21), P_n will assume different values, substituting specific values of P_n^μ that have been found for the compact testing technique in question into (10.20) and choosing $\max P_n^\mu = P_n^m$, we obtain

$$P_n = P_n^m \sum_{\mu=1}^{l} \alpha_\mu P_\nu^\mu,$$

where $\alpha_\mu = P_n^\mu / P_n^m \leq 1$. From the analysis of the latter relation one can see that P_n is the highest when the maximum sum

$$\alpha_1 P_\nu^1 + \alpha_2 P_\nu^2 + \ldots + \alpha_l P_\nu^l \quad (10.22)$$

does not exceed one since the condition (10.21) is met for P_ν^μ and $\alpha_\mu \leq 1$. Among the possible values of P_ν^μ causing the sum (10.22) to be one is the case when all terms of (10.21) save P_v^μ are zero and only $P_v^m = 1$. Then P_n will assume the maximum value of P_n^μ.

Theorem 10.4. The maximum value of probability P_n of failing to detect an error sequence by signature analysis is determined by the expression

$$P_n = \frac{1}{(2^m - 3)}.$$

Proof. For signature analysis, the following recurrence

$$P_n^1 = P_n^2 = 0,$$
$$P_n^\mu = \frac{1}{2^m - \mu}[1 - P_n^{\mu-1} - (\mu-1)P_n^{\mu-2}], \quad \mu = \overline{3, 2^m - 1},$$

holds true for any P_n^i.

Using mathematical induction, we may readily demonstrate that

$$P_n^1 = P_n^2 = 0,$$
$$P_n^\mu = P_n^{\mu-1} = \frac{1}{2^m - \mu}(1 - \mu P_n^{\mu-2}), \quad \mu = 2k+1, \quad k = \overline{1, 2^{m-1} - 2}.$$

Let $P_n^{\mu-2} = \frac{1}{2^m} - \xi$. Then $P_n^\mu = \frac{1}{2^m} + [\frac{\mu}{(2^m - \mu)}]\xi$. Factor $\frac{\mu}{(2^m - \mu)}$ is less than one for $\mu \leq 2^{m-1}$ and greater than one for $\mu > 2^{m-1}$. Therefore P_n^μ decreases monotonically with $\mu < 2^{m-1}$ and increases with $\mu > 2^{m-1}$. Hence the value of P_n^μ will be maximum with minimal and maximal values of μ. And finally we obtain

$$P_n^3 = P_n^{2^m - 4} = \frac{1}{(2^m - 3)}.$$

Theorem 10.5. The maximum value of probability P_n of failing to detect an error sequence by ones counting is determined by the expression

$$P_n = \frac{(2^{m-1}-1)}{(2^m-3)}.$$

Proof. For ones counting, the probability P_n^μ is maximum when the value of numerator in the expression

$$P_n^\mu = \frac{C_s^{\mu/2} C_{l-s}^{\mu/2}}{C_l^\mu}, \quad \mu = 2k, \ k = 1, 2, ..., l/2 \quad (10.23)$$

is maximum which is attained with $s=l/2$ where s is the number of ones in the reference output sequence. For an arbitrary s, the denominator of (10.23) increases faster then the numerator with μ and hence the probability P_n^μ assumes the maximum value with $\mu=2$ and $s=l/2$. Therefore

$$\max P_n^\mu = \frac{C_{l/2}^1 C_{l/2}^1}{C_l^2} = \frac{l}{2(l-1)} = \frac{2^{m-1}-1}{2^m-3}.$$

Subject to Lemma 10.1, the maximum value of P_n can be determined by

$$P_n = \frac{(2^{m-1}-1)}{(2^m-3)}.$$

Theorem 10.6. The maximum value of P_n of failing to detect an error sequence by trivial method based on selection of any m of l tested bits in a l-long sequence as a signature is determined by

$$P_n = 1 - \frac{m}{(2^m-2)}.$$

Proof. For the trivial method based on selection of any m of l tested bits in an output sequence, the probability P_n^μ is determined by

$$P_n^\mu = \frac{C_{l-m}^\mu}{C_l^\mu} = \frac{(l-m)!\,(l-\mu)!}{l!\,(l-m-\mu)!}, \quad \mu = \overline{1, l-m}, \qquad (10.24)$$

where $C_{l-m}^\mu = Q^\mu$ is the number of undetectable μ-multiple errors. Here $P_n^\mu = 0$ for $\mu \in \{l-m+1,\ldots,l\}$.

In the expression (10.24), the denominator decreases faster than the numerator with the increase of μ therefore the maximum value of P_n is attained with the minimum μ, i.e. $\mu = 1$. Hence the maximum value of P_n equal to max P_n^μ will be determined by

$$P_n = \frac{(l-m)!\,(l-1)!}{l!\,(l-m-1)!} = 1 - \frac{m}{2^m - 2}.$$

Thus all the three theorems discussed above resulted in the maximum values of P_n for the three compact testing methods.

Table 10.3

Technique	Probability estimate		
	min P_n	max P_n	lim max P_n
Signature analysis	0	$1/(2^m-3)$	0
Ones counting	0	$(2^{m-1}-1)(2^m-3)$	1/2
Trivial	0	$1-m/(2^m-2)$	1

The general table of estimates (Table 10.3) for P_n contains the limiting probability values for each technique. Analysis of the results obtained shows that signature analysis turns out to be more efficient in practically every case when data stream length is large, since its upper estimate P_n has the minimum value $1/(2^m-3)$ which does not depend on specific form of distribution P_v^μ and tends to zero for $m \to \infty$.

Chapter 11

EVALUATION TECHNIQUES FOR REGULAR BINARY SEQUENCE SIGNATURES

11.1. Evaluation of Regular Sequence Signatures

The signature analysis-based techniques for diagnosing digital circuits are finding ever-widening application in manufacturing and usage of computer facilities. This fact is stimulatory to further development and investigation of signature analysis theory and practices. One of the signature analysis problems that remained practically unsolved is evaluation of regular sequence signatures.

In the general case, signature analysis is implemented by an analyzer described by a primitive polynomial

$$\varphi(x) = 1 \oplus \alpha_1 x^1 \oplus \alpha_2 x^2 \oplus \ldots \oplus \alpha_{m-1} x^{m-1} \oplus \alpha_m x^m,$$

where $\alpha_i \in \{0,1\}$ are the constant coefficients, $m = deg\varphi(x)$ is the high degree of polynomial $\phi(x)$. By way of example let us consider the signature analyzer of Fig.11.1 described by the polynomial $\phi(x) = 1 \oplus x^1 \oplus x^4$. Note that the signature analyzer described by $\phi(x)$ is basically similar to a M-sequence generator described by the same polynomial $\phi(x)$ which permits the use of M-sequence theory for studying signature analysis properties.

An arbitrary regular sequence is a collection of symbols that are most often binary 0s or 1s. To describe the sequence, we shall use the notation adopted for specifying characteristic M-sequences, where 0^n or 1^n is respectively a sequence of $n>0$ of repeated 0s or 1s. For $n=1$, we have $0^1=0$ and $1^1=1$. Thus the sequence 000000 can be represented as 0^6, 111111111 as 1^9. At the same time, the sequence 101000111010111100000 can be represented as $1010^31^30101^30^5$. More compact representation may be obtained for cyclic-type regular sequences. For example the sequence 000110001100011000110001 00011 can be written as $(0^31^2)^5$.

188   Chapter 11

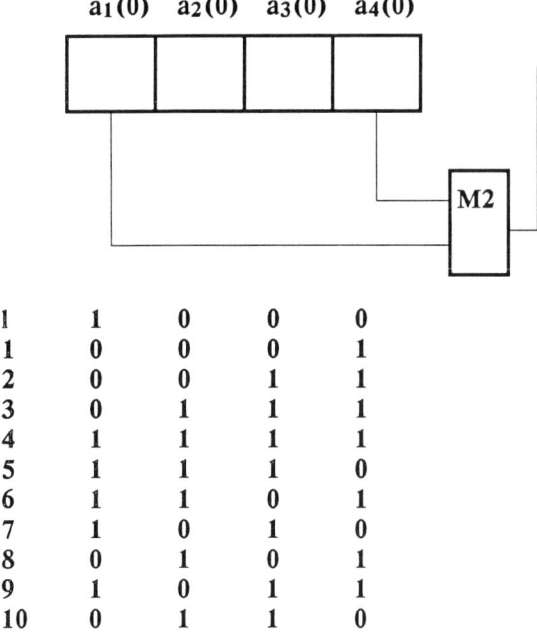

	a₁(0)	a₂(0)	a₃(0)	a₄(0)
1	1	0	0	0
1	0	0	0	1
2	0	0	1	1
3	0	1	1	1
4	1	1	1	1
5	1	1	1	0
6	1	1	0	1
7	1	0	1	0
8	0	1	0	1
9	1	0	1	1
10	0	1	1	0

Fig. 11.1. Signature analyzer

To describe the signature analyzer behavior, we use matrix notation. Then a signature $S(l)$ of l-long sequence $\{y(k)\}$, $k=\overline{1,l}$, obtained at the signature analyzer described by polynomial $\phi(x)$ will be expressed as

$$S(l) = V^{\,l} S(0) \oplus \sum_{k=1}^{l} V^{\,l-k} R(k) \qquad (11.1)$$

where

$$V = \begin{vmatrix} 0 & 1 & 0 & 0 & \dots & 0 & 0 \\ 0 & 0 & 1 & 0 & \dots & 0 & 0 \\ 0 & 0 & 0 & 1 & \dots & 0 & 0 \\ \dots & \dots & \dots & \dots & \dots & \dots & \dots \\ 0 & 0 & 0 & 0 & \dots & 0 & 1 \\ \alpha_m & \alpha_{m-1} & \alpha_{m-2} & \alpha_{m-3} & \dots & \alpha_2 & \alpha_1 \end{vmatrix} ;$$

α_i, $i=\overline{1,m}$, are coefficients of polynomial $\phi(x)$;

$$S(0) = \begin{vmatrix} a_1(0) \\ a_2(0) \\ a_3(0) \\ \ldots \\ a_m(0) \end{vmatrix}$$

is the column vector of initial states of signature analyzer storage elements; $R(k)$ is the vector

$$\begin{vmatrix} y(k) \\ 0 \\ 0 \\ \ldots \\ 0 \end{vmatrix},$$

defined by the symbol of sequence to be compressed.

By way of regular sequence example let us consider a zero sequence $\{y(k)\}$ and a sequence of unity symbols.

For a zero sequence $\{y(k)\}=\{000...0\}$ consisting of l zero symbols, the relation (11.1) is transformed into

$$S(l) = V^l S(0) \qquad (11.2)$$

This relation describes the behavior of M-sequence generators thereby allowing the use of the M-sequence theory for evaluation of $S(l)$. For $S(0)=0$, i.e., for zero initial states, $S(l)=0$ as follows from relation (11.2). The situation becomes more complicated for $S(0) \neq 0$ and large values of l. Then determination of $S(l)$ becomes very time-consuming.

Now we consider some approaches that can make evaluation of regular sequence signatures less time-consuming.

For a cyclic-type sequence $\{y(k)\}$, the following theorem holds true.

Theorem 11.1. If the period r of l-long sequence to be compressed meets the condition $L \bmod r = 0$, where $L=2^m-1$ and m is the high degree of generating polinomial $\phi(x)$, then for $l \geq L$ the signature $S(l)$ will meet the relation

$$S(l) = S(l \bmod L) \oplus [(l - l \bmod L) \bmod 2] V^{l \bmod L} S(L), \qquad (11.3)$$

where $S(l)$ is the signature of sequence to be compressed with the period r and length of L binary symbols under the zero initial settings.

Proof. For $l \geq L$, the relation (11.1) is transformed into

$$S(l) = V^{gL} V^{l \bmod L} S(0) \oplus [V^{(g-1)L + l \bmod L} \sum_{k=1}^{L} \oplus V^{L-k} R(k) \oplus$$

$$V^{(g-2)L + l \bmod L} \sum_{k=L+1}^{2L} \oplus V^{2L-k} R(k) \oplus \ldots \oplus \qquad (11.4)$$

$$V^{l \bmod L} \sum_{k=gL+1}^{gL + l \bmod L} \oplus V^{gL-k} R(k)],$$

where $g = (l - l \bmod L)/L$, then $gL + l \bmod L = l$. Since $L \bmod r = 0$, the equality $R(n) = R(n+iL)$, $n = \overline{1,L}$, $i = \overline{1,g}$, holds true for $R(k)$, $k = \overline{1,L}$. In view of the equality and $V^L = V^0$, we can transform the relation (11.4) into

$$S(l) = V^{l \bmod L} S(0) \oplus [V^{l \bmod L} [\sum_{n=1}^{L} \oplus V^{L-n} R(n) \oplus$$

$$\sum_{n=1}^{L} \oplus V^{L-n} R(n) \oplus \ldots \oplus \sum_{n=1}^{l \bmod L} \oplus V^{0-n} R(n)]] = V^{l \bmod L} S(0) \oplus$$

$$\sum_{n=1}^{l \bmod L} \oplus V^{l \bmod L - n} R(n) \oplus V^{l \bmod L} (l - l \bmod L) \bmod 2 \sum_{n=1}^{L} \oplus V^{L-n} R(n) =$$

$$= S(l \bmod L) \oplus (l - l \bmod L) \bmod 2 \, V^{l \bmod L} S(L),$$

where $S(L)$ is the signature of sequence of period r and the length of L binary symbols under the zero initial settings, i.e. $S(0) = 0$. For L, the equality $L \bmod r = 0$ holds true.

For a zero and a unity sequence of period $r = 1$, we can write the relation (11.3), in view of $S(L) = 0$, as

$$S(l) = S(l \bmod L). \qquad (11.5)$$

By using the theorem (11.1) we can make evaluation of signatures less time-consuming for $l \gg L=2^m-1$. Thus, for instance by formula (11.2) for $S(0)=0$ and $l=100$, on the analyzer of Fig.11.1, we must perform only 10 iterations. For $m=4$, $S(100)=S(100 \bmod 2^4-1)=S(100 \bmod 15)=S(10)$. At the same time, for comparable values l and $L=2^m-1$, evaluation of signature $S(l)$ may be rather time-consuming. The below discussed methods are given for $l < 2^m-1$.

As noted above, the relation (11.2) is the primary expression describing the M-sequence generator behavior. Therefore various algorithms can be used for efficient evaluation of $S(l)$. An algorithm based on decimation by $q=2$ turns out to be less time-consuming. The algorithm consists of 7 steps.

Algorithm 11.1

1. Represent the value l is in binary notation as a sum

$$l = C_{m-1} 2^{m-1} + C_{m-2} 2^{m-2} + \ldots + C_1 2^1 + C_0 2^0,$$

where $C_j \in \{0,1\}$, $j=\overline{0,m-1}$.

2. Set the initial m-bit code $a_0 a_1 a_2 \ldots a_{m-1}$ of the original M-sequence $\{a_i\}$ associated with vector $S(0)$, i.e. $a_0 = a_1(0)$, $a_1 = a_2(0), \ldots, a_{m-1} = a_m(0)$.

3. Set $j=0$.

4. Determine $C_j + m-1$ successive symbols of sequence $\{a_i^j\}$.

5. Using $C_j + m - 1 + m$ successive symbols of $\{a_i^j\}$, generate m symbols of sequence $\{a_i^{j+1}\}$ performing decimation by $q=2$ on $\{a_i^j\}$, i.e. $a_r^{j+1} = a_{2r}^j$. Thus symbol a_0^{j+1} in $\{a_i^{j+1}\}$, equals to the C_j th symbol in $\{a_i^j\}$.

6. Increment j by one.

7. Test for $j > m-1$. If the condition is met, the code $a_0 a_1 a_2 \ldots a_{m-1}$ is the signature $S(l)$ of the zero sequence of length l for $S(0) \neq 0$. Otherwise, go to step 4.

By way of example, let us consider evaluation of a signature of zero sequence 0^{10} of length $l=10$ and $S(0)=1000$ on the signature analyzer of Fig.11.1

1. $l=1010$, where $C_0=C_2=0$ and $C_1=C_3=1$.

2. The initial setting is $a_0^0 a_1^0 a_2^0 a_3^0 = 1000$.

3. $j=0$.

4.1. Calculate $0+4-1=3$ successive symbols of $\{a_i^0\}$. This yields $a_4^0 a_5^0 a_6^0 = 111$.

5.1. Since $C_0=0$, then $a_0^1 = a_0^0, a_1^1 = a_2^0, a_2^1 = a_4^0, a_3^1 = a_6^0$, hence $a_0^1 a_1^1 a_2^1 a_3^1 = 1011$.

6.1. $j=1<3$.

4.2. $a_4^1 = 0, a_5^1 = 0, a_6^1 = 1, a_7^1 = 0$.

5.2. $C_1=1$, hence $a_0^2 = a_1^1, a_1^2 = a_3^1, a_2^2 = a_5^1, a_3^2 = a_7^1$, i.e. $a_0^2 a_1^2 a_2^2 a_3^2 = 0100$.

6.2. $j=2<3$.

4.3. $a_4^2 = 0, a_5^2 = 1, a_6^2 = 1$.

5.3. $C_2=0$, hence $a_0^3 = a_0^2, a_1^3 = a_2^2, a_2^3 = a_4^2, a_3^3 = a_6^2$, i.e. $a_0^3 a_1^3 a_2^3 a_3^3 = 0001$.

6.3. $j=3=3$.

4.4. $a_4^3 = 1, a_5^3 = 1, a_6^3 = 1, a_7^3 = 0$.

5.4. $C_3=1$, hence $a_0^4 a_1^4 a_2^4 a_3^4 = 0110$.

6.4. $j=4>3$, hence code 0110 is the signature $S(0^{10})$.

The obtained result is supported by the timing chart for the analyzer of Fig.11.1.

Let us estimate the labor-consuming Q of calculating signature $S(l)$ by the above algorithm. For the worst case of $C_j=1$, m steps are needed to calculate the M-sequence symbols in each cycle. The total number of cycles is also m, hence $Q \leq m^2$. Note that the most time-consuming step of the algorithm in question is finding the symbols of M-sequence.

The corollary to the below theorem is most important for calculating the signature of unity sequence.

Theorem 11.2. Compression of a unity sequence of length l on storage elements of a signature analyzer described be a primitive polynomial $\phi(x)$ results in symbols of a sequence that is the inverse of the M-sequence described by $\phi(x)$.

Proof. When a zero sequence is compressed on the signature analyzer a successive symbol produced matches a M-sequence symbol and may be calculated by the formula

$$a_i = \sum_{k=1}^{m} {}^{\oplus} \alpha_k a_{i-k}, \qquad i=0,1,2,\dots . \qquad (11.6)$$

At the same time, the value of a successive symbol produced by compressing a unity sequence is determined by the formula

$$1 \oplus \sum_{k=1}^{m} {}^{\oplus} \alpha_k \bar{a}_{i-k}.$$

Hence, in view of $\overline{x} = x \oplus 1$ and the fact that for the primitive polynomials only the even number of coefficients α_k equals 1, we obtain that

$$1 \oplus \sum_{k=1}^{m} {}^{\oplus} \alpha_k \bar{a}_{i-k} = 1 \oplus \sum_{k=1}^{m} {}^{\oplus} \alpha_k a_{i-k} = \bar{a}_i.$$

It follows that the signature analyzer produces the symbols of inverse sequence.

Based on the above theorem, we can represent calculation of signature for the unity sequence 1^l as an algorithm.

Algorithm 11.2.

1. Represent the initial setting $S(0)$ of signature analyzer as inverse code $\overline{S(0)}$.

2. Using the algorithm 11.1, determine the signature of sequence 0^l for $\overline{S(0)}$.

3. Invert the signature obtained in step 2 to produce the signature of unity sequence 1^l.

By way of example, let us consider calculation of signature for sequence 1^{10} in the signature analyzer described by polynomial $\phi(x) = 1 \oplus x \oplus x^4$ for its initial zero setting.

1. Based on $S(0)=0000$, we obtain $\overline{S(0)} = 1111$.

2. Using the algorithm 11.1, we obtain $l=1010$ and $a_0^0 a_1^0 a_2^0 a_3^0 a_4^0 a_5^0 a_6^0 = 1111010$. Calculation results by the above algorithm are summarized in Table 11.1.

Table 11.1

J	a_0^j	a_1^j	a_2^j	a_3^j	a_4^j	a_5^j	a_6^j	a_7^j
0	1	1	1	1	0	1	0	
1	1	1	0	0	1	0	0	0
2	1	0	0	0	1	1	1	
3	1	0	1	1	0	0	1	0
4	0	1	0	0				

3. By inverting $a_0^4\ a_1^4\ a_2^4\ a_3^4$, we obtain $S(1^{10})=1011$.

To test the result for validity, calculate $S(1^{10})$ in a normal way by the analyzer of Fig11.1. The timing chart (Table 11.2) shows its performance.

11.2. A Technique for Calculating Periodic Sequence Signatures

Calculation of signatures for periodic sequences of length l and the period of r binary digits is of prime interest. The following theorem holds true for such sequences.

Table 11.2

k	$a_1(k)$	$a_2(k)$	$a_3(k)$	$a_4(k)$	k	$a_1(k)$	$a_2(k)$	$a_3(k)$	$a_4(k)$
0	0	0	0	0	6	1	0	0	1
1	0	0	0	1	7	0	0	0	1
2	0	0	1	0	8	0	1	1	0
3	0	1	0	1	9	1	1	0	1
4	1	0	1	0	10	1	0	1	1
5	0	1	0	0					

Theorem 11.3. Signature $S(l)$ of a periodic sequence of length l and period r is determined by

$$S(l) = V^l S(0) \oplus \sum_{i=1}^{l_v-1} {}^{\oplus} V^{ir} S(r), \qquad (11.7)$$

where $S(r)$ is a signature for one period of an initial sequence.

Proof. The initial sequence $y=(z)^{l/r}$, where z is the aggregate of zeros and ones that form a single period of the sequence, can be represented as a modulo-2 sum of sequences $z0^{l-r}, 0^r z0^{l-2r}, 0^{2r} z0^{l-3r},..., 0^{l-r} z$, i.e.

$$y = (z)^{l_r} = z0^{l-r} \oplus \sum_{j=1}^{l_r-2} {}^{\oplus} 0^{jr} z 0^{l-(j+1)r} \oplus 0^{l-r} z. \qquad (11.8)$$

Using the linearity property of signature analysis, we can demonstrate that signature $S(l)$ of an initial sequence equals to a bitwise modulo sum of signatures of sequences that are the terms of signatures of sequences of expression (11.8) with the initial analyzer state $S(0)$. Note that $S(0)$ can assume any value, including zero. The signature of sequence $0^{l-r}z$ equals to signature $S(r)$ of one period z of length r. This follows from the relation (11.1). On the basis of the relation we obtain that $V^r S(r)$ is a signature of sequence $0^{l-2r} z 0^r$, $V^{2r} S(r) - 0^{l-3r} z 0^{2r},..., V^{l-r} S(r) - z^{l-r}$. By the relation (11.1), the signature of zero sequence is $V^l S(0)$ under initial conditions of $S(0)$. The resulting signature $S(l)$, in view of the above linearity property, is determined as

$$S(l) = V^l S(0) \oplus \sum_{i=0}^{l_r-1} {}^{\oplus} V^{ir} S(r).$$

Based on the above theorem, we can develop an algorithm for calculating signature $S(l)$ for a periodic sequence.

Algorithm 11.3.

1. Determine signature $S(r)$ for one period of sequence under test by relation (11.1)

2. Specify $i=0$ and assign a zero to vector $S(l)$.

3. Calculate the new value of $S(l)$ by $S(l)=S(l) \oplus S(r)$.

4. Determine $i=i+1$.

5. For $i<l/r$ proceed to step 6, otherwise proceed to step 8.

6. Generate the new value of $S(r)$ by $S(r)=V^r S(r)$.

7. Execute step 3.

8. Calculate the value $V^l S(0)$.

9. Find the final value of $S(l)$ by bitwise modulo-2 summing of the current value of $S(l)$ with that obtained in step 8.

By using the algorithm of 11.3, calculate the value of $S(l)$ of sequence $y=00110010011$ for the signature analyzer described by polynomial $\phi(x)=1 \oplus x \oplus x^4$ under the initial condition $S(0)=1000$. The above sequence consists of $l=12$ symbols and three periods of $r=4$ in length, and the matrix V that describes the analyzer's behavior has the form

$$V = \begin{vmatrix} 0 & 1 & 0 & 0 \\ 0 & 0 & 1 & 0 \\ 0 & 0 & 0 & 1 \\ 1 & 0 & 0 & 1 \end{vmatrix}.$$

1. By the formula (11.1) calculate the signature $S(r)$ for one period of the sequence

$$S(r)=0010.$$

2. $i=0$, $S(l)=0000$.

3.1. $S(l)=0000 \oplus 0010 = 0010$.

4.1. $i=1$.

5.1. $i<3$.

6.1. Determine a successive value

$$S(r) = V^r S(r) = \begin{vmatrix} 0 & 1 & 0 & 0 \\ 0 & 0 & 1 & 0 \\ 0 & 0 & 0 & 1 \\ 1 & 0 & 0 & 1 \end{vmatrix}^4 \begin{vmatrix} 0 \\ 0 \\ 1 \\ 0 \end{vmatrix} = \begin{vmatrix} 1 & 0 & 0 & 1 \\ 1 & 1 & 0 & 1 \\ 1 & 1 & 1 & 1 \\ 1 & 1 & 1 & 0 \end{vmatrix} \begin{vmatrix} 0 \\ 0 \\ 1 \\ 0 \end{vmatrix} = \begin{vmatrix} 0 \\ 0 \\ 1 \\ 1 \end{vmatrix}.$$

7.1. Proceed to step 3.

3.2. $S(l) = 0010 \oplus 0011 = 0001$.

4.2. $i=2$.

5.2. $i<3$.

6.2. Calculate a successive value

$$S(r) = \begin{vmatrix} 1&0&0&1\\1&1&0&1\\1&1&1&1\\1&1&1&0 \end{vmatrix} \begin{vmatrix} 0\\0\\1\\1 \end{vmatrix} = \begin{vmatrix} 1\\1\\0\\1 \end{vmatrix}.$$

7.2. Proceed to step 3.

3.3. $S(l) = 0001 \oplus 1101 = 1100$.

4.3. $i=3$.

5.3. $i=3$.

8. Calculate

$$V^l S(0) = \begin{vmatrix} 0&1&0&0\\0&0&1&0\\0&0&0&1\\1&0&0&1 \end{vmatrix}^{12} \begin{vmatrix} 1\\0\\0\\0 \end{vmatrix} = \begin{vmatrix} 1&1&0&0\\0&1&1&0\\0&0&1&1\\1&0&0&0 \end{vmatrix} \begin{vmatrix} 1\\0\\0\\0 \end{vmatrix} = \begin{vmatrix} 1\\0\\0\\1 \end{vmatrix}.$$

9. $S(l) = 1100 \oplus 1001 = 0101$.

Let us estimate time required to calculate the signature by the above algorithm for the most probable case when $S(0) = 000...0$.

As noted above, the most time-consuming operation is finding the product of vector into matrix which is equivalent for M-sequences to the time consumed by the signature analyzer to get a successive symbol or perform a single compression cycle. By the above algorithm, time consumption Q for obtaining the resulting signature is determined by the number of multiplications of a vector by a matrix according to the relation

$$Q = r + rm + (V_r - 1) \qquad (11.9)$$

where the first summand specifies time consumption for calculating $S(r)$, the second for V^r and the third for the intermediate values of $S(r)$ with $m = \deg \phi(x)$. Q may assume different values depending on the r and l relation. The minimum value of Q is obtained for $r = [l/(m+1)]^{1/2}$, then

$$Q_{\min} = 2\,[l\,(m+1)]^{1/2} - 1.$$

In this case the efficiency K of the above algorithm 11.3 as compared to the classic calculation method is determined by

$$K = \frac{l}{2\,[l\,(m+1)]^{1/2}} = \frac{l^{1/2}}{2\,(m+1)^{1/2}} = \frac{1}{2}\left(\frac{l}{m+1}\right)^{1/2} \quad (11.10)$$

Note that time consumed by signature calculation by the classic method is determined by l. Thus the value K indicates the extent of classic method efficiency over the signature calculation algorithm 11.3.

By the formula (11.10) K is *121* for $m=16$ and $l=10^6$.

11.3. Fast Signature Calculation Algorithm

Signature evaluation offers few advantages over the conventional technique for obtaining signature values on a real digital device. First, signature evaluation programs allow the signature analyzer to be automatically prepared for the use at the debugging stage when the know-good device does not exist yet.

Fast signature evaluation procedures for periodic sequences are based on a theory of residues that is closely related to that of M-sequences.

To describe the periodic sequences, we shall use the notation adopted for specifying characteristic M-sequences.

Preparatory to discussing the algorithms for signature evaluation, we will give some auxiliary definitions and statements.

To describe the behavior of signature analyzer, we will use the matrix notation in the form of relation (11.1) and the value of $S(0)$ is assumed to be zero. We may note, however, that nonzero initial settings can always be accounted for by relation (11.1).

For nonzero initial settings, the following statements hold true.

Lemma 11.1. Signature $S(y)$ of a zero sequence $y=0^l$ will be zero for zero initial settings.

Lemma 11.2. Signature $S(y)$ of a sequence $y=0^n z$ equals to the signature of sequence z for zero initial settings, i.e.

$$S(y) = S(0^n z) = S(z). \quad (11.11)$$

Lemma 11.3. Signature $S(y)$ of a sequence $y=z0^n$ is defined for zero initial settings by the relation

$$S(y) = S(z0^n) = V^n S(z). \qquad (11.12)$$

Proof of the above three lemmas follows from definition of a binary sequence signature described by (11.1).

Lemma 11.4. Signature $S(y)$ of a sequence $y=z_1 \oplus z_2 \oplus ... \oplus z_n$ that is a bit-by-bit modulo-2 sum of sequences z_i, $i=\overline{1,n}$, equals to a bit-by-bit modulo-2 sum of original sequences for zero initial settings, i.e.

$$S(y) = S(z_1 \oplus z_2 ... \oplus z_n) = \sum_{i=1}^{n} {}^{\oplus} S(z_i) \qquad (11.13)$$

This lemma is based on a fundamental signature analysis property, i.e. linearity property.

As opposed to lemmas 11.1 to 11.4 that follow from signature analysis definition, the below statement calls for more complete proof based on Theorem 11.2.

Theorem 11.4. Signature $S(y)$ of a unit sequence $y=1^l$ is determined for zero initial settings by

$$S(y) = \bar{s}_1 \bar{s}_2 \bar{s}_3 ... \bar{s}_m,$$

where vector $S = s_1 s_2 s_3 ... s_m$ is determined as a product

$$V^l \begin{vmatrix} 1 \\ 1 \\ 1 \\ ... \\ 1 \end{vmatrix}.$$

<u>Proof.</u> According the Theorem 11.2 and Algorithm 11.2. we get the expression

$$\begin{vmatrix} s_1 \\ s_2 \\ s_3 \\ \dots \\ s_m \end{vmatrix} = V^l \begin{vmatrix} \bar{a}_1(0) \\ \bar{a}_2(0) \\ a_3(0) \\ \dots \\ \bar{a}_m(0) \end{vmatrix} = V^l \begin{vmatrix} 1 \\ 1 \\ 1 \\ \dots \\ 1 \end{vmatrix}.$$

When deriving the latter expression we took into consideration that $S(0)$, according to the theorem, assumes a zero value.

A major problem arising at signature evaluation is to estimate computational complexity of the procedure. For a unit sequence, signature evaluation is equivalent to finding the replica of a M-sequence shifted by l cycles as follows from Theorem 11.4. Accordingly, computational complexity for the signature of 1^l will equal to that of producing M-sequence replicas. For the purpose, we can use fast algorithms for shifted-replica generation whose computational complexity Q may be estimated by m^2, where $m=deg\phi(x)$. Note that computational complexity Q is estimated by the number of vector-by-matrix multiplications which is equivalent to generation of a single M-sequence symbol or to a single cycle of signature analyzer operation. It is evident that computational complexity of signature generation for a unit sequence 1^l by a classical method is idefined as $Q=l$.

In what follows we denote by $R^{(n)}$ raising the binary vector R to the n th power, and by universally adopted notation R^n raising binary polynomials into power.

Theorem 11.5. Under zero initial settings and $n=2^d$, where d is a positive integer obtained at the signature analyzer with external modulo-2 adders, signature $S(y)$ of a periodic sequence $y=(10^{n-1})^r 1$ is determined by

$$S(y) = S[(A\,S(z))^{(n)}] \qquad (11.16)$$

where $S(z)$ is the signature of sequence $z=1^{r+1}$, $A^{-1}A = E$; E is a unitary matrix, and the matrix

$$A = \begin{vmatrix} 1 & 0 & 0 & \dots & 0 & 0 \\ \alpha_{m-1} & 1 & 0 & \dots & 0 & 0 \\ \alpha_{m-2} & \alpha_{m-1} & 1 & \dots & 0 & 0 \\ \dots & \dots & \dots & \dots & \dots & \dots \\ \alpha_2 & \alpha_3 & \alpha_4 & \dots & 1 & 0 \\ \alpha_1 & \alpha_2 & \alpha_3 & \dots & \alpha_{m-1} & 1 \end{vmatrix}$$

is constructed on the basis of coefficients of a primitive polynomial $\varphi^{-1}(x)$ that is the reciprocal of a generating polynomial $\phi(x)$.

Proof. The original sequence $y=(10^{n-1})^r1$ can be represented as a binary polynomial $y(x)$ of the form

$$y(k) = \sum_{k=0}^{r} \oplus x^{kn} = [\sum_{k=0}^{r} \oplus x^k]^n = z^n(x),$$

where $z(x)$ is a binary polynomial describing the unit sequence $z=1^{r+1}$. The above relation has been obtained by using the Schoenemann property determined by the equality

$$[a(x)]^{p^d} = a(x^{p^d})$$

for polynomial $a(x)$ over GF(p). With $p=2$, the property holds true for $d=2^k$.

By using the mathematical definition of signature evaluation as a binary polynomial division which is equivalent to compression by the analyzer with internal modulo-2 adders, we can demonstrate that

$$C[y(x)] = C^n[z(x)] \bmod \varphi^{-1}(x), \qquad (11.7)$$

where $\varphi^{-1}(x)$ is the divisor or, what is the same, the polynomial describing the internal modulo-2 analyzer.

The values of signatures produced by analyzers with external and internal modulo-2 adders are related as

$$C(x) = A\,S(x). \qquad (11.18)$$

Substituting the value $C(x)$ from relation (11.18) into (11.17) we obtain

$$S[y(x)] = A^{-1}[[A\,S[z(x)]]^n \bmod \varphi^{-1}(x).$$

Changing from polynomial representation to binary sequences and taking account that the operation $\bmod\,\varphi^{-1}(x)$ on polynomial $\{AS[z(x)]\}^n$ is equivalent to multiplying the result $S\{[AS(z)]^{(n)}\}$ of compressing the binary sequence $[AS(z)]^{(n)}$ by the matrix, we can perform compression at an analyzer with external modulo-2 adders described by polynomial $\phi(x)$. We shall finally have

$$S(y) = S[[AS(z)]^{(n)}].$$

Based on Theorem 11.5 we can develop a fast signature evaluation algorithm. Thus for periodic sequences satisfying the condition of Theorem 11.5, we can use the following algorithm.

Algorithm 11.4

Evaluation of signature $S(y)$ for a sequence $y=(10^{n-1})^r 1$ by the signature analyzer described by $\phi(x)$.

1. Evaluate the signature of unit sequence $z=1^{r+1}$ by a fast algorithm to obtain the value of $S(z)$.

2. Multiply the vector $S(z)$ by matrix

$$A = \begin{vmatrix} 1 & 0 & 0 & \ldots & 0 & 0 \\ \alpha_{m-1} & 1 & 0 & \ldots & 0 & 0 \\ \alpha_{m-2} & \alpha_{m-1} & 1 & \ldots & 0 & 0 \\ \ldots & \ldots & \ldots & \ldots & \ldots & \ldots \\ \alpha_2 & \alpha_3 & \alpha_4 & \ldots & 1 & 0 \\ \alpha_1 & \alpha_2 & \alpha_3 & \ldots & \alpha_{m-1} & 1 \end{vmatrix}$$

where $\alpha_i \in \{0,1\}$, $i=\overline{1,m}$, are coefficients of the generating polynomial $\phi(x)$. This produces an intermediate result

$$R = AS(z).$$

3. Raise vector R to the nth power.

4. Compress the binary sequence associated with vector $R^{(n)}$ at the analyzer described by polynomial $\phi(x)$. The procedure result will be the desired signature value.

Consider an example of signature evaluation for the sequence $y=(10^3)^7 1$ by the analyzer described by polynomial $\phi(x)=1 \oplus x \oplus x^4$.

1. Determine the value of signature for $z=1^8$. We obtain $S(z)=0110$.
2. Calculate the vector

$$R = \begin{vmatrix} 1&0&0&0 \\ 1&1&0&0 \\ 0&1&1&0 \\ 0&0&1&1 \end{vmatrix} \begin{vmatrix} 0 \\ 1 \\ 1 \\ 0 \end{vmatrix} = \begin{vmatrix} 0 \\ 1 \\ 0 \\ 1 \end{vmatrix}.$$

3. Generate the sequence $R^{(4)}=(0101)^{(4)}=100000001$.

4. Compress the sequence 100000001 obtained in step 3 by the signature analyzer; we obtain $S(y)=1010$.

Let us estimate computational complexity of signature $S(y)$ by algorithm 11.4. We must initially note that computational complexity of finding the signature of $y=(10^{n-1})^r 1$ by a classical method is

$$Q = l = nr + 1. \qquad (11.19)$$

To find the value of $S(y)$ by the above algorithm, we must perform $r+1$ cycles of analyzer at the first step and a single cycle at the second step. Computational complexity of the third step is negligible since the Shoenemann relation can be used for raising to the power $n=2^d$. At the fourth step, the complexity will be $(m-1)n+1$ cycles. Thus computational complexity of the algorithm is determined by

$$Q_1 = (m-1)n + r + 3. \qquad (11.20)$$

It is evident that the efficiency of a fast signature evaluation algorithm depends to a large measure on the relationship between m, n and l. The minimum value Q_l is obtained for

$$n = \left(\frac{l-1}{m-1}\right)^{1/2}$$

as

$$Q_{1\ min} = 2\,[(l-1)(m-1)]^{1/2} + 3 \approx 2\,(lm)^{1/2} \qquad (11.21)$$

Whence, in particular, follows that, for $l=10^6$, $Q_{1\,min} = 4000$, i.e. the complexity is 250 times less than for $m=16$ calculating a signature at the analyzer by a classical method. The efficiency can be even greater by using fast signature evaluation procedures for a unit sequence. The maximum value $Q_{1\,max}$ for $l>>m$ is twice less than that of Q_1 obtained with $n=2$.

By using the above statements we can significantly decrease the computational complexity for the signatures of regular sequences produced by different counter structures.

Consider an example of signature computation for binary sequences $y_1 = 0101010101010101 = 0(10)^7 1$ and $y_2 = 001100110011 = (0^2 1^2)^4$ produced at the outputs from the low-order positions of a binary counter for polynomial $\phi(x) = 1 \oplus x \oplus x^4$.

Let us first find the value $S(y_1)$ which is defined by Lemma 11.2 as $S(z_1)$, where $z_1 = (10)^7 1$. The form of sequence z_1 satisfies the Theorem 11.4. Therefore we can use the algorithm 11.4 to calculate $S(z_1)$. Calculation of $S(z_1)$ differs from the preceding example in implementation of steps 3 and 4. Thus in step 3 obtain $R^{(2)} = (0101)^{(2)} = 10001$ and $S(y)$ is 1111. The sequence $y_2 = (0^2 1^2)^4$ can be represented as a bit-by-bit modulo-2 sum of two sequences $y_3 = 0^3 (10^3)^3 1$ and $y_4 = 0^2 (10^3)^3 10$. Among sequences y_3 and y_4 there is a sequence $z_2 = (10^3)^3 1$ which satisfies Theorem 11.4. Accordingly, using the algorithm 11.4, we obtain $S(z_2) = 0110$. Then in view of Lemma 11.2, we obtain $S(y_3) = S(z_2) = 0110$, and using Lemmas 11.2 and 11.3, we obtain

$$S(y_4) = V S(z_2) = \begin{vmatrix} 0 & 1 & 0 & 0 \\ 0 & 0 & 1 & 0 \\ 0 & 0 & 0 & 1 \\ 1 & 0 & 0 & 1 \end{vmatrix} \begin{vmatrix} 0 \\ 1 \\ 1 \\ 0 \end{vmatrix} = \begin{vmatrix} 1 \\ 1 \\ 0 \\ 0 \end{vmatrix}.$$

Eventually, by using Lemma 11.4, we obtain $S(y_2) = S(y_3) \oplus S(y_4) = 0110 \oplus 1100 = 1010$.

By using Lemmas and Theorem 11.4, it turns out possible to develop signature evaluation algorithms whose computational complexity closely approximates to the minimum value obtained by relation (11.21). Considering the fact that for $n=2$, the value of Q reaches its maximum, we can use the relation.

$$0^{n-1}(10^{n-1})^r 1 = 0^{n-1}(10^{2n-1})^{(r-1)/2} 10^n \oplus 0^{2n-1}(10^{2n-1})^{(r-1)/2} 1,$$

which allows to obtain the signature of $0^{n-1}(10^{n-1})^r 1$ by using the signatures of two sequences for which n is increased twice thereby decreasing the complexity Q_1.

Chapter 12

MULTI-LINE COMPRESSION SCHEMES

12.1. Design of Parallel Signature Analyzers

The problem of analyzing multi-output digital circuits in the course of their testing consists of detecting a fault by the output responses of the circuit. A characteristic property of such analysis is the need of testing a large amount of output responses of the circuit (it may run into hundreds). Therefore, in this case, conventional compact testing methods used for single-output digital circuits would not do the job. In fact, any attempt to analyse a n-input digital circuit by a serial signature analyzer either increases the circuit testing time by n or requires hardware for implementing n signature analyzers. Besides, the length of signature may increase n times. However, the use of serial signature analyzers allows one to find a compromise solution which is based on the use of only $u < n$ analyzers with hardware complexity and testing time increased equally.

Therefore, the use of serial signature analyzers for multi-output digital circuits causes either the increase of testing time or the complexity of analyzing section implementation as well as the increase of signature length. To overcome the difficulty, other techniques for compressing the responses of digital circuits under test have been developed.

Among them we can distinguish the method based on multiplexing sequences to be compressed, where n responses of a digital circuit are compressed through the multiplexer at the serial signature analyzer. Specific organization of compression by this method has its advantages and drawbacks. So if n responses are compressed successively in time the test experiment implementation will increase significantly. When n symbols of all sequences are being compressed with the constant test stimuli the frequency of applying test patterns will be made n time less than that of analyzer by changing addresses at the multiplexer inputs. The efficiency of diagnosing declines since the required test dynamics is not provided, and many faults in the circuit may escape detection.

The use of serial signature analyzers and multiplexing can be placed into practical techniques for solving the problem of multi-output circuit analysis.

Here exist, however, more systematic approaches to the problem based on deep theoretic grounds. Consider one of them, that is compression in space and time.

12.2. Compression in Space and Time

Compact testing methods are directed towards solving the problem of compressing long output responses of a digital circuit into short keywords. The possibility of such a procedure is based on the hypothesis that the final number of symbols in the output sequence of the circuit under test might be distorted. The hypothesis allows one to estimate the occurrence of a fault in the circuit by an integral characteristic of its output response. In this case, the length l of sequence being analyzed normally far exceeds the length m of keyword; however, the test coverage obtained will be high.

The procedure of compressing output responses of digital circuits is by its very nature the procedure of compression in time. An example of the procedure is syndrome testing implementation, which consists of adding the current symbol of the sequence under test to the sum of symbols obtained at previous instants of time. When used in practice, compression in time allows one to gain significant advantages from reduction in test data volume, application of long test sequences, etc. Every compact testing technique is oriented to a single-output digital circuit, therefore, a multi-output circuit is normally tested at each output by a unified scheme of compression method implementation. However, for a v-output digital circuit, we may suppose that the circuit faults will be observable by distortion of $w \leq v$ output responses. This is supported by the example of a v-output digital circuit whose outputs are associated with the specific set of its gates and inputs. The specified sets and the associated sets for other circuit outputs are disjoint. In this case any single error in the circuit will cause only one output sequence to be distorted. In the general case, the number of output responses with error characters will depend on the number of circuit faults, the type of test sequence and the level of functional dependence between the sequences produced at its outputs. We may always assume that any fault(s) in the circuit with v outputs will always be observable at least at $w \leq v$ outputs of the circuit. This makes it possible for the character values to be tested by an integral characteristic at v circuit outputs at a time. An example of such a test is the case of a v-output digital circuit whose faults manifest themselves at $w=1$ outputs only. Then their appearance will be observable by testing only one sequence produced as a modulo-2 sum of v output responses of the circuit. In practice, the implementation of the circuit that generates the summed sequence will consist of the use of a v-input modulo-2 adder.

The example just discussed has demonstrated the procedure of compression in space that consists of reducing the amount of output responses to be tested in the circuit, which are then compressed in time to obtain the final characteristic in the form of a keyword. Compact testing that realizes two-stage compression is performed by a special device. Such a device may be a circuit for compression in space, which allows one to reduce the number of tested sequences to u, to be further compressed in time to obtain a m-bit keyword. Then the value u, which defines the number of outputs in a circuit producing the integral characteristic, is a function of v and w, i.e. $u = f(v, w)$. If the value v is uniquely characterized by the number of circuit outputs, it will be rather difficult to estimate w. For the purpose we must either simulate all possible circuit faults or analyse the circuit structure. We may practically always estimate the maximum value of w, which is known to be greater than its actual value, but less than the number of circuit outputs v.

Thus the problem arises of designing a device that converts v output sequences of the digital circuit with a fault manifesting itself on w outputs at least into $u = f(v, w)$ sequences containing data on the fault that has occurred. Then the primary requirement imposed on the device is its ease of implementation, which resides in the fact that the number of elements to be used for obtaining the minimal value of u should be small. Any of the discussed compact testing methods whose hardware implementation allows compression in space may be a candidate for designing such a device. However, the scheme of linear conversion of v input sequences into u output sequences, which is based on coding theory concepts, is more suitable. In this case, the scheme realizing the procedure of compression in space will perform conversion associated with testing matrix H of linear code $(v, v-u, w+1)$. The basic characteristics of such a scheme are determined by the following lemma.

Lemma 12.1. The data compression scheme that realizes the testing of matrix H of linear code $(v, v-u, w+1)$ with v inputs and u outputs allows one to detect all errors whose multiplicity does not exceed w.

The proof of the above Lemma follows uniquely from the property of the linear code $(v, v-u, w+1)$, which includes $(v-u)$ data characters, total code word length of v and code distance $w+1$, to detect all possible errors of multiplicity less than $w+1$.

An example of a linear code that allows one to detect any single error is the code $(7, 6, 2)$ described by the testing matrix

$$H = |1 1 1 1 1 1 1|.$$

Implementation of the compression scheme based on matrix H has a seven-input modulo-2 adder. A more complex example is the use of code *(7, 3, 3)* whose testing matrix appears as

$$H = \begin{vmatrix} 0 & 0 & 0 & 1 & 1 & 1 & 1 \\ 0 & 1 & 1 & 0 & 0 & 1 & 1 \\ 1 & 0 & 1 & 0 & 1 & 0 & 1 \end{vmatrix} \quad (12.1)$$

The scheme of a device that performs compression in space and is based on matrix (12.1) has *7* inputs and *3* outputs. It provides for detecting all errors whose multiplicity does not exceed 2.

The maximum and minimum estimates for $u=f(v, w)$ are summarized in Table 12.1 for the most commonly used codes.

Table 12.1

v	w	u=f(v,w)		Code applied
		min	max	
v	1	1	1	Parity checking
2^m-1	2	m	m	Hamming Code (2^m-1, 2^m-1-m,3)
v	2	int[$\log_2(m+1)$]	int[$\log_2(m+1)$]	Reduced Hamming Code
v	3	1+int($\log_2 m$)	1+int[$\log_2(m+1)$]	External Hamming Code
23	6	11	11	Golay Code (23, 12, 7)

As seen from Table 12.1, the Golay code allows one to construct compression schemes with *23* inputs and *11* outputs. In so doing, the volume of test data is decreased by *23-11=12* output responses of a digital circuit of length *l*. At the same time, provision is made for detecting all faults that manifest themselves at no more than six outputs from the circuits.

The proposed procedure of compressing data in space can be built into test equipment or VLSI chip. By this means the amount of test data to be analyzed in the course of a test experiment may be significantly reduced. However, the efficiency of such an approach is not high without the procedure of compression in time.

Therefore, the most commonly used practical solutions are based on the use of multiline time compression schemes, which perform data compression in space and time concurrently. We next consider the basic methods of constructing such schemes.

12.3. State Count Testing

The principle of the technique is isolation and check (against the reference pattern) the count S of occurrences of a certain prespecified binary vector within the set time interval on n outputs from a digital circuit under test. When the count of occurrences matches the reference value S_s, i.e. $S=S_s$, the circuit under test is considered fault free.

The probability P_d of an error sequence to escape detection by the technique under the assumption that all reference and error sequences are equiprobable to appear on the outputs from the fault-free device under test can be found by the formula

$$P_n = \sum_{r=1}^{l} \frac{C_l^r(2^n-1)^{l-r}-1}{2^{nl}-1} \cdot \frac{C_l^r(2^n-1)^{l-r}}{2^{nl}} \qquad (12.2)$$

where C_l^r is the number of different placements of $r = S$ controllable vectors in l positions of a n-dimensional sequence under test; $(2^n-1)^{l-r}$ is the number of all possible placements of any uncontrollable vectors in the vacant positions of the output sequence; 2^{nl} is the number of all possible n-dimensional sequences of length l.

Based on the expression (12.2) we can demonstrate that with the count of controllable vector occurrences $r \geq 3$ and the sequence length $l \geq 2^n$ the efficiency of transition count testing is higher that of signature analysis which is equal in complexity.

When the reference value $r=S$ of a controllable vector occurrence is known a priori, the expression (12.2) assumes the form

$$P_n = \frac{C_l^r(2^n-1)^{l-r}-1}{2^{nl}-1}. \qquad (12.3)$$

Of fundamental importance resulting from expression (12.3) is the relationship between P_n and the number of controllable vectors r. The value of r to implement the state count testing with the specified validity which defines the probability P_n can be selected from the family of curves $P_n = f(r)$ that has been plotted for different r by (12.3). Fig. 12.1 shows an example of the family of curves P_n for $l=8$ and $n=1, 2, 3$. From the plot for the number of outputs $n=2$ it will be obvious that to obtain the test validity not worse than $P_n=0.2$ by state counting, one must select the value of r from the set $\{0, 4, 5, 6, 7, 8\}$.

A logical follow-on of state count testing is the modified technique based on checking the number r of the selected vector occurrences in the specified positions of the output sequence being examined. The modified state count testing provides for detecting a fault in a digital circuit by the first controllable vector occurrence in unexpected time steps as well as by the occurrence of any other vector in certain time steps. Owing to this property the proposed technique has some advantages over other compact testing techniques since they detect a fault only upon the analysis of the resulting signature.

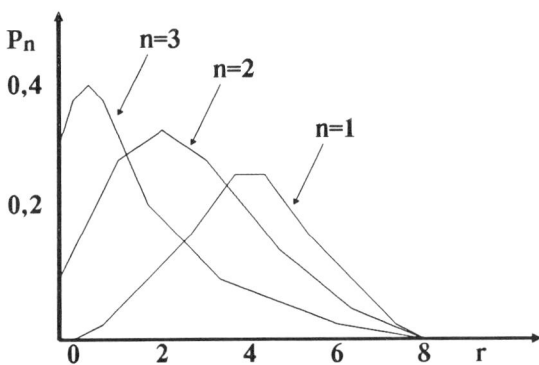

Fig.12.1. The family of curves Pn

The validity of the modified transition count testing is defined by the probability

$$P_n = \frac{(2^n - 1)^{l-r} - 1}{2^{nl} - 1}, \qquad (12.4)$$

where $(2^n - 1)^{l-r}$ is the number of different placements of any uncontrollable n-dimensional vectors in the analyzed sequence of length l in $l-r$ uncontrollable positions.

Multi-Line Compression Schemes

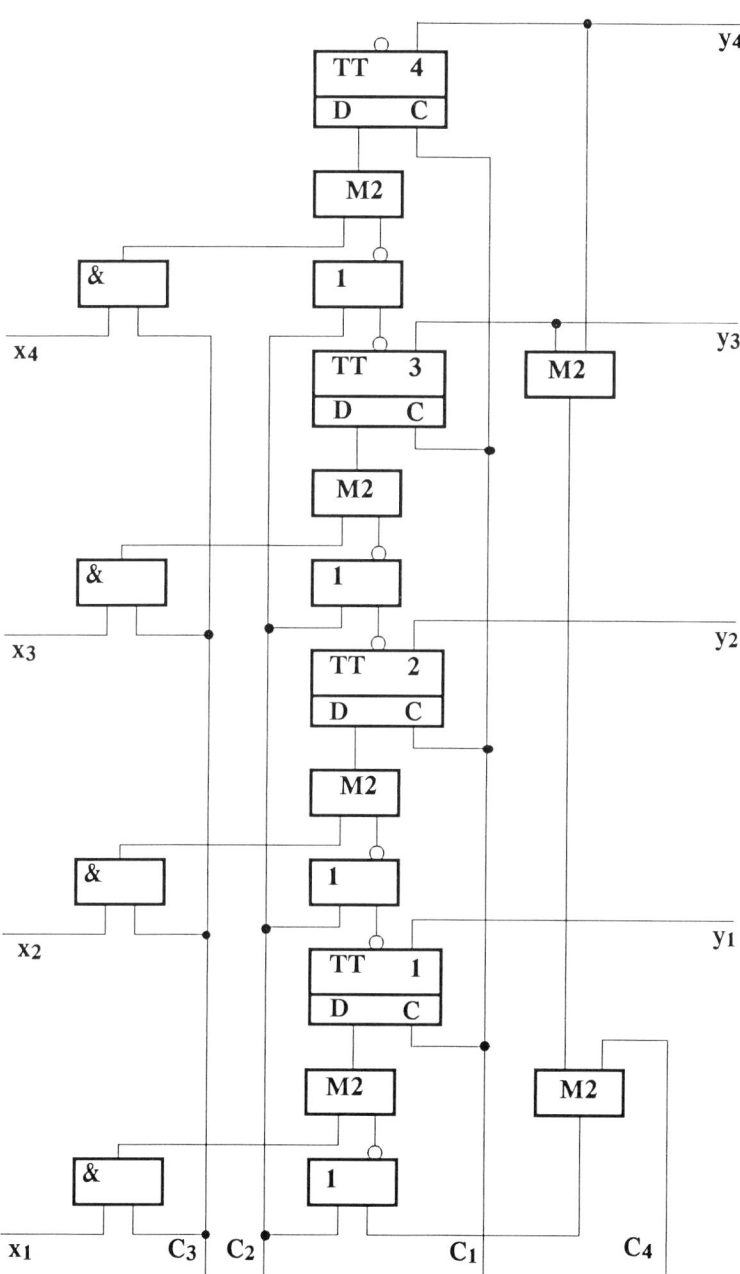

Fig. 12.2. BILBO

12.4. Universal Module BILBO

Various approaches for compressing response data into short signatures (keywords) are used to obtain compressed VLSI self-test data. Any difference in the actual and the expected signature indicates that the VLSI chip may be faulty whereas a match between the actual and expected signatures indicates a highly probable good condition of the chip.

A self-test approach using signature analyzers based on polynomial division as the facility for test response compressions is currently being widely used.

In the majority of cases, the BILBO structure which allows one to realize a number of functions required for the use of self-test techniques, is examined. Fig.12.2 shows the logic diagram of a BILBO for a generating polynomial $\phi(x)=1 \oplus x^3 \oplus x^4$. Below are the basic functions that can be performed by the BILBO.

1. Every storage element is set to 0 by applying control signals $C_3=0$ and $C_2=1$. Under these conditions, the input of each flip-flop becomes a logic 0 that is stored by applying a clock to input C_1.

2. With control signals $C_2=1$ and $C_3=1$ applied, the storage elements operate as expected in the VLSI chip under consideration. Each storage element can be loaded with input data applied to inputs x_i, $i=\overline{1,4}$, by a clock C_1 and unloaded through the outputs.

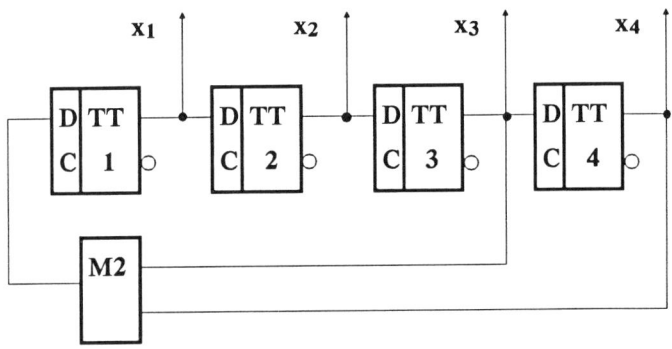

Fig. 12.3. Pseudorandom test pattern generator

3. For conditions $C_2=0$, $C_3=0$ and $C_4=1$, the structure under consideration is converted into a pseudorandom test pattern generator whose equivalent circuit is shown in Fig.12.3. The resulting structure allows one to generate a pseudorandom test pattern applied to a combinational portion of the VLSI chip.

4. By inverting control signal C_3, the circuit is converted into a multiple-input signature analyzer whose input patterns will be given by x_i, $i=\overline{1,4}$. Compression

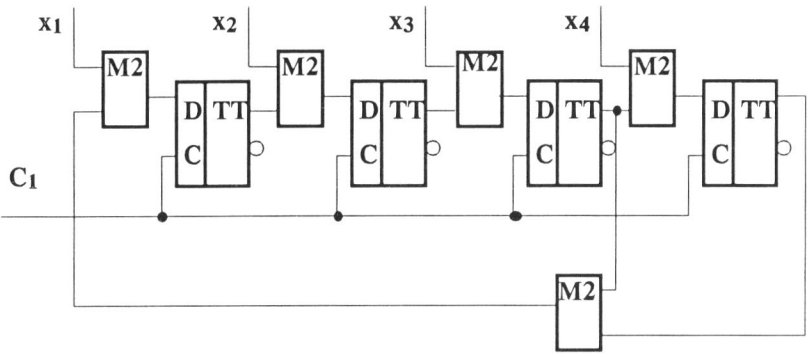

Fig. 12.4. Multi-input signature analyzer

of the above patterns will result in a signature (the contents of storage elements in a structure of Fig.12.4).

Signature analyzers of BILBO design have some advantages. There are low hardware cost, regular structure and short testing time. However parallel analyzers apart from the risk of losing error information at compression on each individual line have one more source of validity degradation, i.e. mutual error compensation by the lines which may heavily affect the efficiency. The chapters that follow will discuss the validity of parallel signature analyzers.

Chapter 13

ANALYSIS OF BILBO-PSA EFFICIENCY

13.1. Equivalent PSA Circuits with Internal and External XOR Gates

The present Section deals with the detectability of most popular parallel signature analyzer (PSA) structures - PSA with internal and PSA with external XOR-gates that is well known as a BILBO-module PSA. These analyzers are compared from the test efficiency point of view.

As opposed to the existing SSA (serial signature analyzer) structures, the structures under study have an additional source reducing the error detectability due to error aliasing on different analyzer channels, therefore we shall attempt to analyse the process of interchannel error aliasing separately in order to estimate their detectability. For the purpose, we shall describe a PSA signature generation in terms of the behavior of two-stage circuits.

A functional scheme of a PSA with internal XOR-gates is given in Fig.13.1 and, for the general case, its operation can be described by the set of equations:

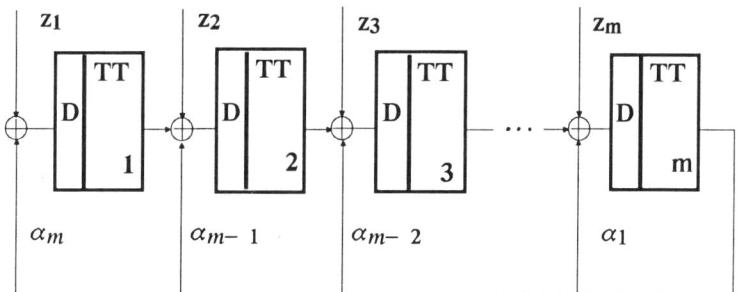

Fig. 13.1 PSA with internal XOR gates

Chapter 13

$$a_i(0) = 0, \quad i = \overline{1,m};$$
$$a_1(k) = z_1(k) \oplus a_m(k-1)$$
$$a_j(k) = z_j(k) \oplus a_{j-1}(k-1) \oplus \alpha_{m-j+1} a_m(k-1), \quad (13.1)$$
$$j = \overline{2,m}, \quad k = \overline{1,l},$$

where m is the PSA length; l is the number of binary vectors to be compressed; $a_i(k) \in \{0,1\}$ is the contents of the i-th storage element of PSA in the k-th cycle; $z_i(k) \in \{0,1\}$ is the value at the i-th input to PSA in the k-th cycle; $\alpha_i \in \{0,1\}$ are the coefficients that determine the PSA feedback structure. Thus for $\alpha_i = 1$ there exists a feedback between the m-th storage element output and the $(m-i+1)$-th storage element input.

When represented in vector notation, the above system appears as

$$A(0) = 0$$
$$A(k) = VA(k-1) \oplus Z(k), \quad k = \overline{1,l}, \quad (13.2)$$

where $A(k) = (a_1(k), a_2(k), \ldots a_m(k))^T$ is the state vector of PSA; $Z(k) = (z_1(k), z_2(k), \ldots z_m(k))^T$ is the input vector of PSA; V is a square matrix of the form:

$$V = \begin{vmatrix} 0 & 0 & 0 & \ldots & 0 & \alpha_m \\ 1 & 0 & 0 & \ldots & 0 & \alpha_{m-1} \\ 0 & 1 & 0 & \ldots & 0 & \alpha_{m-2} \\ 0 & 0 & 1 & \ldots & 0 & \alpha_{m-3} \\ \ldots & & & & & \ldots \\ 0 & 0 & 0 & \ldots & 1 & \alpha_1 \end{vmatrix}$$

For the given PSA, its characteristic polynomial can be represented as

$$\varphi(x) = x^m \oplus \alpha_1 x^{m-1} \oplus \alpha_2 x^{m-2} \oplus \ldots \oplus \alpha_{m-1} x \oplus \alpha_m$$

The contents of PSA storage elements obtained for l clocks

$$A(l) = \sum_{k=1}^{l} \oplus V^{l-k} Z(k)$$

is a resulting signature.

Signal $z_i(k)$ applied to the i-th input of a PSA with internal XOR-gates in the k-th cycle is equivalent to the same signal applied to the $(i-1)$-th input in the $(k-1)$-th cycle, where $i=\overline{2,m}$, from the resulting signature value standpoint.

Since the input signal values are sequentially transferred to the first PSA input from the remaining input nodes, we may conclude that application of signals $z_i(k)$ to the i-th input in the k-th cycle is equivalent to application of the same value to the first PSA input in the $(k-i+1)$ cycle.

Considering that modulo-2 addition is linear in nature, we shall use the superposition principle to conclude that application of an input vector $z_1(k)$, $z_2(k)$, ... , $z_m(k)$ to the first, second, ... , m-th PSA inputs, respectively, in the k-th instant is equivalent to application of values

$$y(k-m+1+i) = z_{m-i}(k) \tag{13.3}$$

to its first input in the $(k-m+1+i)$-th cycle, $i=\overline{1,m-1}$.

Therefore, application of an input vector $z_1(k)$, $z_2(k)$, ... , $z_m(k)$ to the 1-st, 2-nd, ..., m-th PSA inputs in the k-th cycle is equivalent to application of binary values of (13.3) to the SSA whose feedback structure is identical to that of the PSA in question. The behavior of such a SSA is described by the set of equations

$$\begin{aligned}
b_i(0) &= 0, \quad i = \overline{1,m}; \\
b_1(k) &= y(k) \oplus b_m(k-1) \\
b_j(k) &= b_{j-1}(k-1) \oplus \alpha_{m-j+1} \, b_m(k-1), \\
& j = \overline{2,m}, \quad k = \overline{1,l},
\end{aligned} \tag{13.4}$$

where $b_i(k) \in \{0,1\}$ is the contents of the i-th storage element of SSA in the k-th cycle; $y(k) \in \{0,1\}$ is the input value of SSA in the k-th cycle; $\alpha_i \in \{0.1\}$ are the coefficients that determine the SSA feedback structure.

Actually, by sequentially substituting the values of $y(k)$ (13.3) into the set of equations (13.4), we obtain that SSA storage elements $1,2,...,m$ will contain $z_1(k), z_2(k), ..., z_m(k)$ by the k-th cycle. This corresponds to the contents of storage elements in the PSA in question with the vector $z_1(k), z_2(k), ..., z_m(k)$ applied to its inputs in the k-th instant. Since the feedback structures of both SSA and PSA are identical, hence, the resulting signatures of SSA and PSA will be identical too.

Having generalized the expression (13.3) for the number of binary vectors compressed on the superposition basis, we may state the following.

Statement 13.1. The resulting signatures obtained by compressing m binary sequences of length l at a PSA described by the set of equations (13.1) (Fig.13.1) and at a SSA (13.4) with the same feedback layout will be identical if binary data application to a SSA input is performed by the rule

$$y(k)=\sum_{j=1}^{m}{}^{\oplus} z_{m-j+1}(k-j+1), \quad k=\overline{1,l+m-1} \qquad (13.5)$$

$$z_q(p)=0 \text{ for } n<p<1$$

where $z_q(p)$ are the values applied to the q-th PSA input in the p-th cycle.

According to Statement 13.1, a PSA with internal XOR-gates can be represented as a two-stage equivalents circuit (Fig.13.2).

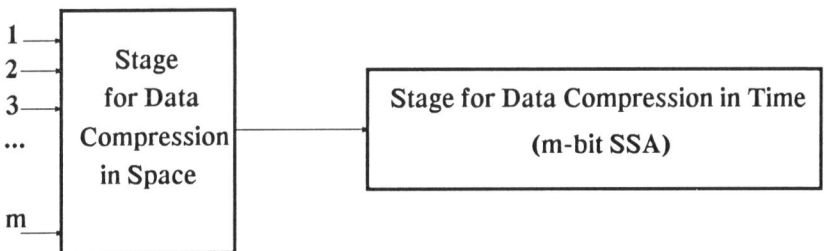

Fig. 13.2 A Two-Stage Equivalent Circuit of PSA

The first stage implements data compression in space (m binary sequences of length l are transformed into a single sequence of length $(l+m-1)$ according to equation (13.5)). The second stage of an equivalent PSA circuit is a SSA which is identical in structure to the PSA under study and compresses data in time by (13.4).

Table 13.1 contains the values of binary characters produced sequentially by the first stage (compression in space) of the circuit for a three-bit PSA with internal XOR gates.

Such representation greatly simplifies the study of PSA detectability which reduces in fact to that of properties of the equivalent circuit stage 1, since error

Analysis of BILBO-PSA Efficiency

detectability of the second stage (SSA) in respect to binary sequences will known by now.

In a similar way, we shall try to represent a PSA with external XOR gates (BILBO) as a two-stage equivalent circuit.

The BILBO-design PSA behavior (Fig.13.3) can be described by the set of equations

$$a_i(0) = 0, \quad i = \overline{1,m};$$

$$a_1(k) = z_1(k) \oplus \sum_{i=1}^{m}{}^{\oplus} \alpha_i\, a_i(k-1) \qquad (13.6)$$

$$a_j(k) = z_j(k) \oplus a_{j-1}(k-1) \quad j=\overline{2,m},\ k=\overline{1,l},$$

where m, l, $a_i(k)$, $z_i(k)$ and $\alpha_i(k)$ have the same meaning as in (13.1).

When described in the vector notation, the PSA with external XOR-gates can be represented by (13.1) where the matrix V has the form

$$V = \begin{vmatrix} \alpha_1 & \alpha_2 & \alpha_3 & \cdots & \alpha_{m-1} & \alpha_m \\ 1 & 0 & 0 & \cdots & 0 & 0 \\ 0 & 1 & 0 & \cdots & 0 & 0 \\ 0 & 0 & 1 & \cdots & 0 & 0 \\ \cdots & \cdots & \cdots & \cdots & \cdots & \cdots \\ 0 & 0 & 0 & \cdots & 1 & 0 \end{vmatrix}$$

In so doing, the PSA of interest like that with internal XOR-gates (13.1) has characteristic polynomial of the form

$$\varphi(x) = x^m \oplus \alpha_1 x^{m-1} \oplus \alpha_2 x^{m-2} \oplus \ldots \oplus \alpha_{m-1} x \oplus \alpha_m$$

and its state vector will be $A(l) = \sum_{k=1}^{l}{}^{\oplus} V^{n-k} Z(k)$ in l cycles to form a resulting signature S.

Application of a $z_i(k)$ to the i-th input of the PSA in the k-th cycle is equivalent to application of a $z_i(k)$ to its *(i-1)*-th input in the *(k-1)*-th cycle and a $\alpha_{i-1} z_i(k)$

Chapter 13 219

to its first input in the k-th cycle from the resulting signature stand point. By successively carrying input values over from the i-th input to the first input of PSA we have found out that application of values $z_i(k)$ to its i-th input in the k-th cycle is equivalent to application of the following to its first input: $\alpha_{i-j} z_i(k)$ in the $(k-j+1)$-th cycle, $j=\overline{1,(i-1)}$; $z_i(k)$ in the $(k-i+1)$-th cycle.

Since the modulo-2 addition is linear we can apply the superposition principle. Hence, application of the input vector $z_1(k), z_2(k), \ldots, z_m(k)$ to the 1-st, 2-nd,..., m-th input of PSA, respectively, in the k-th cycle is equivalent to application of the following values to its first input:

Table 13.1

Number of sequence section	Cycle number	A PSA equivalent circuit first stage compression result
1	1	$z_3(1)$
	2	$z_3(2) \oplus z_2(1)$
2	3	$z_3(3) \oplus z_2(2) \oplus z_1(1)$
	4	$z_3(4) \oplus z_2(3) \oplus z_1(2)$
	5	$z_3(5) \oplus z_2(4) \oplus z_1(3)$
	...	...
	n	$z_3(n) \oplus z_2(n-1) \oplus z_1(n-2)$
3	n+1	$z_2(n) \oplus z_1(n-1)$
	n+2	$z_1(n)$

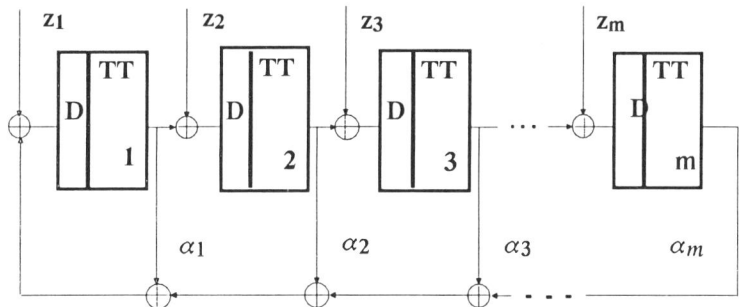

Fig. 13.3 PSA with external XOR gates (BILBO)

$$y(k-m+1) = z_m(k) \qquad - \text{ in the } (k-m+1)-th \text{ cycle}$$

$$y(k-m+1+i) = z_{m-i}(k) \oplus \sum_{j=1}^{i} {}^{\oplus} \alpha_j z_{m-i+j}(k) \qquad - \text{ in the} \quad (13.7)$$

$$i = \overline{1, m-1} \qquad (k-m+1+i)-th \text{ cycle}$$

Therefore application of the input vector $z_1(k), z_2(k), ..., z_m(k)$ to the 1st, 2nd, ..., m-th input of PSA, respectively, in the k-th cycle is equivalent to application of binary values by eq. (13.7) to an input of a SSA whose feedback layout is identical to that of PSA in question. The behavior of such SSA is described by the set of equations

$$b_i(0) = 0, \quad i = \overline{1, m};$$

$$b_1(k) = y(k) \oplus \sum_{i=1}^{m} {}^{\oplus} \alpha_i b_i(k-1) \qquad (13.8)$$

$$b_j(k) = b_{j-1}(k-1) \quad j = \overline{2, m},$$

where $b_i(k)$, $y(k)$, $\alpha_i(k)$ have the same meaning as in system (13.4).

Actually, by successively substituting the values $y(k)$ (13.7) into the system of equations (13.8) we obtain the values $z_1(k), z_2(k), ..., z_m(k)$ in SSA storage elements $1, 2, ..., m$ by the k-th cycle. This corresponds to the contents of PSA storage elements with the vector applied to its inputs in the k-th instant. Since the feedback layouts of SSA and PSA are identical, the resulting signatures of PSA and SSA are also identical.

Having generalized the expression (13.7) for l binary vectors compressed on the superposition basis, we may state the following.

Statement 13.2. The resulting signatures obtained by compressing m binary sequence of length l at a PSA based on XOR-gates (13.6) and a SSA (13.8) having the same feedback layout will be identical if binary data is applied to a SSA input by the rule

$$y(k) = \sum_{j=1}^{m} {}^{\oplus} z_{m-j+1}(k-j+1) \oplus \sum_{i=1}^{m-1} {}^{\oplus} \alpha_i \sum_{j=1}^{m-i} {}^{\oplus} z_{m-j+1}(k-i-j+1), \quad (13.9)$$

$$k = \overline{1, l+m-1}, \quad z_q(p) = 0 \text{ for } l < p < 1$$

where $z_q(p)$ are the values applied to the q-th PSA input in the p-th cycle.

Table 13.2

Number of sequence section	Cycle number	A PSA BILBO equivalent circuit first stage compression result		
		Always	If there is a feedback	
			α_1	α_2
1	1	$z_3(1)$		
	2	$z_3(2) \oplus z_2(1)$	$\oplus z_3(1)$	
2	3	$z_3(3) \oplus z_2(2) \oplus z_1(1)$	$\oplus z_3(2) \oplus z_2(1)$	$\oplus z_3(1)$
	4	$z_3(4) \oplus z_2(3) \oplus z_1(2)$	$\oplus z_3(3) \oplus z_2(2)$	$\oplus z_3(2)$
	5	$z_3(5) \oplus z_2(4) \oplus z_1(3)$	$\oplus z_3(4) \oplus z_2(3)$	$\oplus z_3(3)$
	...	...	...	
	n	$z_3(n) \oplus z_2(n-1) \oplus z_1(n-2)$	$\oplus z_3(n-1) \oplus z_2(n-2)$	$\oplus z_3(n-2)$
3	n+1	$z_2(n) \oplus z_1(n-1)$	$\oplus z_3(n) \oplus z_2(n-1)$	$\oplus z_3(n-1)$
	n+2	$z_1(n)$	$\oplus z_2(n)$	$\oplus z_3(n)$

According to Statement 13.2, a PSA based on XOR-gates can be represented as a two-stage equivalent circuit (Fig.13.2). The first stage compresses data in space (m binary sequences of length l are transformed into a single sequence of length (l+m-1) according to the equation (13.9). The second stage of PSA equivalent circuit is a SSA which has an identical feedback layout and compresses data in time by (13.8). Table 13.2 contains the values of binary characters for a three-bit PSA with internal XOR gates which are sequentially produced by the first stage (compression in space) of its equivalent circuit.

13.2. Estimating PSA Detectability on a Two-Stage Equivalent Circuit

As a criterion of compression efficiency we have selected probability distribution P_n^μ for a μ-multiple error to escape detection that is widely used for analyzing compression techniques. The value of P_n^μ is determined as a ratio of undetected μ-multiple errors to all possible errors of μ corrupted characters.

If only one of m binary sequences compressed is corrupted for the PSA under investigation, the escape probability for errors of different multiplicity will match those for its appropriate SSA.

This class, however, is only a small subset of possible errors in the output binary sequences of digital circuits under test.

Let us examine the PSA structure for the efficiency of detecting random errors of multiplicity μ by using the obtained equivalent circuit.

Assuming that error escape events are independent at the first and second stages of the PSA equivalent circuit which sequentially compresses binary data, we can determine a μ-multiple error detection probability in view of equiprobable distribution of SSA detectability by the formula:

$$P_d^\mu = 1 - P_n^\mu = (1 - P_{n1}^\mu)(1 - P_{n2})$$

where P_n^μ is the probability of a μ-multiple error to escape detection by an equivalent circuit.; P_{n1}^μ and P_{n2} are the probabilities of a μ-multiple error to escape detection by the space and time compression stages, respectively.

From the above formula it follows that the probability for a μ-multiple error to escape detection is

$$P_n^\mu = P_{n1}^\mu + P_{n2} - P_{n1}^\mu P_{n2} \qquad (13.10)$$

This formula defines the value P_n^μ for PSA accurately enough since P_{n1}^μ reflects error aliasing on different PSA channels (interchanel error aliasing) and P_{n2} reflects error aliasing in the shift register during compression in time.

Therefore the problem of estimating the efficiency of a PSA based on a linear feedback shift register reduces to estimation of the first stages of equivalent circuits that compress in space m l-bit sequences into a single sequence of length $(m+l-1)$ according to the expressions (13.5) and (13.9).

13.3. Examining Equivalent PSA circuits with Internal XOR gates

Let us examine the efficiency of the first stage of a PSA with internal XOR-gates.

Escape probability P_{n1}^{μ} (13.10) for an error of multiplicity $\mu \in \{1, 2, \dots ,ml\}$ that has occurred when m sequences $\{z_i(k)\}$ of length l were transformed by eq. (13.5) can be detected by the formula:

$$P_{n1}^{\mu} = \frac{V_n^{\mu}}{V_o^{\mu}} \qquad (13.11)$$

where V_n^{μ} is the number of μ-multiple errors that escaped detection; V_o^{μ} is the total number of μ-multiple errors.

In the sequences of bits produced by a space-compression circuit in accordance with eq. (13.5), we can isolate three sections with different data compression rules (Table 13.1). The length of sections 1 and 3 is determined by the PSA length only and equals to $(m-1)$ in both cases, whereas the length of section 2 depends on the number of test vectors and equals to $(l-m+1)$. When testing the circuit, the number of test patterns exceeds the PSA length by far, hence, the effect of sections 1 and 3 on a resulting value P_{n1}^{μ} is negligible. Taking account of the fact, we can approximately describe data compression process in the first stage of an equivalent circuit of PSA with internal XOR gates as

$$W(k) = \sum_{i=1}^{m} {}^{\oplus} z_i(k), \quad k = \overline{1,l} \qquad (13.12)$$

Let us find the escape probability P_{n1}^{μ} for an error of multiplicity $\mu \in \{1,2,3, \dots ,ml\}$ that has occurred in m sequences $\{z_i(k)\}$ of length l as a result of their transformation by (13.12).

Any single error that manifested itself in corruption of a bit in any of m output sequences will cause one bit to be inverted in the compression result. Therefore, we may deduce that any error of multiplicity $\mu=1$ will be detectable by the compression technique in question.

For $\mu=3$, the same as for any odd value of μ, $P_{n1}^{\mu}=0$, since corruption of odd number of bits in sequences $\{z_i(k)\}$ will always cause a difference between actual sequence $\{W^*(k)\}$ and its estimated form $\{w(k)\}$, as (13.12) suggests.

At the same time, the above relation demonstrates that for any μ assuming an even value, $P_{n1}^{\mu} \neq 0$.

In fact, when actual values $z_j^*(k)$ and $z_d^*(k)$, where $j \neq d \in \{1,2,3,...,m\}$, differ from their estimated values $z_j(k)$ and $z_d(k)$, the actual value of $w^*(k)$ bit in sequence $\{W(k)\}$ will coincide with its estimated value $w(k)$. For $\mu=2$, the number of such situations is determined by

$$V_m^2 = C_l^1 C_m^2,$$

and the total number V_o^2 of possible errors of multiplicity $m=2$ equals to C_{ml}^2. Thus the probability P_{n1}^{μ} can be calculated as

$$P_{n1}^2 = \frac{C_l^1 C_m^2}{C_{ml}^2} = \frac{m-1}{ml-1}$$

In view of the fact that $l \gg m$ and $m > 1$, the escape probability for double errors can be estimated as

$$\tilde{P}_{n1}^2 = \frac{m-1}{m} \frac{1}{l} \qquad (13.13)$$

For $\mu=4$, the value P_{n1}^4 will be calculated by

$$P_{n1}^4 = \begin{cases} \dfrac{C_m^2 C_m^2 C_l^2}{C_{ml}^4}, & m < 4 \\ \dfrac{C_m^2 C_m^2 C_l^2}{C_{ml}^4} + \dfrac{C_l^1 C_m^4}{C_{ml}^4}, & m \geq 4 \end{cases}$$

Having performed the transformations and substituting $(ml-1)$ and $(l-1)$ for ml and l, in view of the above assumptions, we obtain

$$\tilde{P}_{n1}^4 = \begin{cases} 3(\frac{m-1}{m})^2 \frac{1}{l^2}, & m<4 \\ 3(\frac{m-1}{m})^2 \frac{1}{l^2} + \frac{(m-1)(m-2)(m-3)}{m^3} \frac{1}{l^3}, & m \geq 4 \end{cases}$$

Having omitted the second term for $m \geq 4$, we can obtain P_{n1}^4 by

$$\tilde{P}_{n1}^4 = 3(\frac{m-1}{m})^2 \frac{1}{l^2} \qquad (13.14)$$

Similarly to (13.14), the escape probability for six-bit errors can be evaluated by

$$\tilde{P}_{n1}^6 = 15(\frac{m-1}{m})^3 \frac{1}{l^3} \qquad (13.15)$$

With equations (13.13), (13.14) and (13.15) generalized, we obtain the resulting value to be used for calculating the probability P_{n1}^μ for even $\mu \ll l$:

$$\tilde{P}_{n1}^\mu = \prod_{j=1}^{\mu/2} (2j-1)(\frac{m-1}{m})^{\mu/2} \frac{1}{l^{\mu/2}} \qquad (13.16)$$

Let us prove that the obtained estimate for the error aliasing probability in binary sequences compressed by the first stage of an equivalent circuit holds true for even μ.

Arbitrary partition of μ into different even terms can be generally represented as

$$2k_2 + 4k_4 + 6k_6 \ldots + Tk_T = \mu$$

where the i-th term occurs in this partition k_i times, where $i=2,4,6,\ldots$ and the maximum term T in each partition does not exceed the PSA length m. It should be noted that some k_i in a specific partition may be zero. Let us denote the entire set of all possible partitions of μ into even terms by Q. Then, in the general form P_{n1}^{μ} can be represented as

$$P_{n1}^{\mu} = \sum_Q \frac{(C_m^2)^{k_2} C_l^{k_2} (C_m^4)^{k_4} C_{l-k_2}^{k_4} \cdots (C_m^T)^{k_T} C_{l-k_2-k_4-\ldots-k_{T-2}}^{k_T}}{C_{ml}^{\mu}}$$

where summation is performed over the entire set Q.

We can write the expression in a more compact form as

$$P_{n1}^{\mu} = \sum_Q \frac{\prod_{i=1}^{T/2} ((C_m^{2i})^{k_{2i}}) \, C_l^{k_2} C_{l-k_2}^{k_4} \cdots C_{l-k_2-k_4-\ldots k_{T-2}}^{k_T}}{C_{ml}^{\mu}}$$

where the combinations in the second portion of the numerator are taken for nonzero coefficients k_i for the appropriate partition of μ.

Let us rewrite the expression as

$$P_{n1}^{\mu} = \sum_Q \prod_{i=1}^{T/2} ((C_m^{2i})^{k_{2i}}) \frac{1}{C_{ml}^{\mu}}$$

$$\frac{l!}{k_2!(l-k_2)!} \frac{(l-k_2)!}{k_4!(l-k_2-k_4)!} \cdots \frac{(l-k_2-k_4-\ldots-k_{T-2})!}{k_T!(l-k_2-k_4-\ldots-k_{T-2}-k_T)!} =$$

$$= \sum_Q \prod_{i=1}^{T/2} ((C_m^{2i})^{k_{2i}}) \frac{\mu!}{\prod_{i=1}^{T/2}(k_{2i}!)}$$

$$\frac{(l-k_2+1)(l-k_2+2)\ldots l \, (l-k_2-k_4+1) \ldots (l-k_2) \prod_{j=1}^{k_T}(l-\sum_{p=1}^{T/2} k_{2p}+j)}{(ml-\mu+1)(ml-\mu+2)\ldots ml}$$

where, for $k_i=0$, the appropriate terms in the latter term numerator will be absent for a certain i.

It is evident that for $\mu \ll l$ and $l \gg 1$, the expression will take the form

$$P_{n1}^{\mu} \approx \sum_Q \prod_{i=1}^{T/2}((C_m^{2i})^{k_{2i}}) \frac{\mu!}{\prod_{i=1}^{T/2}(k_{2i}!)} \frac{l^{\sum_{i=1}^{T/2}k_{2i}}}{(ml)^{\mu}} =$$

$$= \sum_Q \prod_{i=1}^{T/2}((C_m^{2i})^{k_{2i}}) \frac{\mu!}{\prod_{i=1}^{T/2}(k_{2i}!) \, m^{\mu}} \frac{1}{l^{\mu - \sum_{i=1}^{T/2}k_{2i}}}$$

The analysis of the latter relation testifies that the factor

$$l^{\mu - \sum_{i=1}^{T/2}k_{2i}}$$

will equal to $\dfrac{1}{l^{\mu/2}}$ in the term that takes account of partitioning μ into multiples of 2. For other terms of the obtained expression, the multiplier assumes the value of $\dfrac{1}{l^{\mu/2 + E}}$, where E is a positive integer, since for all partitions of μ into even terms, except for the above, the inequality

$$\sum_{i=1}^{T/2} k_{2i} < \frac{\mu}{2}$$

holds true.

Therefore, in view of the fact that $l \gg 1$, we can restrict ourselves with the first term of the above relation, since any subsequent term will be a second or greater order infinitesimal. In such a case, the escape probability for an error of multiplicity μ can be estimated, with the above assumptions, as

$$\widetilde{P}_{n1}^{\mu} = \frac{(C_m^2)^{\mu/2} C_l^{\mu/2}}{C_{ml}^{\mu}}$$

Let us prove that for the same assumptions $l \gg 1$ and $\mu \ll l$, the probability for an error of even multiplicity μ to escape detection by the first stage of PSA equivalent circuit may be estimated by equation (13.16).

To prove the statement, we shall use mathematical induction for all even μ.

1. Prove the validity of expression for error multiplicity $\mu=2$.

$$\frac{C_m^2 C_l^1}{C_{ml}^2} = \frac{m!}{2!(m-2)!} \frac{l!}{(l-1)!} \frac{2!(ml-2)!}{(ml)!} = \frac{(m-1)ml}{(ml-1)ml} =$$

$$= \frac{m-1}{lm-1} \approx \frac{m-1}{m} \frac{1}{l}$$

where approximation is carried out on assumption of $l \gg 1$

$$\prod_{j=1}^{1}(2j-1)(\frac{m-1}{m})^1 \frac{1}{l^1} = \frac{m-1}{m} \frac{1}{l}$$

2. Suppose that for an arbitrary even $\mu > 2$, the equality of interest

$$\frac{(C_m^2)^{\mu/2} C_l^{\mu/2}}{C_{ml}^\mu} = \prod_{j=1}^{\mu/2}(2j-1)(\frac{m-1}{m})^{\mu/2} \frac{1}{l^{\mu/2}}$$

holds true.

3. Prove the validity of the above equation for error multiplicity $\mu=2$.

$$\frac{(C_m^2)^{\mu/2+1} C_l^{\mu/2+1}}{C_{ml}^{\mu+2}} = \frac{(C_m^2)^{\mu/2} C_l^{\mu/2}}{C_{ml}^\mu} \frac{m(m-1)(l-\mu/2)(\mu+2)(\mu+1)}{2!(\mu/2+1)(ml-\mu-1)(ml-\mu)} \approx$$

$$\approx \frac{(C_m^2)^{\mu/2} C_l^{\mu/2}}{C_{ml}^\mu} \frac{m(m-1) l(\mu+1)}{m^2 l^2} =$$

$$= \frac{(C_m^2)^{\mu/2} C_l^{\mu/2}}{C_{ml}^\mu} \frac{(m-1)}{m} \frac{1}{l} (\mu+1) ,$$

where the approximation has been carried out in assumption of $l \gg \mu$.

$$\prod_{j=1}^{\mu/2+1} (2j-1)(\frac{m-1}{m})^{\mu/2+1} \frac{1}{l^{\mu/2+1}} =$$

$$= \prod_{j=1}^{\mu/2} (2j-1)(\frac{m-1}{m})^{\mu/2} \frac{1}{l^{\mu/2}} (\mu+1) \frac{(m-1)}{m} \frac{1}{l},$$

Thus we have proved the validity of estimate (13.16) obtained for calculating the probability of an even-multiplicity error to escape detection by the first stage of an equivalent circuit of PSA with internal XOR gates.

The analysis of the obtained probability of an data loss about error in the first stage of an equivalent circuit of PSA with internal XOR gates (13.10) reveals the following rules of test-response compression:

- the values of P_{n1}^{μ} may be fairly different for various error multiplicity μ;

- the difference is especially noticeable for small error multiplicity, whereas with the increase of m, P_{n1}^{μ} decreases for even multiplicity, and the differences mentioned become negligible;

- the difference of P_{n1}^{μ} values for various error multiplicity is highly pronounced for relatively small lengths of binary sequences l being compressed, and with the increase of l, P_{n1}^{μ} approaches zero for even multiplicity;

The plots of Fig.13.4 demonstrate how the error escape probability changes with the multiplicity and the length of binary sequences compressed for a 4-bit PSA with internal XOR gates.

The efficiency of the second compression stage in the circuit of Fig.13.2 which is a single-input signature analyzer, can be expressed for the compressed binary sequence length $2^{\omega} - 1$ as:

$$P_{n2}^1 = P_{n2}^2 = 0$$
$$P_{n2}^{\mu} = \frac{1}{2^{\omega} - \mu} [1 - P_{n2}^{\mu-1} - (\mu-1) P_{n2}^{\mu-2}], \mu = \overline{3, 2^{\omega} - 1}$$

and, for $l \gg \omega$, it can be estimated as

$$\tilde{P}_{n2}^{\mu} = \frac{1}{2^{\omega}},$$

where ω is the signature length.

These expressions allow to evaluate the efficiency of a two-stage compression circuit that corresponds to a PSA of the discussed structure.

Actually, for $\mu=3$ $P_n^3 = P_{n2}^\mu = 1/2^\omega$, since $P_{n1}^3 = 0$, and, for $\mu=4$, P_n^4 will be determined as

$$P_n^4 \approx 3(\frac{m-1}{m})^2 \frac{1}{l^2} + \frac{1}{2^\omega} - 3(\frac{m-1}{m})^2 \frac{1}{l^2} \frac{1}{2^\omega}$$

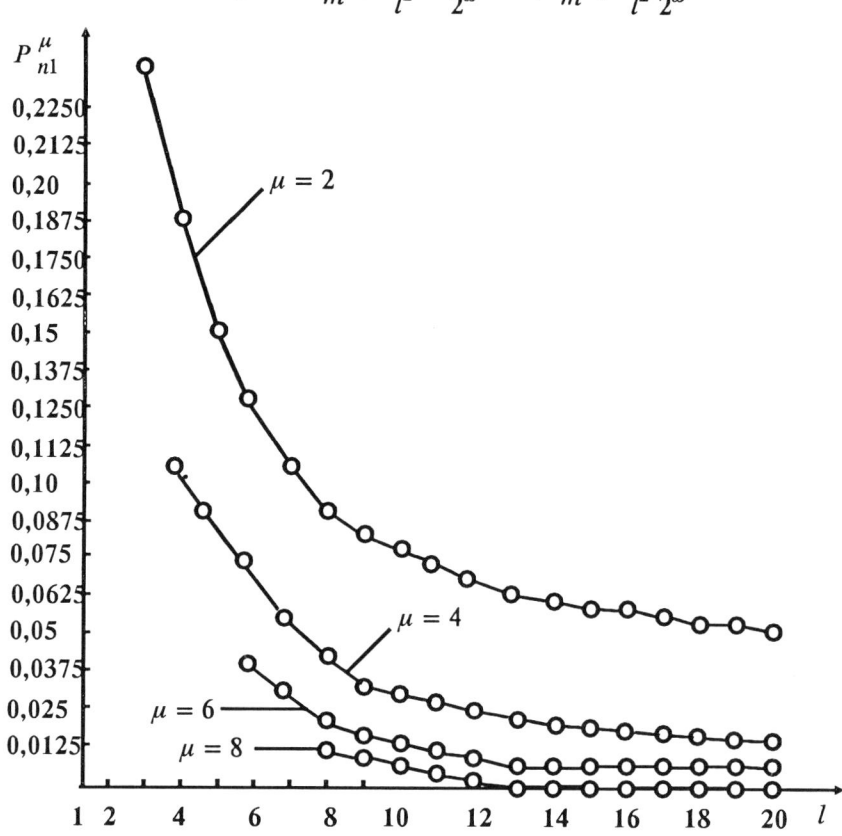

Fig. 13.4

where the final value of P_n^4 depends on the relationship between m, l and 2^ω, then for the case of $m=16$ and $l=256$, we shall have $P_n^4 \approx 4/2^\omega$ which is more than P_n^3 by a factor of 4.

Let us analyse an equivalent circuit of a PSA with external XOR gates (BILBO).

For $\mu=1$, any error caused by bit corruption in any of m binary sequences compressed inverts at least one bit in the compression result. Hence, any error of multiplicity $\mu=1$ will be detected by this data compression circuit.

For $\mu=2$, there may be cases when errors escape detection, as exemplified by (13.9). Thus for a BILBO PSA with $\alpha_i=0$, $i=\overline{1,m-1}$ (ring signature register), the number of aliasing errors for sequence section 2 will be $(l-m+1)C_m^2$ whereas for sections 1 and 2 C_i^1. Therefore, the number of undetectable errors of multiplicity $\mu=2$ will be

$$V_n^2 = 2\sum_{i=1}^{m-1} C_i^2 + (l-m+1) C_m^2. \qquad (13.17)$$

for a ring register scheme.

If there exists a feedback path from the i-th storage element $(\alpha_i=1)$ in the PSA, the number of undetectable errors for section 2 will be

$$V_{n2}^2 = (l-m+1)(C_{m-i}^2 + C_i^2), \quad i = \overline{1,m-1}$$

and for sections 1 and 3

$$V_{n1}^2 = \sum_{j=1}^{m-i} C_j^2 + (i-1) C_{m-i}^2 + \sum_{p=1}^{i-1} C_p^2$$
$$V_{n3}^2 = \sum_{j=1}^{m-i-1} C_j^2 + (m-i-1) C_i^2 + \sum_{p=1}^{i} C_p^2$$
$$i=\overline{1,m-1}.$$

respectively, by the compression rule (13.9).

Therefore, for the BILBO-design PSA with a single feedback path (the feedback from the last storage element is known to exist for sure), the number of undetectable errors is determined by

$$V_1^2 = V_{01}^2 + V_{02}^2 + V_{03}^2 \qquad (13.18)$$

With $l \gg m$, sections 1 and 3 exert very slight effect on the total number of undetectable errors and the expression (13.18) takes the form

$$V_i^2 = V_{n2}^2 = (l-m+1)(C_{m-i}^2 + C_i^2), \quad i = \overline{1, m-1}, \qquad (13.19)$$

The total possible number of two-bit errors depends on only m and l and equals to $V_a^2 = C_{ml}^\mu = C_{ml}^2$.

One can see from the obtained equations that for a BILBO-design PSA the probability of a $(\mu=2)$-multiple error to escape detection depends on location of a feedback path.

For example, for a 16-bit PSA with a feedback path from the first storage element $(\alpha_i=1)$, the number of $(\mu=2)$-multiple errors undetectable by the first stage of an equivalent circuit is

$$V_1^2 = (l-15) C_{15}^2 = 15 \times 7 (l-15),$$

with $l \gg m$, and for a PSA with a feedback path from the eighth storage element, it is

$$V_8^2 = (l-15) 2 C_8^2 = 8 \times 7 (l-15),$$

Besides, for the actual values n, the probability of a $(\mu=2)$—multiple error to escape detection by the circuit under test is far in excess of that for the 16-bit SSA (the second stage). Hence the probability of a $(\mu=2)$—multiple error to escape detection by PSA totally depends on the value P_{n1}^μ (13.10).

Thus, for a PSA with a feedback path from the first storage element, the probability of a $(\mu=2)$-multiple error to escape detection is almost twice as higher as that for a PSA with a feedback path from the eighth storage element.

Consider a BILBO-design PSA with several feedback paths $\alpha_i = \alpha_j = ... = \alpha_q = \alpha_p = 1$. Then, for the multiplicity $\mu=2$ and $l \gg m$, the number of errors undetectable by a space-compression circuit is the PSA equivalent structure, will be determined by the formula:

Chapter 13   233

$$V_n^2 = (l-m+1)(C_{m-i}^2 + C_{i-j}^2 + \ldots + C_{q-p}^2 + C_p^2), \quad (13.20)$$

where $i > j > \ldots > q > p$.

The analysis of relations (13.19) and (13.20) reveals that as the number of feedback paths increases, the probability of a $(\mu=2)$—multiple error to escape detection by the discussed compression circuit will decrease and even become zero in the case of feedback paths from all storage elements.

A comprehensive investigation of the process of data compression by the first stage of the equivalent circuit of PSA with external XOR gates (13.9) reveals that it has a pronounced effect on the escape probability P_n^μ (13.10) for errors of low multiplicity $(\mu < m)$, and the efficiency of detection for these errors increases with the increase in the number of feedback paths in an original structure (Fig.13.3).

13.4. PSA Efficiency Testing by Software Simulation

To examine the efficiency of parallel signature analyzers in detecting errors in binary sequences being compressed, they were simulated.

The technique used for generating a program model of PSA is based on the linearity property of modulo-2 addition. Let us think of data compressed as a binary sequence $Z(i)$, $i=\overline{1, lm}$. Then, in the general case, the resulting binary sequence compressed at the output of circuit under test can be expressed as

$$Z_a(i) = Z_f(i) \oplus Z_e(i), \quad i=\overline{1,lm},$$

where $Z_a(i)$ is the actual binary sequence to be compressed; $Z_f(i)$ is the expected (error-free) output sequence of circuit under test; $Z_e(i)$ is the so called error binary sequence whose unit values check for corrupted bits in $Z_a(i)$ against $Z_f(i)$.

Let $S_a(j), S_f(j), S_e(j), j=\overline{1,m}$, represent the values of signatures produced as a result of compressing binary sequences $Z_a(i), Z_f(i), Z_e(i), i=\overline{1, lm}$, at the signature analyzer.

Analysis of BILBO-PSA Efficiency

From the linearity property of modulo-2 addition it follows that the resulting actual signature may be expressed as

$$S_a(j) = S_f(j) \oplus S_e(j), \qquad j = \overline{1,m},$$

Therefore, the actual erroneous binary sequence Z_a, when compressed at PSA, will produce other than the expected signature S_f if signature S_e is nonzero.

Based on the property described, we implemented the program experiment on estimating the efficiency of various PSA structures. The program in Pascal consists of the following modules: - error sequence generator, - various PSA structure modules; - S_e analyzer.

The program module of error sequence Z_e generator can generate any binary patter of different length with the specified number of unit characters (error multiplicity). The generated binary sequences are compressed at the PSA program model and the resulting signatures S_e are processed by the signature analysis module. By varying the settings we can select the module associated with the required analyzer structure that is set to specific length m and feedback

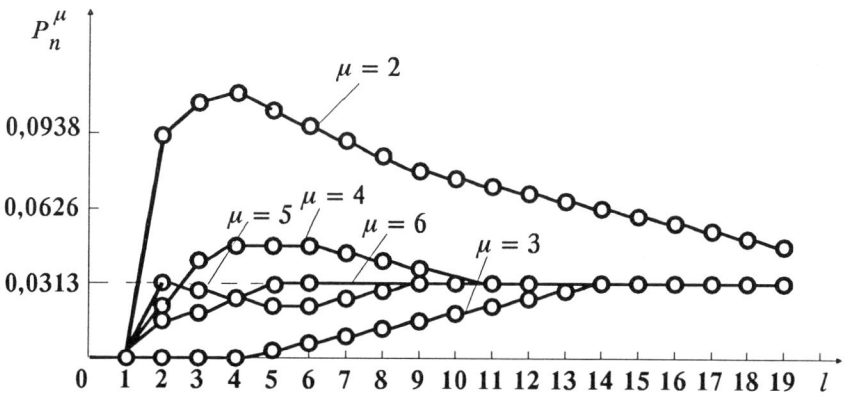

Fig. 13.5

structure.

Software simulation of PSA results in the values of error escape probability depending on the multiplicity of errors for different lengths of binary sequences being compressed.

By using the above described model, we have estimated the efficiency of PSA with internal XOR gates for two optional lengths and feedback structures.

Fig. 13.5 shows the plots representing variation in the probability of errors of multiplicity $\mu=2,3,4,5,6$ to escape detection by a five-bit PSA with internal XOR gates described by characteristic polynomial $\varphi(x)=x^5\oplus x^4\oplus x^3\oplus x^2\oplus 1$. Similar plots are given in Figs. 13.6, 13.7 and 13.8 for PSAs based on characteristic polynomials $\varphi(x)=x^6\oplus x^5\oplus x^2\oplus x\oplus 1, \varphi(x)=x^7\oplus x^6\oplus x^5\oplus x^4\oplus x^2\oplus x\oplus 1$,

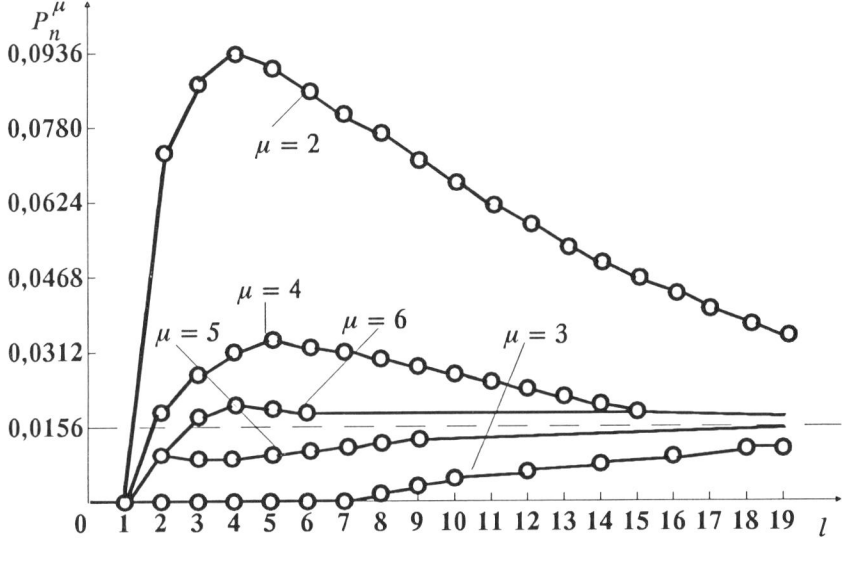

Fig. 13.6

and $\varphi(x)=x^9\oplus x^8\oplus x^7\oplus x^6\oplus x^5\oplus x^4\oplus x^3\oplus x\oplus 1$ respectively.

The analysis of these plots reveals that the structure of a PSA with internal XOR gates is characterized by a non-uniform rule of error escape probability variation with the multiplicity and length of sequence compressed. The effect of this property is most pronounced for low multiplicity of an error and for small lengths of binary sequence compressed.

Using the discussed program model, we tested the BILBO-design PSAs of different lengths and based on various characteristic polynomials.

Figs. 13.9 and 13.10 show the obtained error escape probabilities depending on multiplicity for 6-bit BILBO-design PSAs with one (polynomial $\varphi(x)=x^6\oplus x\oplus 1$), and three (polynomial $\varphi(x)=x^6\oplus x^5\oplus x^2\oplus x\oplus 1$) feedback paths, respectively, excluding the path from the last storage element. It is

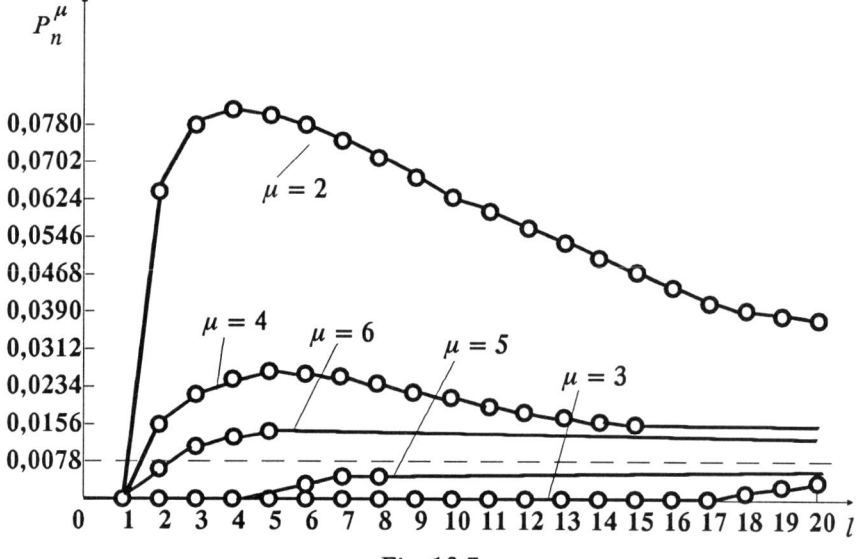

Fig. 13.7

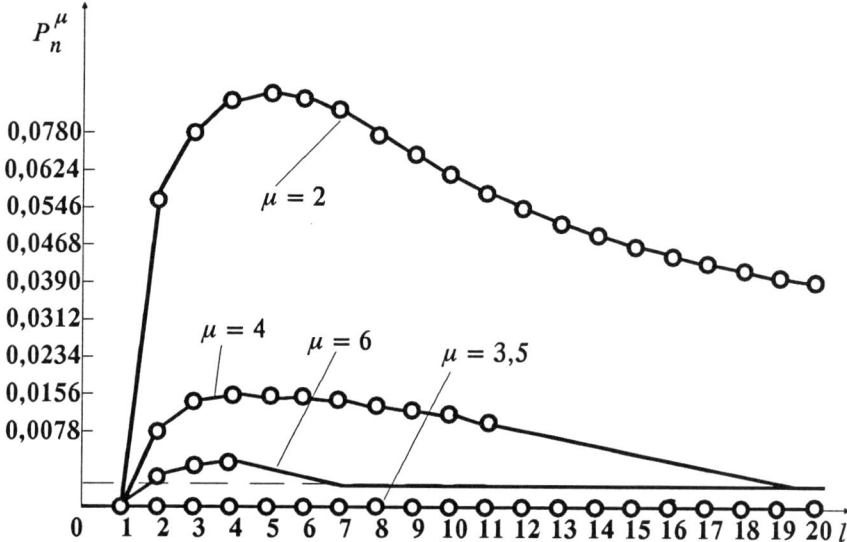

Fig. 13.8

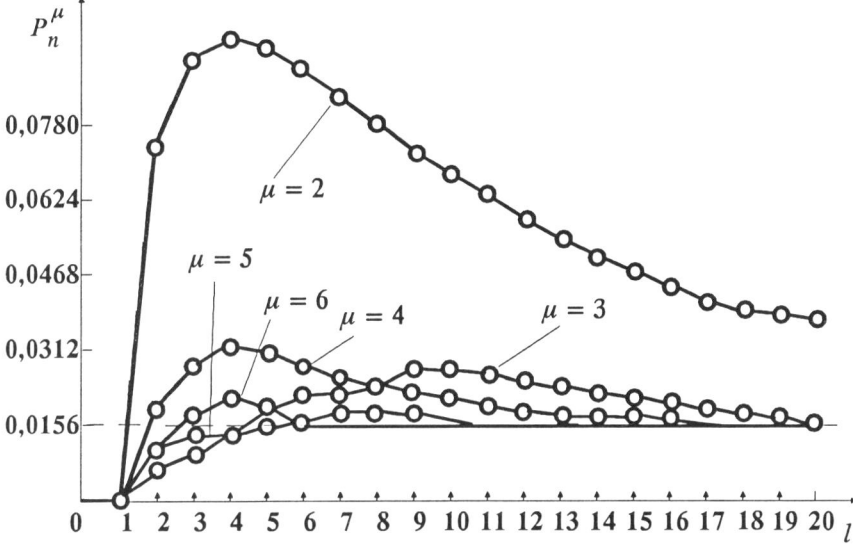

Fig. 13.9

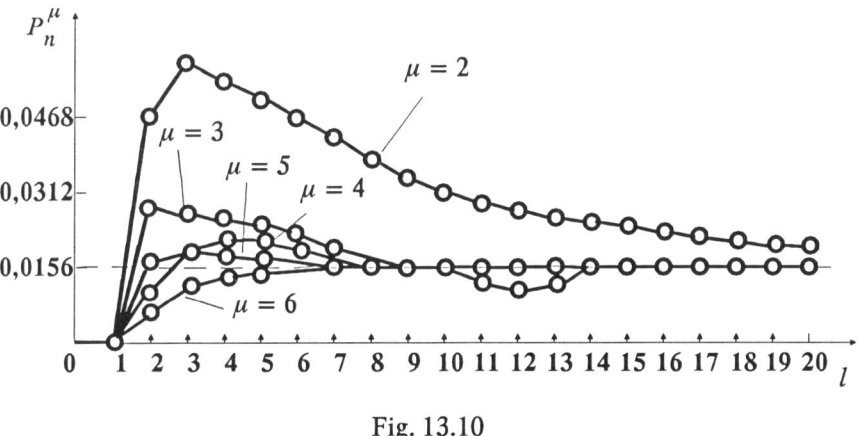

Fig. 13.10

evident from the plots that the efficiency of PSA with a great number of feedback paths (polynomial $\varphi(x)=x^6\oplus x^5\oplus x^2\oplus x\oplus 1$) is higher for even multiplicity. This regularity is more pronounced for $\mu=2$. As the Figures suggest, the

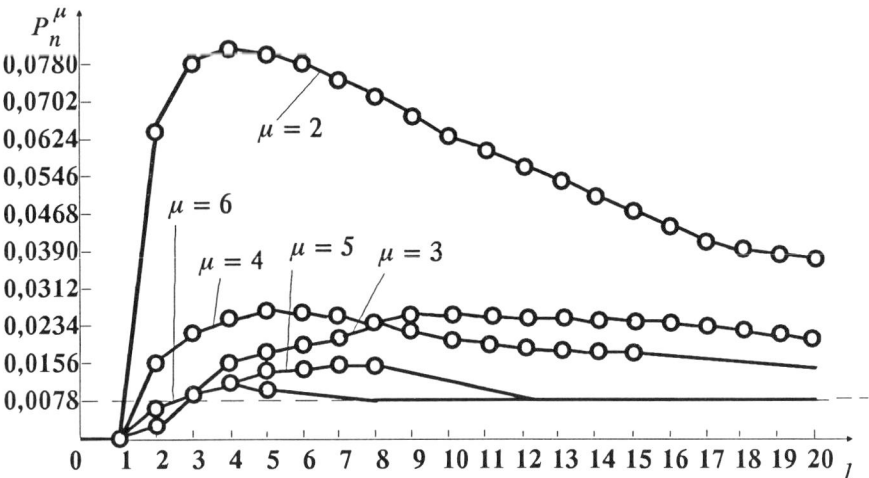

Fig.13.11

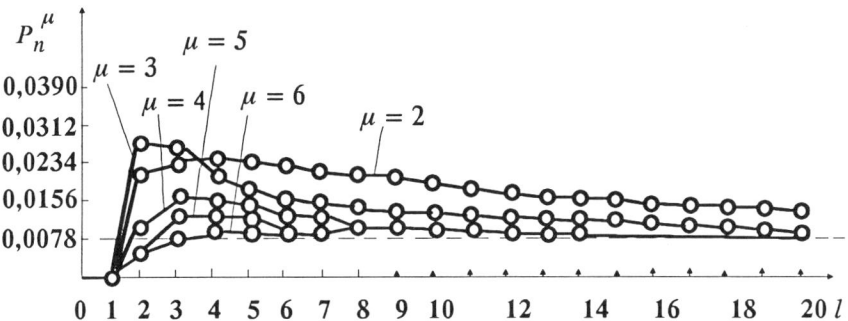

Fig.13.12

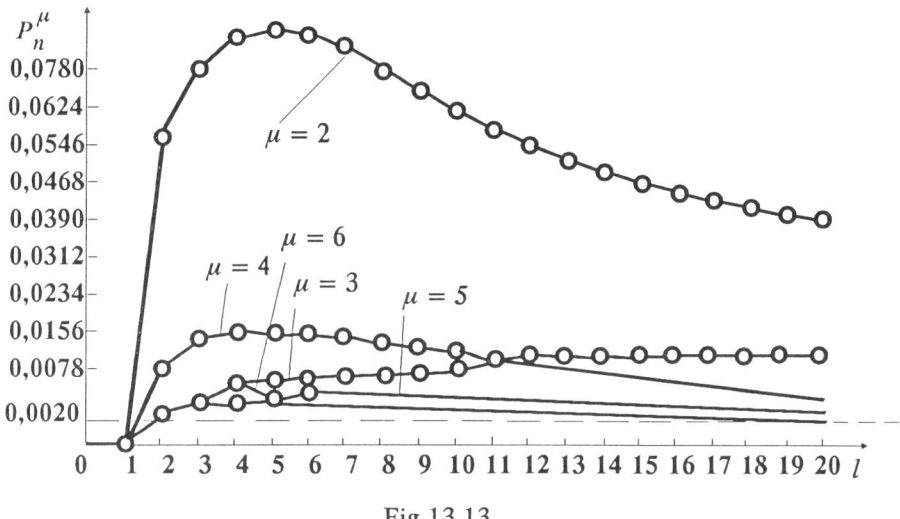

Fig.13.13

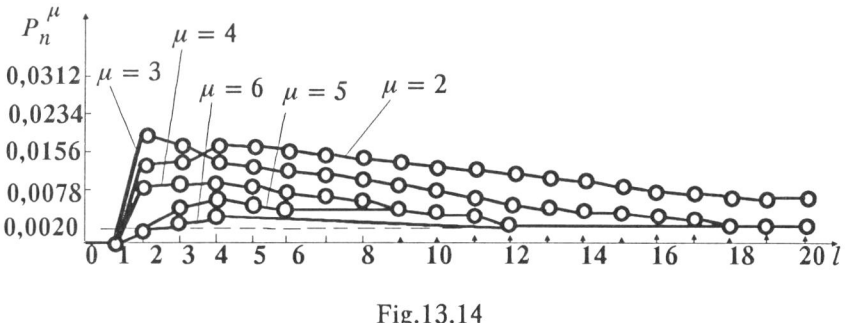

Fig.13.14

tendency is also preserved, although to a lesser degree, for successive even multiplicity values.

The plots of error escape probability variation versus multiplicity and the length of sequence compressed for seven-bit (characteristic polynomials $\varphi(x)=x^7\oplus x^3\oplus 1$ and $\varphi(x)=x^7\oplus x^6\oplus x^5\oplus x^4\oplus x^2\oplus x\oplus 1$) and nine-bit (polynomials $\varphi(x)=x^9\oplus x^4\oplus 1$ and $\varphi(x)=x^9\oplus x^8\oplus x^7\oplus x^6\oplus x^5\oplus x^4\oplus x^3\oplus x\oplus 1$) PSAs represented in Figs. 13.11, 13.12, 13.13 and 13.14, respectively, totally support the pointed out features of the analyzer structure in question. Also, one can see that with the increase in PSA length, the difference in PSA efficiency becomes more pronounced depending on the number of feedback paths.

Thus, the theoretical analysis and the simulation results allow us to state that in the same cases a BILBO-design PSA is inefficient at testing. However, to improve the efficiency, structures with the maximum number of feedback paths described by a primitive polynomial must be selected.

Chapter 14

PSA DESIGN FOR VLSI SELF TEST

14.1. An Efficient Parallel Signature Analyzer

When using the self-test strategy to test a modern digital device we are often forced to solve the problem of designing a parallel signature analyzer with fairly large number of inputs. This is due to an ever increasing number of nodes in a digital circuit under test to be analyzed during test experiment and is aggravated by relatively long tests which prevents from expanding the process in time by testing small groups of test points alternatively.

A widely-used PSA referred to as BILBO that is characterized by low hardware overhead for relatively small number of its inputs is seemed impracticable for solving the problem for the following reasons. First, any extra input to a BILBO-design PSA adds an extra bit to he signature but the excessive signature length has practically no effect on the efficiency of test compressed for the appropriate number of PSA inputs. Moreover it complicates its processing at testing digital devices. Second, for the BILBO-design PSA, its hardware overhead linearly depends on the number of inputs thereby making hardware overheads significant for implementation of parallel signature analyzers.

Of great interest in solving the problem is a two-stage PSA structure where the first stage is a XOR gates and the second is a classical SA. This implementation is characterized by the lack of dependency between signature length, which in this instance is only determined by the requirements for compressed data validly, and the number of PSA inputs. This structure calls for less hardware overhead per each extra PSA input, one XOR gate input. However, the above PSA implementation has a grave drawback, i.e. low validity of compressed data.

Similar drawback is also inherent in the space and time compression technique, which also produces a two-stage SA structure. Here the first and the second stages are designed on the basis of the requirement for detecting specific error patterns. However, this does not ensure high efficiency of detecting other types of error sequences. Besides, the popular assumption that data compression processes run independently in the first and second PSA stages has not been supported by theory and practice.

Therefore, to solve the problem of interest we must design a data compressor characterized by ease of implementation, guaranteed error detection efficiency and low hardware overhead for each extra input.

Let us consider the behavior of a m-bit BILBO-design PSA (Fig.13.3). This analyzer based on characteristic polynomial

$$\varphi(x) = x^m \oplus \alpha_1 x^{m-1} \oplus \alpha_2 x^{m-2} \oplus \ldots \oplus \alpha_{m-1} x \oplus \alpha_m$$

is described by the set of equations (13.6). As it has been noted in Chapter 13, the behavior of such a PSA can be represented by an equivalent two-stage circuit (Fig.13.2), the first stage of which implements data compression in space by the equation (13.9), and the second is a classical SA described by the set of equations (13.8).

By way of example Fig.14.1 shows an implementation of an equivalent BILBO-design PSA described by primitive characteristic polynomial $\varphi(x) = x^4 \oplus x \oplus 1$. It is easy to verify that with a binary pattern $z_1(1)$, $z_2(1)$, $z_3(1)$, $z_4(1)$ applied to the first stage inputs, storage elements of the second PSA stage will contain $z_1(1)$, $z_2(1)$, $z_3(1)$, $z_4(1)$ in D-flipflops 1 to 4, respectively, after four cycles. Since matrices V in sets (13.6) and (13.8) that describe signature analyzer behavior are identical, the second stage of the equivalent scheme will repeat the BILBO-design PSA state in each successive cycle and hence the appropriate signature will be obtained at the end of test experiment.

It is evident that with two binary patterns $z_i(1)$ and $z_i(2)$, $i=\overline{1,4}$, applied to the inputs of the first stage of an equivalent circuit, for $j=\overline{1,5}$, the values $y(j)$ at its output will be determined by (13.9) as

$$\begin{aligned}
y(1) &= z_4(1) \\
y(2) &= z_3(1) \oplus z_4(2) \\
y(3) &= z_2(1) \oplus z_3(2) \\
y(4) &= z_1(1) \oplus z_2(2) \oplus z_4(1) \\
y(5) &= \phantom{z_1(1) \oplus{}} z_1(2) \oplus z_4(2)
\end{aligned}$$

The same binary sequence $y(j)$, $j=\overline{1,5}$, can be obtained at the output of circuit of Fig.14.2 for input patterns $z_i(1)$ and $z_i(2)$, $i=\overline{1,4}$, by applying the patterns to the circuit inputs in the same cycle. By substituting we obtain

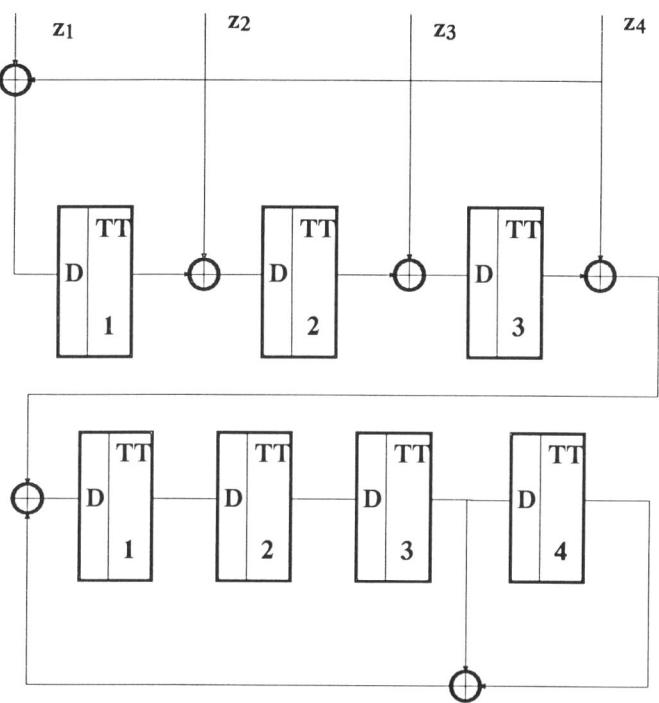

Fig. 14.1. The equivalent BILBO-design PSA

$$z_5(1) = z_1(2)$$
$$z_6(1) = z_2(2)$$
$$z_7(1) = z_3(2)$$
$$z_8(1) = z_4(2)$$

Therefore the circuit of Fig.14.2 has twice as many inputs as the circuit of Fig.14.1. In so doing, the resulting signature obtained by using the circuit for an input pattern $z(i)$, $i=\overline{1,8}$, will be the same as the resulting signature produced by the BILBO PSA described by characteristic polynomial $\varphi(x) = x^4 \oplus x \oplus 1$ when two patterns $z(i)$ and $z(i+1)$, $i=\overline{1,4}$, are compressed alternatively. Specific hardware overhead per any extra input forms *1/4* of D-flipflop and *5/4* of XOR gate in average.

Fig. 14.3 shows a functional diagram of a two-stage parallel signature analyzer based on the above considerations. In so doing, the length of signature (number of storage elements in a PSA second stage) equals to *r* and does not

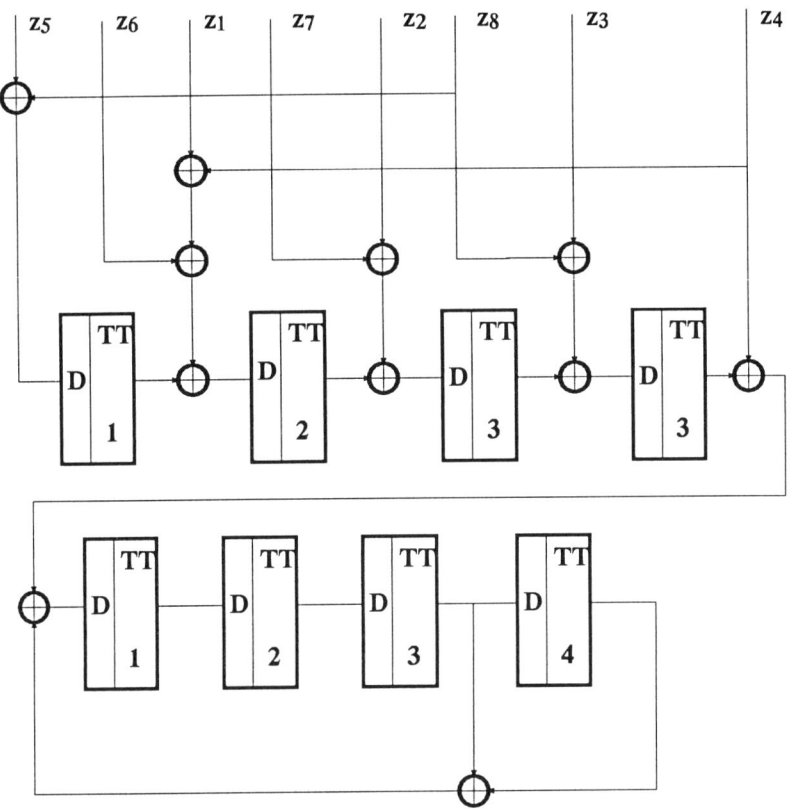

Fig. 14.2. Two stage signature analyzer

depend on the number of inputs to signature analyzer. The number of first-stage storage elements is $h=(\lceil N/r \rceil + r - 1)$. All data inputs to PSA, the total number of which is N, are divided into $\lceil N/r \rceil$ groups of inputs each, where $\lceil N/r \rceil$ is the nearest integer no less that N/r (the last group will contain less than r inputs if N is not divisible by r). Then the i-th input of j-th group is connected to an input to the $(i+j-1)$-th XOR gate and in accordance with the SA (second-stage) characteristic polynomial from is connected to one of the inputs to the $(i+j)$, $(i+j+1)$,...,$(j+r-1)$ XOR gates (with the characteristic polynomial coefficient $\alpha_q = 1$, the ith input of the jth group is connected to an input of the $(j+q)$-th XOR gate, with $\alpha_q = 0$, the connection does not exist), $i=\overline{1,r}$, within each group j, j from 1 to $\lceil N/r \rceil$

z_1 -z_n

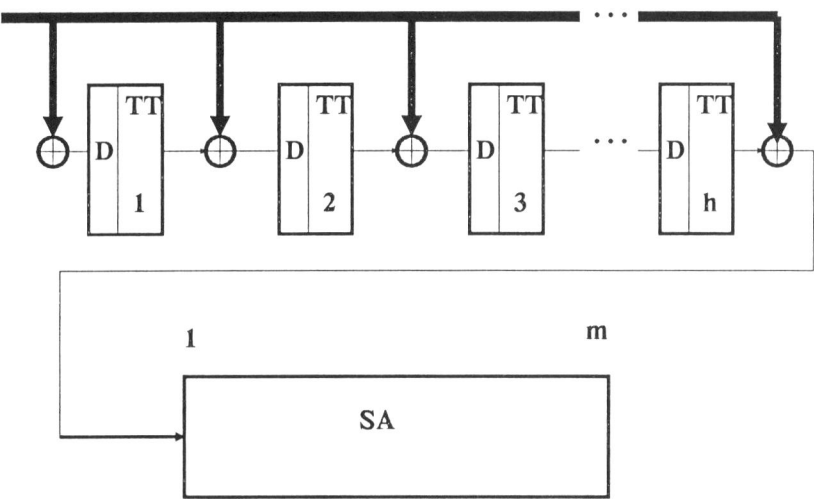

Fig. 14.3. The functional diagram of the signature analyzer

Therefore, we have obtained a PSA scheme that satisfies the requirements formulated. Each extra input to the circuit produces hardware overhead that averages to *1/r*-th part of D-flipflop and *(r+1)/2* inputs to XOR gates.

It is evident that the second component of hardware overheads can be significantly reduced by selecting a proper characteristic polynomial for the second PSA stage with preference being given to polynomials with the minimum number of terms. Thus, for a characteristic trinomial, specific hardware costs per any extra input will average to *1/r* of D-flipflop and *3/2* input to a XOR gate, for a trinomial appearing as $\varphi(x) = x^r \oplus x \oplus 1$, these values will be *1/r* and *((r+1)/r)*, respectively.

By way of example Fig.14.4 shows a functional diagram of a 16-channel 4-bit PSA that has been based on primitive polynomial $\varphi(x) = x^4 \oplus x \oplus 1$

14.2. An Efficient Two-Stage Parallel Signature Analyzer

In the previous section we attempted to solve the problem of designing a parallel signature analyzer with low hardware overhead per each extra input.

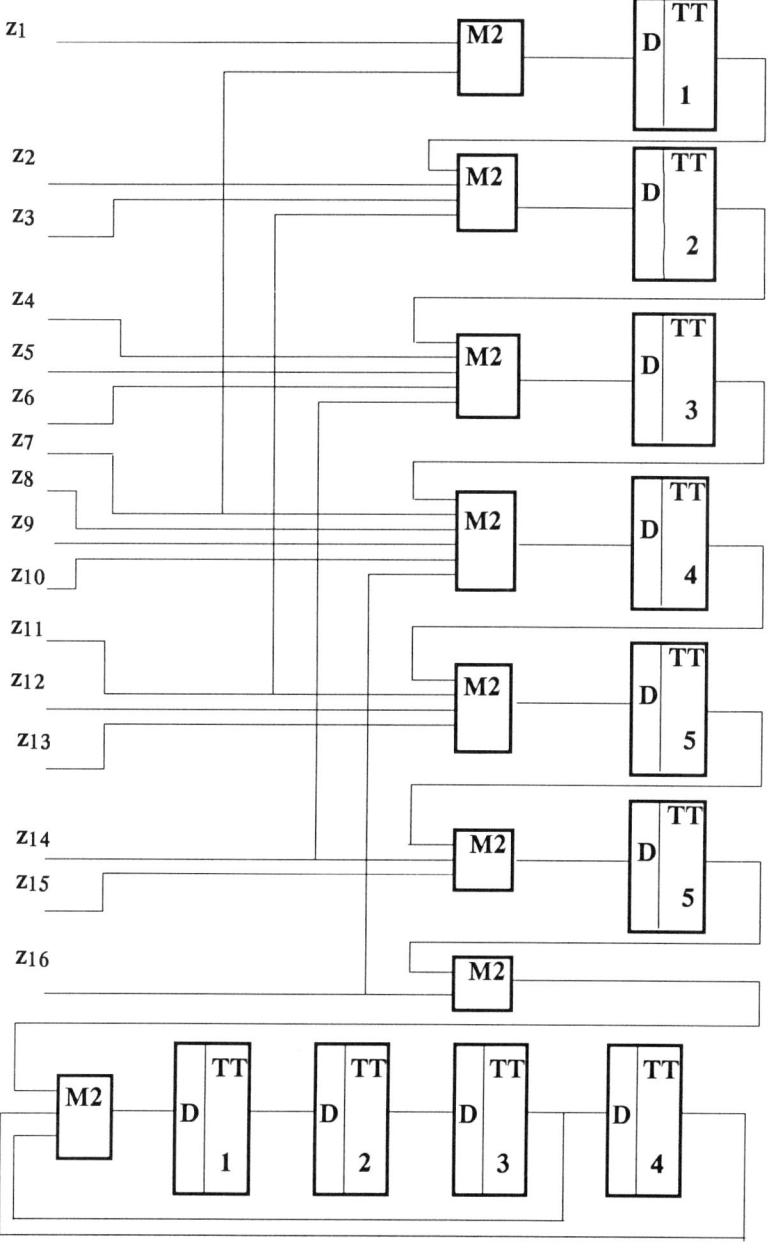

Fig. 14.4. The functional diagram of a 16-channel 4-bit PSA

We have found a PSA scheme with the number of inputs not restricted by the signature length and binary data compression rule close to that of BILBO-design PSA. However, as we have demonstrated it in Section 13, the data compression rule prevailing in the BILBO-design PSA is characterized by nonuniform probability distribution for an error to escape detection depending on its multiplicity.

In this Section we present the design technique for a PSA with the number of inputs not restricted by the number of storage elements which demonstrates high test efficiency characteristic of a classical SA described by a primitive characteristic polynomial.

Lets consider the behavior of a SA described by primitive characteristic polynomial $\varphi(x) = x^m \oplus \alpha_1 x^{m-1} \oplus \alpha_2 x^{m-2} \oplus \ldots \oplus \alpha_{m-1} x \oplus \alpha_m$. In view of the characteristic polynomial form, we can represent the behavior of such SA in matrix notation:

$$A(k+1) = V A(k) \oplus Y(k) \qquad (14.1)$$

where $A(k) = (a_1(k)\, a_2(k) \ldots a_m(k))^T$ is the state vector of SA; $Y(k) = (y(k)\, 0\, 0 \ldots 0)^T$ is the input vector of SA; V is a square matrix which assumes the form:

$$V_I = \begin{vmatrix} 0 & 0 & 0 & \ldots & 0 & \alpha_m \\ 1 & 0 & 0 & \ldots & 0 & \alpha_{m-1} \\ 0 & 1 & 0 & \ldots & 0 & \alpha_{m-2} \\ 0 & 0 & 1 & \ldots & 0 & \alpha_{m-3} \\ \ldots & & & & & \ldots \\ 0 & 0 & 0 & \ldots & 1 & \alpha_1 \end{vmatrix} \qquad (14.2)$$

for a SA structure with internal XOR gates and

$$V_E = \begin{vmatrix} \alpha_1 & \alpha_2 & \alpha_3 & \ldots & \alpha_{m-1} & \alpha_m \\ 1 & 0 & 0 & \ldots & 0 & 0 \\ 0 & 1 & 0 & \ldots & 0 & 0 \\ 0 & 0 & 1 & \ldots & 0 & 0 \\ \ldots & \ldots & \ldots & \ldots & \ldots & \ldots \\ 0 & 0 & 0 & \ldots & 1 & 0 \end{vmatrix} \qquad (14.3)$$

for a SA with external XOR gates, where α_i are coefficients of primitive characteristic polynomial $\varphi(x)$.

Let us denote the initial state vector for SA by $A(0)$. Then its states in the subsequent cycles will be determined as

$$A(1) = V A(0) \oplus Y(0)$$
$$A(2) = V(V A(0) \oplus Y(0)) \oplus Y(1) = V^2 A(0) \oplus V Y(0) \oplus Y(1)$$
$$A(3) = V^3 A(0) \oplus V^2 Y(0) \oplus V Y(1) \oplus Y(2)$$

and so on and, in the general case, for the k-th instant:

$$A(k) = V^k A(0) \oplus \sum_{j=1}^{k} {}^{\oplus} V^{k-j} Y(j-1) \qquad (14.4)$$

Let us consider binary data compression from an n-output digitial circuit by sequentially applying its test responses from each of its n test to the SA input for all l test stimuli, which is equivalent to implementation of a circuit with a multiplexer and SA.

Thus n-bit output patterns $y_1(k), y_2(k), \ldots, y_n(k), k=\overline{0,l-1}$, are transformed into a single binary sequence of the form:

$$y_1(0)\, y_2(0) \ldots y_n(0)\ y_1(1)\, y_2(1) \ldots y_n(1) \ldots y_1(l-1)\, y_2(l-1) \ldots y_n(l-1)$$

which can be represented in a polynomial form as

$$q(x) = \sum_{k=1}^{l} {}^{\oplus} \sum_{v=1}^{n} {}^{\oplus} y_v(k-1) x^{(k-1)n+v-1} \qquad (14.5)$$

According to the equation (14.4), the state of SA after compression of the first n bits of the sequence will be defined by

$$A(n) = V^n A(0) \oplus \sum_{j=1}^{n} \oplus V^{n-j} Y_j(0) \qquad (14.6)$$

where $Y_j(k)=(y_j(k)\ 0\ 0\ \ldots\ 0)^T$, $j\in\{1,2,\ldots,n\}$, $y_j(k)$ are the characters of sequence (14.5).

It is evident that the same SA's state $A(n)$ can be attained by using a certain PSA scheme, which performs the (14.6) transformation for a single cycle, i.e. described by the expression

$$A_{PSA}(1) = V^n A_{PSA}(0) \oplus \sum_{j=1}^{n} \oplus V^{n-j} Y_j(0)$$

where $A_{PSA}(1)=A(n)$.

It is evident then that the result of sequential compression of each successive n-bit word $y_1(k), y_2(k), \ldots, y_n(k)$ at SA or for a single cycle at PSA will be the same. For such a PSA, the state in the kth instant will be determined by the formula

$$A_{PSA}(k) = V^n A_{PSA}(k-1) \oplus \sum_{j=1}^{n} \oplus V^{n-j} Y_j(k-1) \qquad (14.7)$$

or

$$A_{PSA}(k) = V_{PSA} A_{PSA}(k-1) \oplus B\, Y_{PSA}(k-1) \qquad (14.8)$$

where $Y_{PSA}(k)=(y_1(k)\ y_2(k) \ldots y_n(k))^T$ is the input vector for PSA; B is the nxm matrix where its i-th column is the first column of the matrix V^{n-i}, $i=\overline{1,n}$,
with $V^0=I$, where I is a unitary matrix; $V_{PSA}=V^n$ and $A_{PSA}(k)=A(nk)$.

Therefore, a PSA that has been implemented in accordance with equation (14.8) is totally equivalent to data compression circuit based on a SA and a multiplexer from the resulting signature point of view. Hence the validity of data analyzed by such a PSA will be in complete agreement with that of the classical SA. In so doing, the number of PSA storage elements (length) m must be determined based on the requirements of obtaining the required error escape probability level P_n in the compressed sequence in accordance with the commonly accepted estimate (10.3).

Let us consider as an example the procedure of designing a 4-bit PSA with inputs $n=4$.

As a SA to be used as a basis for building a PSA, we select a structure with external XOR gates described by primitive characteristic polynomial $\varphi(x) = x^4 \oplus x \oplus 1$.

According to the equation (14.7), we can find the set of equations that describes the behavior of the sought-for PSA:

$$a_1(k+1) = a_1(k) \oplus a_3(k) \oplus a_4(k) \oplus y_1(k) \oplus y_4(k)$$
$$a_2(k+1) = a_1(k) \oplus a_2(k) \oplus y_3(k)$$
$$a_3(k+1) = a_2(k) \oplus a_3(k) \oplus y_2(k)$$
$$a_4(k+1) = a_3(k) \oplus a_4(k) \oplus y_1(k)$$

which in the form of (14.8) will appear as:

$$\begin{vmatrix} a_1(k+1) \\ a_2(k+1) \\ a_3(k+1) \\ a_4(k+1) \end{vmatrix} = \begin{vmatrix} 1 & 0 & 1 & 1 \\ 1 & 1 & 0 & 0 \\ 0 & 1 & 1 & 0 \\ 0 & 0 & 1 & 1 \end{vmatrix} \begin{vmatrix} a_1(k) \\ a_2(k) \\ a_3(k) \\ a_4(k) \end{vmatrix} \oplus \begin{vmatrix} 1 & 0 & 0 & 1 \\ 0 & 0 & 1 & 0 \\ 0 & 1 & 0 & 0 \\ 1 & 0 & 0 & 0 \end{vmatrix} \begin{vmatrix} y_1(k) \\ y_2(k) \\ y_3(k) \\ y_4(k) \end{vmatrix},$$

The structural diagram for the found PSA is given in Fig.14.5. It can be readily verified that the signatures obtained by compressing any binary sequence on the thus designed PSA (Fig.14.5) and the SA with internal XOR gates described by primitive polynomial $\varphi(x) = x^4 \oplus x \oplus 1$ will be equal.

14.3. Efficiency of a Parallel Signature Analyzer

Since the described PSA design technique has been based on a algorithm for binary data compression by a classical SA, we can estimate its efficiency by completely relying on the findings of multiple studies which testify to the advantage of SA described by primitive characteristic polynomial over other analyzer types. The expressions used for counting numeric values of different SA efficiency estimates will be true for the entire binary sequence compressed at this PSA.

However, when a PSA is used to analyze data on only one channel and only one channel is in error when operating in parallel mode, the efficiency may degrade as compared with the classical SA described by primitive polynomial. Thus for a 5- input signature analyzer described by primitive polynomial $\varphi(x) = x^4 \oplus x \oplus 1$ and implemented by the above described technique, the resulting set of equations representing its behavior will appear as

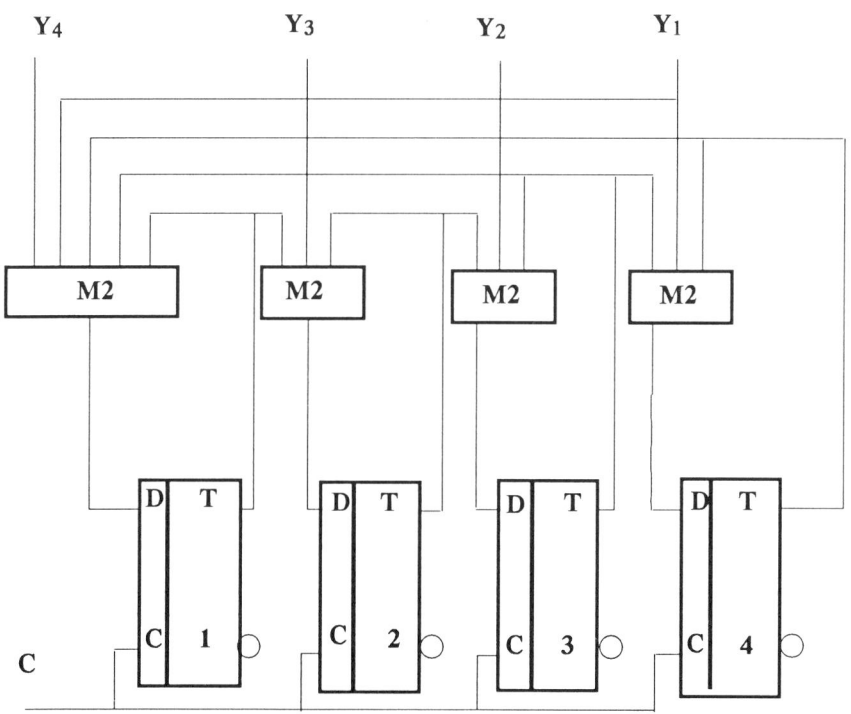

Fig. 14.5. The structural diagram of the PSA

$$a_1(k+1) = a_1(k) \oplus a_2(k) \oplus y_1(k) \oplus y_2(k) \oplus y_5(k)$$
$$a_2(k+1) = a_1(k) \oplus a_3(k) \oplus a_4(k) \oplus y_1(k) \oplus y_4(k)$$
$$a_3(k+1) = a_1(k) \oplus a_2(k) \oplus y_3(k)$$
$$a_4(k+1) = a_2(k) \oplus a_3(k) \oplus y_2(k)$$

Let us consider the case when an error occurred in a binary sequence compressed on the first line $\{y_1(k)\}$, $k=\overline{1,15}$, of a thus designed PSA and corrupted the first and the fourth bits there. In the test response $\{y_1(k)\}$ we can represent the error as a sequence $\{e_1(k)\}$, $k=\overline{1,15}$, where $e_1(1)=e_1(4)=1$, i.e. 100100000000000. In other words, the values of an error sequence $\{y_1^*(k)\}$ are determined as

$$y_1^*(k) = y_1(k) \oplus e_1(k), \quad k=\overline{1,15,}$$

It can be readity verified that the error signature $S(e)$ obtained at this PSA from compressing binary patterns $\{e_i(k)\}$, $i=\overline{1,5}$, $k=\overline{1,15}$, will be zero. By virtue of equality $S(y^*)=S(y) \oplus S(e)$ that is true due to the linearity property of modulo-2 addition, this testifies that the error cannot be detected by the PSA although any aleasing error in this sequence is detectable by a classical SA.

This fact suggests that the efficiency of binary data compression may lower on a single line of the above PSA structure than that obtained by the classical SA.

Besides, a drawback of the PSA designed by the above method is its irregular structure that manifests as complexity of internal links in the analyzer.

Let us define the conditions under which a PSA that has been designed by the proposed technique will be equal to a classical SA from the test efficiency standpoint, with binary data being compressed in a single-line mode. At first we prove the following Lemma.

Lemma 14.1. The sets of detectable (undetectable) errors of a binary sequence $e(k)$ are identical when compressed on a single input to a signature analyzer based on a linear feedback shift register and described by a primitive characteristic polynomial and on any subset of its inputs.

Proof. Consider a signature analyzer based on a shift register with internal XOR gates, i.e. having the matrix V of system (14.8) that describes its behavior in the form of V_I (14.2), with B=I, where I is a unitary matrix, for system (14.8).

Without loss of generality, assume that the sequence $e(k)$ when compressed on the first input to the signature analyzer in question, has resulted in a nonzero signature S. With sequence $e(k)$ compressed on inputs to the signature analyzer, the signature

$$S_{PSA}=S\oplus SV\oplus SV^2\oplus\ldots\oplus SV^{m-1}=S(I\oplus V\oplus V^2\oplus\ldots\oplus V^{m-1})$$

is produced.

With $e(k)$ compressed on a certain subset of signature analyzer inputs, the associated terms (but not all at a time) will be non-existent.

All the elements of Galois field $GF(2^m)$ as extensions of a simple field $GF(2)$ may be represented as a vector space based on a set $\{1, \alpha, \alpha^2, \alpha^3, \ldots, \alpha^{m-1}\}$, where α is a primitive element of field $GF(2^m)$. As a way of expressing elements of $GF(2^m)$, we may use matrix notation, where the role of irreducible polynomial root (field extension generating element) plays its associated matrix V. For a primitive polynomial, its associated matrix is a generating polynomial for the multiplicative group in $GF(2^m)$ and matrix $(I\oplus V\oplus V^2\oplus\ldots\oplus V^{m-1})$ will not be singular since all terms between brackets form the basis for a vector space of $GF(2^m)$ elements as a simple extension field of $GF(2)$. It is evident that the property retains for the case when any subset of terms (but not all at a time) is absent. Hence signature S_{PSA} is nonzero.

One can also see from the above relation that the values of signatures produced at compression of an error sequence $e(k)$ by the discussed signature analyzer on distinct inputs differ by a factor that equals to a degree of nonsingular matrix V. Therefore, if a zero signature results from compression of an error sequence $e(k)$ on a single input to a signature analyzer, the same sequence compressed on any other channel of the same signature analyzer and, hence, on any subset of its inputs will also produce a zero signature.

For a signature analyzer with external XOR gates, i.e. the one with the associated matrix V of system (14.8) in the from V_E (14.3), the above Lemma also holds true. This follows from the fact that by virtue of characteristic polynomial primitivity we may express its behavior in terms of behavior of an identical structure whose output vectors are identical and the state is related to that of PSA with internal XOR gates by $A_E(k)=S^{-1}A_I(k)$, where nonsingular matrix S is defined by equations $V_E=S^{-1}V_I S$. QED.

Suppose we have a PSA that has been designed by the above technique for n inputs on the basis of primitive characteristic polynomial $\varphi(x)$ of degree m. The behavior of such PSA is described by the set of equations (14.8). If the characteristic polynomial $\psi(x)$ of V_{PSA} is primitive, the following theorem will hold true.

Theorem 14.1. A Set of error detectable (undetectable) at compressing a binary sequence on a single channel of PSA that has been designed by the proposed technique (14.8) is identical to the one detectable (undetectable) by a classical single-channel signature analyzer SA based on a linear feedback shift register defined by the same primitive characteristic polynomial.

Proof. Since the signature analyzers are defined by he same primitive characteristic polynomial, there exists a transposing matrix S such that

$$V_{PSA} = S^{-1} V S.$$

where V is a matrix describing the behavior of a signature analyzer based on a linear feedback shift register of the form V_I (14.2) or V_E (14.3).

Substituting V_{PSA} in the set of equations (14.8), we obtain

$$S A_{PSA}(k+1) = V S A_{PSA}(k) \oplus S B Y(k)$$
$$Z(k) = S^{-1} S A_{PSA}(k),$$

where $Z(k)$ is an output vector of PSA, i.e. the PSA behavior can be expressed in terms of an equivalent signature analyzer with matrix V and state $A_{SA}(k) = S A_{PSA}(k)$.

It is evident that with data compression on only one PSA line, the sequence will be applied to a certain nonzero subset of inputs to a signature analyzer described by matrix V depending on the values of matrices B and S. From the nonsingularity of S and the proved Lemma 14.1 it follows that the sets of errors detectable (undetectable) by compressing a binary sequence on a single line of these signature analyzers are identical as was to be shown.

Thus if the characteristic polynomial of matrix V^n in (14.7) is primitive, the PSA reliability in a single-line mode will be the same as that of a classical SA.

Let us further define the conditions for selecting a primitive polynomial and the number of PSA inputs for which matrix V^n will also have a primitive characterictic polynomial.

Theorem 14.2. Sets of errors detectable (undetectable) in binary sequence being compressed on a single input to the $n=2^d$-channel signature analyzer, where d is a natural number, which has been designed by the above technique on the basis of a primitive polynomial $\varphi(k)$, $m = deg(\varphi(x))$, and those detectable (undetectable) by a classical SA defined by the characteristic polynomial $\varphi(x)$ are identical.

Proof. As it is well known from the theory of finite fields, if an element α in a Galois field $GF(q^m)$ is primitive over the field, its conjugate elements

$$\alpha, \alpha^q, \alpha^{q^2}, \ldots, \alpha^{q^{m-1}}$$

will also be primitive. What's more, they all be roots of the same primitive polynomial. It is also evident that

$$\alpha^{q^m} = \alpha$$

Hence from the matrix notation of $GF(q^m)$, for matrix V having the form of V_I (14.2) or V_E (14.3) as a primitive field element, we may deduce that the matrix

$$V^n = V^{2^d},$$

where d is any natural number, is the root of the same characteristic polynomial and in the sense of linear algebra has the same characteristic polynomial as V. From the above and Theorem 14.1 it follows that the statement holds true.

Theorem 14.3. A set of errors detectable (undetectable) in a binary sequence being compressed on a single input of n-input signature analyzer that has been designed on the basis of a primitive polynomial $\varphi(x)$, $m=deg(\varphi(x))$, maximal divisor of (2^m-1) and n equal to 1, $n \neq 2^d$, where d is a natural number, will be equal to a set of errors detectable (undetectable) at its compression by a classical single-input signature analyzer based on a primitive polynomial $(\psi(x)=det(xI-V^n))$ other than $\varphi(x)$. Also, characteristic polynomial $\varphi(x)$ will be used to describe any similar PSA with the number of inputs equal to $n2^i$ where i is a natural number.

Proof. From the Galois field theory it is known that primitive elements of $GF(q^m)$ (a generating element of its multiplicative group consisting of nonzero values of $GF(q^m)$ are solely those degrees r of a certain primitive element for which a maximal divisor of r and (q^m-1) equals to 1. It is also known that for α as a primitive element of $GF(q^m)$ the elements

$$\alpha, \alpha^q, \alpha^{q^2}, \ldots, \alpha^{q^{m-1}}$$

are also primitive ones of the same field. Moreover, they are roots of the same primitive polynomial of degree m over $GF(q)$. Hence if n is reciprocally prime with 2^m-1, but not equal to 2^i, where i is a natural number, then a primitive polynomial $\psi(x)$ other than $\varphi(x)$ will be a characteristic polynomial $\varphi(x)$, of element α^n. Besides this polynomial will also be characteristic for the elements defined as

$$(\alpha^n)^{2^i} = \alpha^{n2^i}, i=\overline{1,m}.$$

Thus, if matrix V for any primitive polynomial is taken as a primitive element of $GF(q^m)$ then the matrix

$$V^{n2^i}, \quad i \in N$$

will also be the primitive element.

Therefore, from Theorem 14.2, the sets of detectable (undetectable) errors will be indentical for such degrees of matrix V and, hence, such PSAs will have the same primitive characteristic polynomial $(\psi(x)=det(xI-V^n))$ as was to be shown.

14.4. Two-Stage PSA Design

For a PSA determined by (14.8) with n inputs meeting the condition $n=2^i$, where i is a natural number, there exists a nonsingular matrix S, as Theorem 14.2 suggests, for which the equation

$$V^n = S^{-1} V S$$

holds true.

Herefrom and from equality (14.8) follows that the behavior of such a PSA can be expressed in terms of an equivalent circuit described by the system:

$$\begin{aligned} A'_{PSA}(k+1) &= V A'_{PSA}(k) \oplus S B \, Y_{PSA}(k) \\ A_{PSA}(k) &= S^{-1} S \, A'_{PSA}(k), \end{aligned} \qquad (14.9)$$

where $A'_{PSA}(k) = SA_{PSA}(k)$; $A_{PSA}(k)$ is an output vector for the system.

What's more, since matrix S is nonsingular the set of errors that may be detected (undetected) by the PSA of (14.9) will be equal to the set of errors detectable (undetectable) by a PSA described as:

$$A'_{PSA}(k+1) = V A'_{PSA}(k) \oplus B' Y_{PSA}(k), \qquad (14.10)$$

where $B' = SB$.

For an PSA, obtained by the equation (14.8) with the number of inputs n and length m meeting the condition - maximal divisor of (2^m-1) and n equal to 1, $n \neq 2^i$, where i is natural, there may exist a nonsingular matrix S for which the equation:

$$V^n = S^{-1} V' S \qquad (14.11)$$

where V' is an associated matrix of polynomial $(\psi(x) = det(xI - V^n))$ in the form V_I (14.2) or V_E (14.3) is holds true.

Therefore, we can in this case too, use the PSA in the form of (14.10), which is equivalent from the error detectivity point of view.

The PSA form thus obtained is a simple two-stage circuit the first stage of which is purely combinational and converts a n-bit sequence into an m-bit one and the second stage is a well-known PSA with either internal or external XORs.

As opposed to familiar structures of data compression in space and time (13), which stages are created separately and, hence, may cause the resulting efficiency of the whole PSA to decrease, the present design technique is based on the requirement of high efficiency for the whole two-stage structure.

Thus the structure of PSA that has been found by the above technique is noted for the following merits:

- the structure, when used, ensures that the efficiency of error detection in test data being compressed both at analyzing all inputs (multi-line mode) and any input separately (single-line mode) equals to that of a classical SA, described by primitive characteristic polynomial;
- the number of inputs to this PSA is not restricted by its length;
- this PSA has low hardware overhead;
- the present PSA is easy to implement.

Thus the algorithm for PSA design consists of the following sequence of steps:

1. The number of PSA storage elements m is determined according to the specified error detection probability in test responses with due account of digital circuit design features by

$$P_n \geq \frac{1}{2^m} \qquad (14.12)$$

2. The obtained PSA length m is then refined by application of the tentatively specified number of inputs n to meet the condition:

Maximale divisor of (2^m-1) and n equal to 1.

3. Choose a primitive polynomial $\varphi(x)$, $m=deg(\varphi(x))$, for the refined value of m.

4. Using the values n, m, $\varphi(x)$ as a base, set up the equation (14.8) which describes the behavior of an intermediate PSA structure.

5. Reduce the obtained equation (14.8) to the form describing the behavior of the desired two-stage PSA (14.10), and use the specific form of matrices V' and B' for designing the structure of its first and second stages.

As an example of its implementation, let us consider the case of PSA design whose intermediate version found by (14.8) is gived in Fig.14.9.

According to equations (14.10) and (14.11), let's evaluate a transposing matrix S:

$$S \begin{vmatrix} 1 & 0 & 1 & 1 \\ 1 & 1 & 0 & 0 \\ 0 & 1 & 1 & 0 \\ 0 & 0 & 1 & 1 \end{vmatrix} = \begin{vmatrix} 0 & 0 & 1 & 1 \\ 1 & 0 & 0 & 0 \\ 0 & 1 & 0 & 0 \\ 0 & 0 & 1 & 0 \end{vmatrix} S$$

One of the solutions for the given system is the following value of matrix S:

$$S = \begin{vmatrix} 1 & 1 & 1 & 1 \\ 0 & 1 & 0 & 1 \\ 0 & 0 & 1 & 1 \\ 0 & 0 & 0 & 1 \end{vmatrix}.$$

Having calculated the value of matrix B' from (14.10), we obtain the equation that describes the behavior of a two-stage PSA structure:

PSA Design for VLSI Self-test

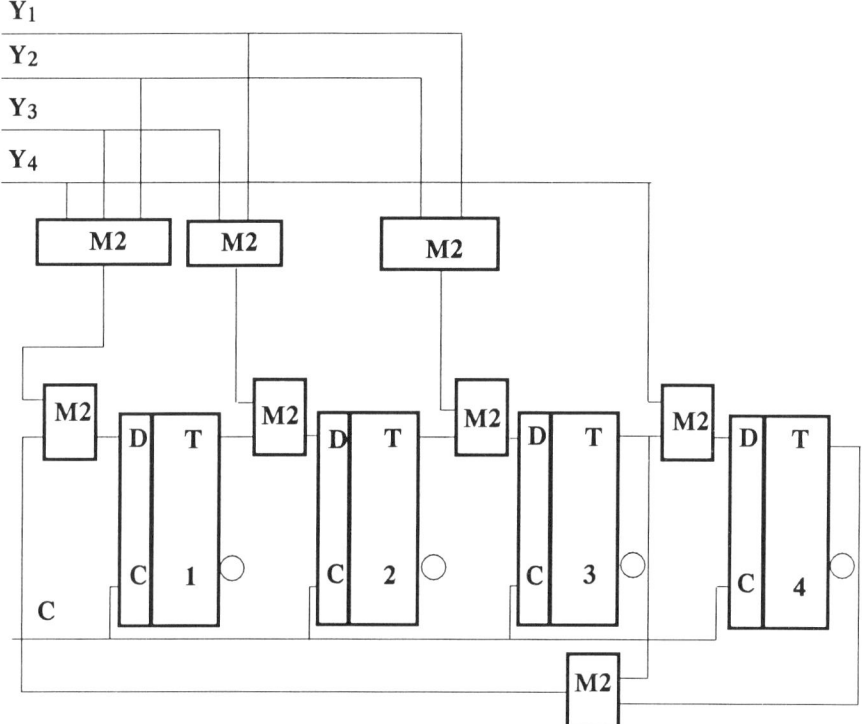

Fig. 14.6. The PSA

$$\begin{vmatrix} a_1(k+1) \\ a_2(k+1) \\ a_3(k+1) \\ a_4(k+1) \end{vmatrix} = \begin{vmatrix} 0 & 0 & 1 & 1 \\ 1 & 0 & 0 & 0 \\ 0 & 1 & 0 & 0 \\ 0 & 0 & 1 & 0 \end{vmatrix} \begin{vmatrix} a_1(k) \\ a_2(k) \\ a_3(k) \\ a_4(k) \end{vmatrix} \oplus \begin{vmatrix} 0 & 1 & 1 & 1 \\ 1 & 0 & 1 & 0 \\ 1 & 1 & 0 & 0 \\ 1 & 0 & 0 & 0 \end{vmatrix} \begin{vmatrix} y_1(k) \\ y_2(k) \\ y_3(k) \\ y_4(k) \end{vmatrix}.$$

Fig. 14.6 shows the PSA fitting the above equation.

Chapter 15

PSA with T-flip-flops Design and Analysis

15.1. Introduction to PSA with T-flip-flops Design

The findings of recent studies testify that a SA described by primitive characteristic polynomial is highly efficient. Hence, to attain minimum aliasing of errors in different binary sequences and thereby ensure that corrupted bits with different multiplicity μ escape detection with similar probability we must use such a signature analyzer scheme which compression rule is identical to the rule of signature generation in a SA described by primitive polynomial. An evident solution to the problem is a PSA based on a SA and a multiplexer which sequentially applies binary data from each test point in the circuit under test on a bitwise basis.

The schemes of single-input signature analyzers (SSA) are given in Fig.9.3, and their behavior is described by the sets of equations (9.4). Such SSAs described by a primitive characteristic polynomial of the form

$$\varphi(x) = x^m \oplus \alpha_1 x^{m-1} \oplus \alpha_2 x^{m-2} \oplus \ldots \oplus \alpha_{m-1} x \oplus \alpha_m$$

are distinguished by high efficiency of error detection in binary sequences compressed. However, although highly efficient in binary data compression from the test efficiency standpoint, the analyzer scheme based on a multiplexer requires that system clock rate be decreased by m thereby increasing the test experiment duration. Besides, with such organization of test process, dynamic fault of a digital device will escape detection.

Thus to attain both high efficiency of testing and short test experiment time, it would be more preferable to use a PSA whose data compression rule is as close as possible to the compression algorithm for a SA described by a primitive characteristic polynomial and is characterized by high speed and low hardware overhead.

There is a well known signature analyzer design method that allows to obtain a PSA whose binary stream compression rule is completely identical to that of data compression by a single-line SA. A major drawback of such a PSA, however, is great complexity and high hardware overhead for implementation which prevent it from being used in self-test digital circuit design.

Consider a SA of Fig.15.1. Its state in a kth instant can be described by a set of equations

$$b_1(k) = y(k) \oplus b_m(k-1)$$
$$b_2(k) = b_1(k-1) \oplus b_m(k-1) \quad (15.1)$$
$$b_j(k) = b_{j-1}(k-1), \quad j = \overline{3,m}$$

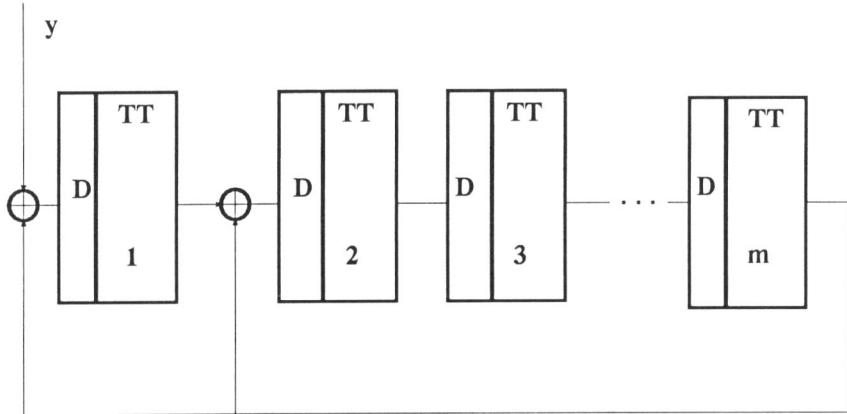

Fig. 15.1. Signature Analyzer

where $b_j(k)$, $j=\overline{1,m}$, is the SSA state in the k-th instant.

This SSA is described by a characteristic polynomial - a trinomial with a single unity coefficient α_{m-1}. Although a primitive polynomial of the like form cannot be found for every SSA length, however, when it does exist (e.g.1 $\varphi(x)=x^2\oplus x\oplus 1$, $\varphi(x)=x^3\oplus x\oplus 1$, $\varphi(x)=x^4\oplus x\oplus 1$, $\varphi(x)=x^6\oplus x\oplus 1$, $\varphi(x)=x^7\oplus x\oplus 1$, etc.) for the required length, the SA scheme distinguished by the simplicity of implementation is more preferable in usage.

From eq. (15.1) it follows that in the $(k+1)$-th clock, the SA state will be

$$b_1(k+1) = y(k+1) \oplus b_m(k) = y(k+1) \oplus b_{m-1}(k-1)$$
$$b_2(k+1) = b_1(k) \oplus b_m(k) = y(k) \oplus b_m(k-1) \oplus b_{m-1}(k-1)$$
$$b_3(k+1) = b_2(k) = b_1(k-1) \oplus b_m(k-1)$$
$$b_j(k+1) = b_{j-1}(k) = b_{j-2}(k-1), \quad j = \overline{4,m}$$

In a successive clock the SSA state will be defined as

$$b_1(k+2) = y(k+2) \oplus b_m(k+1) = y(k+2) \oplus b_{m-2}(k-1)$$
$$b_2(k+2) = b_1(k+1) \oplus b_m(k+1) = y(k+1) \oplus b_{m-1}(k-1) \oplus b_{m-2}(k-1)$$
$$b_3(k+2) = b_2(k+1) = y(k) \oplus b_m(k-1) \oplus b_{m-1}(k-1)$$
$$b_4(k+2) = b_3(k+1) = b_1(k-1) \oplus b_m(k-1)$$
$$b_j(k+2) = b_{j-1}(k+1) = b_{j-3}(k-1), \quad j = \overline{5,m}$$

Thus, in the *(k+m-1)*th clock, the state of SSA in question will be

$$b_1(k+m-1) = y(k+m-1) \oplus b_m(k+m-2) = y(k+m-1) \oplus b_1(k-1) \oplus$$
$$\oplus b_m(k-1);$$
$$b_2(k+m-1) = b_1(k+m-2) \oplus b_m(k+m-2) = y(k+m-2) \oplus$$
$$\oplus b_1(k-1) \oplus b_2(k-1) \oplus b_m(k-1);$$
$$b_3(k+m-1) = b_2(k+m-2) = y(k+m-3) \oplus b_2(k-1) \oplus b_3(k-1); \quad (15.2)$$
$$b_4(k+m-1) = b_3(k+m-2) = y(k+m-4) \oplus b_3(k-1) \oplus b_4(k-1);$$
$$\ldots$$
$$b_m(k+m-1) = b_{m-1}(k+m-2) = y(k) \oplus b_{m-1}(k-1) \oplus b_m(k-1).$$

It is evident that such state can be obtained for a single clock by a PSA based on T flip-flops as storage elements with the same feedback structure as that of equations (15.3) by applying m binary patterns to be compressed in parallel. Fig.15.2 chows the scheme of such a PSA.

$$a_1(k) = z_1(k) \oplus a_1(k-1) \oplus a_m(k-1)$$
$$a_2(k) = z_2(k) \oplus a_1(k-1) \oplus a_m(k-1) \quad (15.3)$$
$$a_j(k) = z_j(k) \oplus a_{j-1}(k-1) \oplus a_j(k-1), \quad j = \overline{3,m}$$

Then, if the following relations

$$a_i(k-1) = b_i(k-1), \quad i = \overline{1,m}$$
$$z_j(k) = y(k+m-j), \quad j = \overline{1,m}$$

are met, the equation

$$a_i(k) = b_i(k+m-1), \quad i = \overline{1,m}$$

will also be met on the basis of the systems of equations (15.2) and (15.3).

Considering that the signature analyzers in question have a linear structure, we may state the following from the superposition standpoint.

Statement 15.1. The resulting signatures obtained by parallel compressing m binary sequences of length l each on a PSA (Fig.15.2) described by the set of equations (15.3) for l clocks and by sequential bitwise compressing l m-dimensional binary vectors on a SSA (Fig.15.1) described by the set of equations (15.1) for lm clocks will be identical for the same initial states of analyzers.

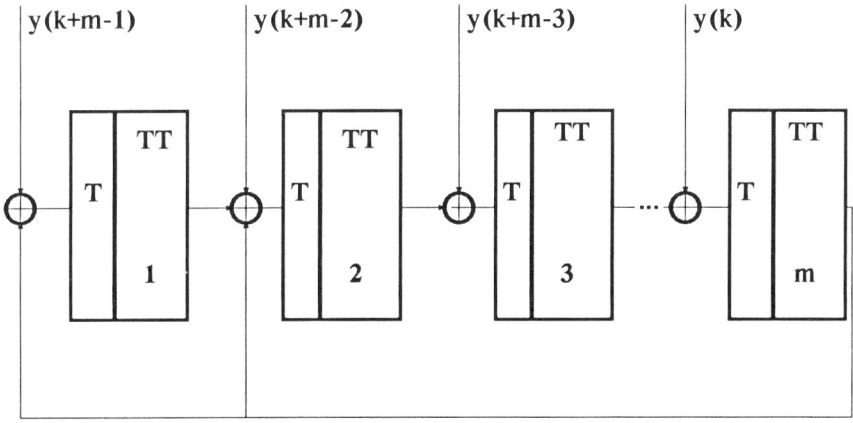

Fig. 15.2. PSA totally identical to the SA of Fig. 15.1

264   PSA with T-flip-flops Design and Analysis

This statement suggests that from the standpoint of error detection probability in binary test sequences, the efficiency of the PSA of Fig.15.2 will be totally identical with that of the SSA of Fig.15.1, which compresses data from several test points on a bitwise basis by sequentially polling (multiplexing) them. To estimate the PSA efficiency, we can use the relations derived for the SSA.

Besides, for the PSA in question, the same as for those discussed in Chapter 13, the test experiment time duration will be m times less than it was for the analyzer employing a SSA and a multiplexer. In so doing, dynamic faults will be detected in the digital circuit at system clock rate.

15.2. Parallel Signature Analyzer with T-flip-flops as Storage Elements

In the preceding Section we have obtained the PSA structure that is characterized by simplicity ease of implementation and high error detection efficiency in binary sequences compressed. The PSA has been designed on the basis of T-flip-flops used as storage elements. As we have mentioned in the Section, a limitation exists for such a PSA thereby making it less versatile. The limitation amounts to the necessity for a primitive polynomial - a trinomial of the form $\varphi(x) = x^m \oplus x \oplus 1$ - for the required PSA of length m. However, the fact that the PSA has been constructed of T-flip-flops makes it advisable to verify whether a structure based on a linear feedback shift register with T-flip-flops as storage elements is suitable for highly efficient PSA design.

It has been known that the basic requirement of any signature analyzer based on a linear feedback shift register is that its characteristic polynomial be primitive, in other words, the possibility of generating a maximal-length pseudorandom sequence off line. This provides for high efficiency of SA when data is compressed in time (on a single line). With this assumption we try to determine the feedback layout for a conventional PSA with T-flip-flops as storage elements.

Let polynomial $\varphi(x) = x^m \oplus \alpha_1 x^{m-1} \oplus \alpha_2 x^{m-2} \oplus \ldots \oplus \alpha_{m-1} x \oplus \alpha_m$ of degree m be irreducible over an elementary Galois field GF(2). It follows that for $\varphi(x)$ being an irreducible polynomial of degree m over GF(p), the field expansion GF(p^m) contains any root γ of polynomial $\varphi(x)$. Therefore GF(2^m) is an expansion of elementary field GF(2), which generating element is the root of polynomial $\varphi(x)$. Hence any element of GF(2^m) can be uniquely represented as a polynomial in x over field GF(2) whose degree does not exceed $m-1$ with

$x=y$, i.e. $\{1, \gamma, \gamma^2, \gamma^3, \ldots, \gamma^{m-1}\}$ is the basis of vector space $GF(2^m)$ over an elementary field $GF(2)$.

Let for a polynomial $\psi(x)$ a $GF(2^m)$ element $\beta=\gamma\oplus1$, where γ is the root of irreducible polynomial $\varphi(x)$, be the root, i.e. $\psi(\beta)=\psi(\gamma\oplus1)=0$. It follows herefrom that with $\psi(x)$ being primitive (for $\gamma\oplus1$, y+1 is a primitive element of $GF(2^m)$), the shift register with T-flip-flops used as storage elements and the feedback layout determined by the coefficientts of irreducible polynomial $\varphi(x)$ (15.4) (Fig.15.3) will generate the entire multiplicative group of $GF(2^m)$ in off-line mode, i.e. produce all possible nonzero patterns of length m:

$$A(k) = V_{TI} A(k-1), \quad k = \overline{1,l}, \qquad (15.4)$$

where $A(k)=(a_1(k), a_2(k), \ldots a_m(k))^T$ is the state vector of shift register and V_{TI} is the square matrix of the form:

$$V_{TI} = \begin{vmatrix} 1 & 0 & 0 & \ldots & 0 & \alpha_m \\ 1 & 1 & 0 & \ldots & 0 & \alpha_{m-1} \\ 0 & 1 & 1 & \ldots & 0 & \alpha_{m-2} \\ 0 & 0 & 1 & \ldots & 0 & \alpha_{m-3} \\ \ldots & \ldots & \ldots & \ldots & \ldots & \ldots \\ 0 & 0 & 0 & \ldots & 1 & \alpha_1\oplus1 \end{vmatrix} \qquad (15.5)$$

Thus to find a polynomial whose coefficients define the feedback layout of the signature analyzer of Fig.15.3, we must substitute x for $x\oplus1$ in an arbitrary primitive polynomial $\psi(x)$ of degree m and then transform it. The resulting irreducible polynomial $\varphi(x)$ is precisely the one which describes the feed back layout for a sought-for PSA (15.6):

$$A(k) = V_{TI} A(k-1) \oplus Z(k), \quad k=\overline{1,l}, \qquad (15.6)$$

where $Z(k)=(z_1(k), z_2(k), \ldots z_m(k))^T$ is the input vector for PSA and V_{TI} is a square matrix of the form (15.5).

From the above considerations it is apparent that such a PSA will be characterized by high efficiency of error detection in binary sequences being compressed on a single line. Let us try now to evaluate the efficiency of its operation in parallel mode.

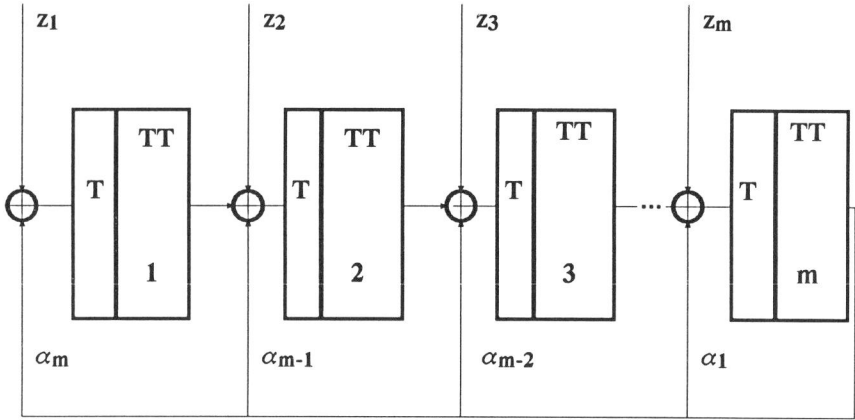

Fig. 15.3. Signature Analyzer

15.3. Efficiency of PSA with T-flip-flops as Storage Elements

Chapter 13 concerned with the analysis of most popular and currently used PSA designs stated that the worst estimates by the test for error escape probability depending on multiplicity, with the occurrence of any fixed multiplicity error being equiprobable, exist for the multiplicity 2. Therefore, we shall find out the probability of failing to detect an error of multiplicity 2 under the assumptions made for the test and for different lengths of binary sequences being compressed.

Let $\varphi(x)$ be an irreducible polynomial which determines the feedback layout for a PSA under test (Fig.15.3) and γ its root. It is evident that in offline mode the design under test is a maximal-length PRP generator (in other words, the generator of multiplicative group elements of $GF(2^m)$). It can be readily seen that element γ is formed not in the cycle following element 1, as happens with PRPG based on a shift register with D-flip-flops and internal XOR gates, but in d cycles of generator operation. Naturally, element γ^2 will also be delayed by d the as the element y of $GF(2^m)$ and so on. It can be easily noticed that the delay d is determined by the following relation:

$$(\gamma \oplus 1)^d = \gamma \ (mod \ \varphi(x))$$

Chapter 15   267

i.e. we must deduct degrees $(\gamma \oplus 1)^i$ until $(\gamma \oplus 1)^i$ equals γ and d equals to the current value of i.

Let us then found the number of undetectable two-bit errors for m binary sequences of length $L \le 2^m-1$, each being compressed on a PSA under test with delay d.

It is evident that undetectable configurations of two corrupted bits in a binary sequence of length $L \le 2^m-1$ that has been compressed on a single line of PSA do not exist since the length of cycle for the generated PRP is 2^m-1 in autonomous mode. Hence the probable undetectable error bits may occur only on different lines of PSA. From the above considerations one can see that an error bit applied to the PSA over the ith line will be aliased by another error bit applied to the $(i+1)$th input of PSA in d cycles or to the $(i+2)$th input of PSA in $2d(mod(2^m-1))$ cycles and so on.

Time delay D between two undetectable error bits can assume the value d, $2d(mod(2^m-1))$, $3d(mod(2^m-1))$, ... , $(m-1)d(mod(2^m-1))$.

Let us consider various situations that may occur for the delay $D<2^{m-1}$. It is evident that for the compressed sequence length $L \le D$, there will be no undetectable errors of multiplicity 2. For any L in the range from D to 2^m-1-D, the number of undetectable two-bit errors will be $(L-D)$. For the compressed sequence length within $2^m-1-D< L \le 2^m-1$, there will be $L-D+L-(2^m-1-D)$ pairs of undetectable bits.

With $D \ge 2^{m-1}$, no undetectable two-bit errors exist for L not exceeding 2^m-1-D. If the length of sequences compressed is within 2^m-1-D to D, the number of all possible undetectable bit pairs will be $L-(2^m-1-D)$. How-

Table 15.1

Delay D	Sequence length L	Number of undetectable two-bit errors
$D < 2^{m-1}$	$L \le D$	0
	$D < L \le 2^m-1-D$	$L-D$
	$2^m-1-D < L \le 2^m-1$	$2L-2^m+1$
$D \ge 2^{m-1}$	$L \le 2^m-1-D$	0
	$2^m-1-D < L \le D$	$L+D-2^m+1$
	$D < L \le 2^m-1$	$2L-2^m+1$

ever, for $D<L\leq 2^m-1$, the number of undetectable two-bit errors will be $L-(2^m-1-D)+L-D$. The resulting expressions can be simplified and written in tabular form (Table 15.1).

It can be readily seen that the number of adjacent line pairs in an m-bit PSA is $m-1$, the number of lines spaced one line apart is $m-2$, etc. The total number of all possible errors of multiplicity 2 equals to the number of combinations of Lm taken at a time. Then we can write a formula to find out the probability of detecting 2-bit errors in the tested binary sequences of length L by using the PSA with delay d:

$$P_n = \frac{\sum_{i=1}^{m-1}(m-i)\,N_{id(mod(2^m-1))}}{C_{Lm}^2}, \qquad (15.7)$$

where $N_{id(mod(2^m-1))}$ is the number of undetectable two-bit errors in a pair of compressed binary sequences as determined by Table 15.1; C_{Lm}^2 is the number of combinations of Lm elements taken 2 at a time.

The resulting formula (15.7) is also appropriate for calculating the escape probability for two-bit errors in a popular PSA with D-flip-flops and internal XOR gates (Fig.13.1). This time, however, the delay d is 1 for any primitive polynomial.

From the analysis of the formula for different multiplicity and compressed sequence length one can see that a PSA based on a shift register with D-flip-flops and internal XOR gates has lower efficiency than the proposed PSA.

Fig.15.4 shows the values of two-bit error escape probability for a 6-bit PSA. Plots 2 and 3 correspond to the PSAs with T-flip-flops (Fig.15.4) whose feedback layout is described by polynomials $\varphi(x)=x^6\oplus x^5\oplus x^2\oplus x\oplus 1$ (characteristic polynomial $\varphi(x)=x^6\oplus x^5\oplus 1$) and $\varphi(x)=x^6\oplus x^4\oplus x^2\oplus x\oplus 1$ (characteristic polynomial $\varphi(x)=x^6\oplus x\oplus 1$), respectively; plot 1 corresponds to a conventional PSA with D-flip-flops (Fig.13.1). The horizontal line is the mean error escape probability for a SA of the same length.

The results obtained from testing a PSA with T-flip-flops as storage elements and internal XOR gates testify that the PSA in question is highly efficient in testing binary data streams. However, there is a need to estimate the probability of errors to escape detection by this analyzer for other multiplicity values. As in the case of PSA with D-flip-flops, there is an alternative design with external XOR gates for the analyzer in question. The behavior of such an analyzer is described by the set of equation

$$A(k) = V_{TE}A(k-1) \oplus Z(k), \quad k=\overline{1,l}, \qquad (15.8)$$

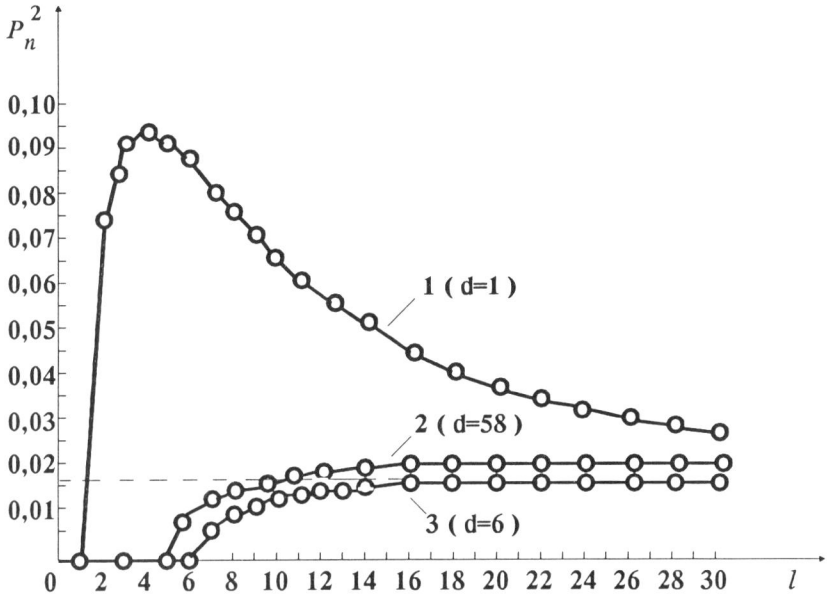

Fig. 15.4

with matrix V appearing as:

$$V_{TE} = \begin{vmatrix} 1 \oplus \alpha_1 & \alpha_2 & \alpha_3 & \ldots & \alpha_{m-1} & \alpha_m \\ 1 & 1 & 0 & \ldots & 0 & 0 \\ 0 & 1 & 1 & \ldots & 0 & 0 \\ 0 & 0 & 1 & \ldots & 0 & 0 \\ \ldots & \ldots & \ldots & \ldots & \ldots & \ldots \\ 0 & 0 & 0 & \ldots & 1 & 1 \end{vmatrix}$$

It is evident that matrices V_{TI} and V_{TE} are similar and have one and the same polynomial $\varphi(x)$, which describes the feedback layout, and characteristic polynomial $\psi(x)$. It follows herefrom that both designs are equivalent as regards the efficiency of their performance in a single-line mode.

For the purpose, we simulated the PSA with T-flip-flops as storage elements by software.

15.4. Simulation of Parallel Signature Analyzers with T-flip-flops

To test for the efficiency of error detection in binary sequences, we simulated a PSA with T—flip-flops. The simulation procedure has been discussed in subsection 13.4.

Based on the above procedure, we tested PSAs with T-flipflops as storage elements and both with internal and external XOR gates of different length and characteristic polynomials.

Fig 15.5 shows the plots of error escape probability versus compresed sequence length for different multiplicity for a 5-bit PSA based on T-flip-flops

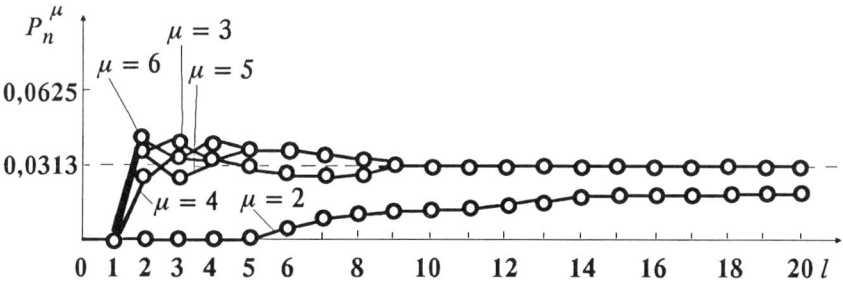

Fig.15.5

as storage elements and having an irreducible polynomial $\varphi(x)=x^5\oplus x^2\oplus 1$ which defines its feedback layout (primitive characteristic polynomial $\psi(x)=x^5\oplus x^4\oplus x^2\oplus x\oplus 1$). Figs.15.6 and 15.7. show similar plots for 6-bit and 7-bit PSAs having the feedback describing polynomials $\psi(x)=x^6\oplus x^4\oplus x^2\oplus x\oplus 1$) (characteristic polynomial $\psi(x)=x^6\oplus x\oplus 1$), $\varphi(x)=x^7\oplus x^6\oplus x^5\oplus x^4\oplus x^2\oplus x\oplus 1$ (characteristic polynomial $\varphi(x)=x^7\oplus x^3\oplus x^2\oplus x\oplus 1$), respectively.

Figs.15.8 to 15.10 show the results of simulation of PSAs with T-flip-flops and external XOR gates which feedback layout has been described by the same polynomials as those for the PSA with internal XOR gates.

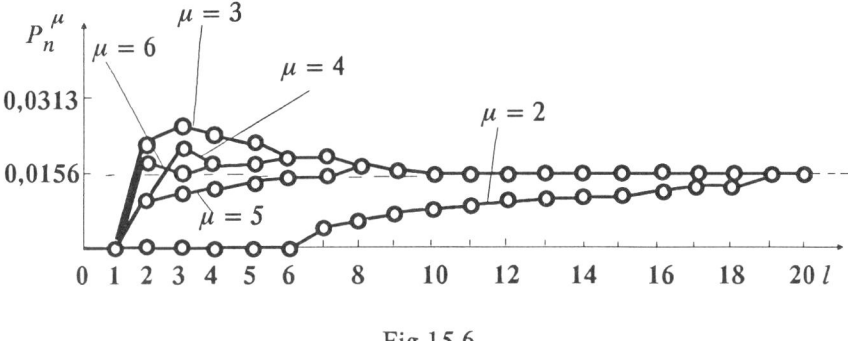

Fig.15.6

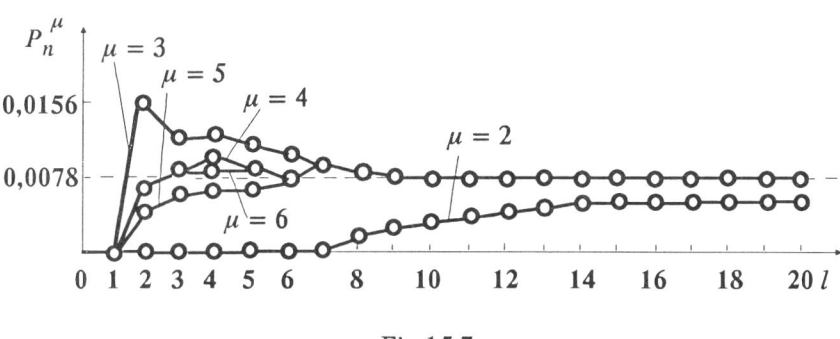

Fig.15.7

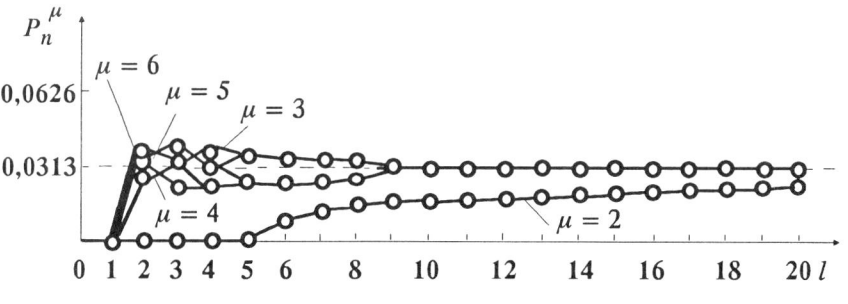

Fig.15.8

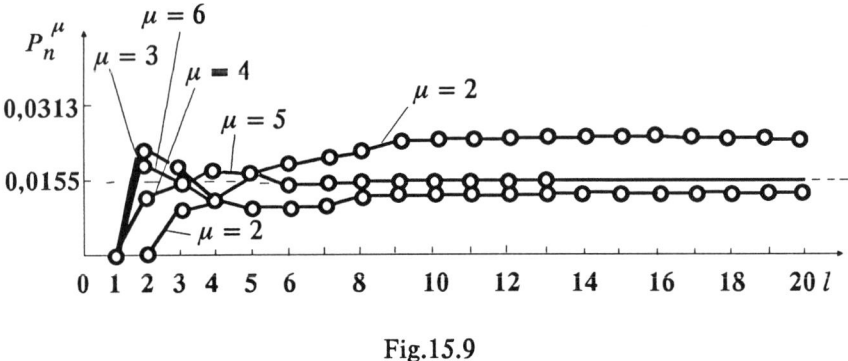

Fig.15.9

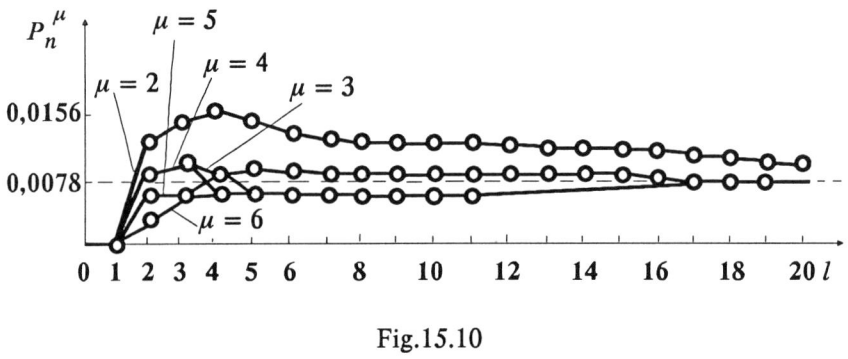

Fig.15.10

The above presented reasoning and simulation results allow us to state that the PSA with T-flip-flops as storage elements that has been designed by one of the techniques discussed in subsections 15.1 and 15.2 is more efficient than the popular PSAs with internal and external XOR gates in detecting errors in binary streams.

Chapter 16

SIGNATURE TESTABILITY

16.1. Signature Testability Analysis

The use of compact testing methods that produce lengthy sequences at the inputs of digital units under test and compress output responses into short keywords called signatures has given rise to the problem of data validity resulted from test experiment. In other words, there is a need of estimating the efficiency of compact testing method over a definite fault class for digital units.

Pseudorandom test sequence generators used for VLSI self-test implementation and compaction networks used for obtaining an integral characteristic of the test experiment result call for efficiency estimation like any other testing scheme. Unlike test generator design providing for the required detectability of the specified fault classes, the use of specific compaction scheme, e.g. signature analyzer, does not provide for the circuit under test to be testable for specific faults. This involves the necessity to estimate circuit fault testability for specific compact testing structures.

To formalize signature analysis efficiency estimation against definite fault classes, let us consider the possible algorithms for calculating the signatures of Boolean functions as well as obtain the generalized conditions for signature testability of two-level digital circuits.

Consider the testability of some digital circuits for some fault classes. For the purpose, we shall first analyse an elementary circuit with four inputs and two outputs that consists of 2OR and 2AND gates. The circuit in question is compact tested as shown in Fig.16.1 where the test generator (TG) is built around a feedback shift register described by the generating polynomial $\psi(z)=1 \oplus z \oplus z^4$. We shall use the signature analyzer (SA) described by polynomial $\phi(z)=1 \oplus z^3 \oplus z^4$ as an analyzing circuit. The test experiment by the above structure consists in applying 2^4-1 test patterns X_k to provide for all possible input combinations for 2OR and 2AND gates to the four inputs to the digital circuit (DC) under test. Output responses of the gates under test are compressed into signatures. Table 16.1 shows the timing chart of signature generation for sequences $\{y_1\}$ and $\{y_2\}$ produced on outputs of a fault-free circuit. Table 16.2 shows sequences $\{y_1'\}$ and $\{y_2'\}$ for a faulty circuit when there is a $\equiv 0$ at the first input to 2OR gate and a $\equiv 1$ at the first input to 2AND gate.

Note that the initial states of signature analyzer and test generator are respectively *0000* and *1000*.

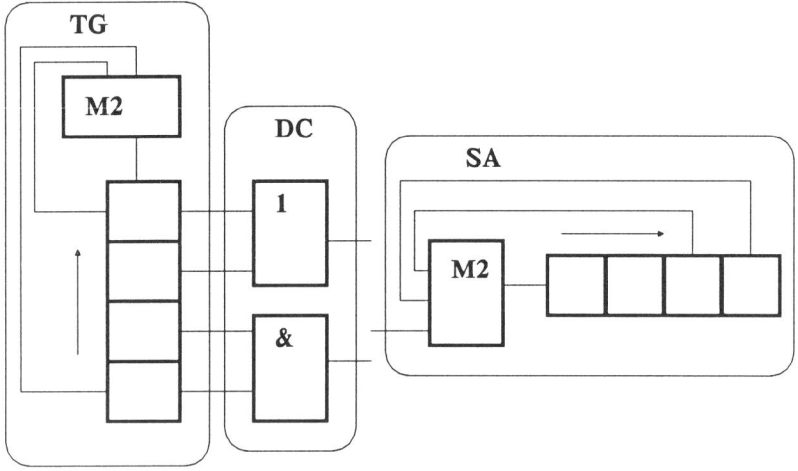

Fig. 16.1. General diagram of compact testing

From the above tables is evident that the values of reference signatures $S(y_1)=0000$, $S(y_2)=0000$ and the values of signatures $S(y_1')$ and $S(y_2')$ obtained for the circuit under test will be *0000* for a faulty state. Hence the mentioned faults cannot be detected by comparing the obtained signatures with the reference values.

A match between the obtained and refernce signatures proves that stuck-at faults $\equiv 0$ and $\equiv 1$ are signature untestable at the primary inputs to 2OR and 2AND gates. Further analysis of whether other stuck-at faults of the circuit in question are detectable produces a sudden result: neither single nor multiple stuck-at fault can be detected by signature analysis. This can be attributed to the fact that the entire set of faulty modifications of the circuit under test in the stuck-at fault class initiates a limited set of output responses of the circuit whose signatures match the reference one within the selected test scheme. The above example demonstrates that prior to using signature analysis for testing the elementary circuit of Fig.16.1 it should be examined for signature testabiluty within the selected compact testing structure.

Table 16.1

k	X_k	$\{y_1\}$	$S(y_1)$	$\{y_2\}$	$S(y_2)$
1	1000	1	1 0 0 0	0	0 0 0 0
2	1100	1	1 1 0 0	0	0 0 0 0
3	1110	1	1 1 1 0	0	0 0 0 0
4	1111	1	0 1 1 1	1	1 0 0 0
5	0111	1	1 0 1 1	1	1 1 0 0
6	1011	1	1 1 0 1	1	1 1 1 0
7	0101	1	0 1 1 0	0	1 1 1 1
8	1010	1	0 0 1 1	0	0 1 1 1
9	1101	1	1 0 0 1	0	0 0 1 1
10	0110	1	0 1 0 0	0	0 0 0 1
11	0011	0	0 0 1 0	1	0 0 0 0
12	1001	1	0 0 0 1	0	0 0 0 0
13	0100	1	0 0 0 0	0	0 0 0 0
14	0010	0	0 0 0 0	0	0 0 0 0
15	0001	0	0 0 0 0	0	0 0 0 0

Analysis of digital circuit signature testability results in on estimate of the selected compact testing scheme efficiency and recommendations as to its modification. Thus for the example of Fig.16.1, the applied compact testing structure is unfit for detecting any stuck-at fault thereby generating a need for its modification or replacement. As alternative compact testing structures we can propose replacements for generating polynomials for the test generator or signature analyzer, increase of their high degree and change the number of test patterns. Any modification in the compact test implementation structure involves a retry of digital circuit analysis for signature testability.

One of the possible solutions for the problem of estimating the circuit faults for signature testability is their physical modelling followed by experimental determination of signatures. The above approach, however, is unfeasable since it consumes much time for conducting test experiment for each fault. Similar result stems from an attempt to solve the problem by programming when the program model of a circuit is used as circuit under test.

Table 16.2

k	X_k	$\{y_1'\}$	$S(y_1')$	$\{y_2'\}$	$S(y_2')$
1	1 0 0 0	0	0 0 0 0	0	0 0 0 0
2	1 1 0 0	1	1 0 0 0	0	0 0 0 0
3	1 1 1 0	1	1 1 0 0	0	0 0 0 0
4	1 1 1 1	1	1 1 1 0	1	1 0 0 0
5	0 1 1 1	1	0 1 1 1	1	1 1 0 0
6	1 0 1 1	0	0 0 1 1	1	1 1 1 0
7	0 1 0 1	1	1 0 0 1	1	0 1 1 1
8	1 0 1 0	0	1 1 0 0	0	0 0 1 1
9	1 1 0 1	1	1 1 1 0	1	1 0 0 1
10	0 1 1 0	1	0 1 1 1	0	1 1 0 0
11	0 0 1 1	0	0 0 1 1	1	1 1 1 0
12	1 0 0 1	0	0 0 0 1	1	0 1 1 1
13	0 1 0 0	1	0 0 0 0	0	0 0 1 1
14	0 0 1 0	0	0 0 0 0	0	0 0 0 1
15	0 0 0 1	0	0 0 0 0	1	0 0 0 0

Let us examine the possibilities of estimating the digital circuit testability analytically. For the purpose, we shall initially define a signature as an analytical function dependent on many variables.

16.2. Signature as a Function of Multiple Variables

Definition of a signfture as a function of sequence to be compressed is general in nature and prevent from taking account of specific features of digital circuits under test. For the signature analysis to be used for VLSI self-test implementation, we must give a more complete definition of a signature to take account of its specific structure. Let us consider signature representation as a function of many variables, including the Boolean expression which describes the circuit under test. For the purpose we shall examine the classical compact testing structure for a single-output combinational circuit of Fig.16.2. The structure consists of a test sequence generator (1), a signature analyzer 2, and a combinational circuit (CC) under test.

The test sequence generator and the signature analyser are the fundamental units providing for the validity of combinational circuit test. Most popular test generator for self- testing VLSI chips is the pseudorandom test sequence generator based on M-sequences. The use of such units makes

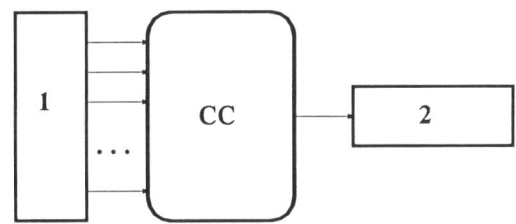

Fig. 16.2. The classical compact testing structure

hardware implementation of block 1 as simple as possible which is the decisive factor in self-test VLSI design. Besides, provision is made for producing all possible input stimuli applied to the combinational circuit inputs. In this case, any fault in the circuit will be observable on the circuit output which is the major advantage of exhaustive testing.

Exhaustive testing assumes the use of a primitive polynomial $\psi(z)$ with the $deg\psi(z) \geq m$, where m is the number of inputs to the circuit under test for constructing a test generator. For $deg\psi(z)=m$ providing for the minimum length N of a test sequence of 2^m-1, exhaustive testing is achieved by applying 2^m-1 m-bit nonzero test patterns to the circuit inputs. A zero test pattern is produced by presetting all storage elements in the test generator to zero. Therefore, in the VLSI self-test mode, the combinational circuit behavior must be tested on all nonzero m-bit patterns. The sequence of test patterns produced is uniquely determined by the form of polynomial $\psi(z)$.

Signature analyzer (2) of Fig.16.2, the same as test generator (1), is based on the primitive polynomial $\phi(z)$. The analyzer operation consists in compressing an output response into a short keyword. Thus, for $deg\phi(z)=deg\psi(z)=m$, 2^m-1 output values are converted into an m-bit signature which, when different from the reference signature, points to a faulty state of CC.

From the above discussed examples we may deduce that the value of signature depends on the form and length of the test sequence, circuit under test described by F, and the analyzer structure, i.e. signature S is the function of many variables. For self-testing VLSI chips, with the test generator and signature analyzer described by primitive polynomials, the function will appear as

$$S = S\;[\psi(z), X_1, L, F, \varphi(z), A_0]\,, \quad (16.1)$$

where $\psi(z)$ and $\phi(z)$ are primitive polynomials used for constructing the test generator and the signature analyzer, respectively; X_1 is the first pattern in the input test sequence of L patterns; $F=F(x_1,...,x_m)$ is the Boolean function describing the circuit under test; A_0 is the initial state value of signature analyzer.

The arguments of function (16.1) that defines the signature may assume any values. However their relationship is somewhat restricted due to the necessity of maximizing the circuit test validity and minimizing the test experiment time. Thus, for polynomials $\psi(z)$ and $\phi(z)$, $deg\psi(z) \geq m$ and $deg\phi(z) \geq m$, and the length L of test sequence must satisfy the inequality $L \geq 2^m - 1$. The first test pattern X_1 and the initial state A_0 of signature analyzer assume specific values of the set of test patterns and the set of analyzer states.

The above discussed argument of signature expressed as (16.1) have normally well-defined values for any circuit under test and never vary with its form. Based on the adopted values of arguments $\psi(z)$, $\phi(z)$, X_1, L, and A_0, and their relationships, signature S will be a function of only Boolean expression $F=F(x_1,...,x_m)$ implemented by the circuit and appear as

$$S = S\,(F) = S\;[F(x_1,...,x_m)]\,, \quad (16.2)$$

Now we shall find the value of signature as a function of $F(x_1,...x_m)$.

Difinition 16.1. Signature $S=S(F)$ of a Boolean function $F=F(x_1,...,x_m)$ is the signature of a sequence produced at the output of circuit realizing F and obtained under the following conditions:

1) The test sequence generator and the signature analyzer are based on primitive polynomials $\psi(z)$ and $\phi(z)$, respectively.

2) The test sequence length is L.

3) The first test pattern produced by the test generator is X_1.

4) The intial state of the signature analyzer is A_0.

16.3. Investigation of Signature Properties

We consider the major properties of signature analysis within the fairly general model of its application (Fig.16.2). For the purpose, we take the definition of signature as a function of many variables expressed by (16.1). The first argument affecting the final value of a signature is a primitive polynomial $\psi(z)$ used for constructing the test generator. The primitive polynomial $\psi(z)$ is needed to provide for all possible input combinations on the inputs to circuit under test, otherwise, the test generator will not test the circuit of Fig.16.2 exhaustively. This will degrade the validity of circuit diagnosis since not all possible input patterns are involved in its testing.

The form of primitive polynomial

$$\psi(z) = 1 \oplus \alpha_1 z^1 \oplus \alpha_2 z^2 \oplus \ldots \oplus \alpha_{m-1} z^{m-1} \oplus \alpha_m z^m$$

is determined by the set of coefficients $\alpha_i \in \{0,1\}$, $i=\overline{1,m}$. Their values as well the value of degree $\psi(z)$, the high degree of polynomial $\psi(z)$, affect the properties of a pseudorandom sequence. If $deg\,\psi(z)$ is the number of inputs m to the circuit under test (Fig.16.2), the test generator will produce all possible binary patterns except for the pattern of m zeros. With $deg\psi(z)=m+1$ each binary pattern other than the all zero pattern will be applied twice to the circuit inputs. The zero pattern will appear only once.

In the general case, for $deg\psi(z)=m+r$, the inputs of circuit under test will be applied with all test patterns 2^r times, except for the zero test pattern which will be produced 2^r-1 times. The order of test pattern generation (their alteration) is uniquely determined by the set of coefficients $\alpha_i \in \{0,1\}$, $i=\overline{1,m}$.

Thus for the same $deg\psi(z)$ variation of coefficients α_i within the primitive polynomials leads to a variation in the alternation of the initial set of test patterns. This is a very important property of the test generator based on a primitive polynomial (refer to Fig.16.1). It should be noted that the example in question demonstrates low efficiency of the selected compact testing scheme for the case of elementary 2OR and 2AND gates. The test generator used in the scheme is based in primitive polynomial $y(z)=1 \oplus z^1 \oplus z^4$ for which $a_1=1$, $a_2=0$, $\alpha_3=0$ and $\alpha_4=1$. Now we consider the effest of variation in test pattern alternation on the result of testing the circuit of Fig.16.1. For the purpose, we modify the form of generating polynomial for the test generator in the compact testing structure for the circuit. With the same high degree of polynomial $\psi(z)$, we find the new values of coefficients α_i associated with the polynomial $\psi(z)=1 \oplus z^3 \oplus z^4$. In this case, the reference signatures assume the following values: $S(y_1)=1011$ and $S(y_2)=0101$. Given one of the possible stuck-at faults, the actual signatures assume the values other than reference as follows from Table 16.3, where x_1, x_2 are input and f_1 is output variables of a 2OR gate; x_3, x_4 are input and f_2-output variables of a 2AND gate (Fig.16.1). Analysis of the content of Table 16.3 demonstrates that all single stuck-at faults are signature testable for 2OR and 2AND gates. This can be attributed to the fact that there is no match between any actual and reference signature, therefore, all the faults are detectable.

Table 16.3

2OR gate		2AND gate	
Fault	Signature	Fault	Signature
$f_1 \equiv 0$	0 0 0 0	$f_2 \equiv 0$	0 0 0 0
$f_1 \equiv 1$	0 0 0 0	$f_2 \equiv 1$	0 0 0 0
$x_1 \equiv 0$	0 1 0 0	$x_3 \equiv 0$	0 0 0 0
$x_1 \equiv 1$	0 0 0 0	$x_3 \equiv 1$	0 0 0 1
$x_2 \equiv 0$	0 0 1 0	$x_4 \equiv 0$	0 0 0 0
$x_2 \equiv 1$	0 0 0 0	$x_4 \equiv 1$	1 0 0 0

Hence by modifying the generation polynomial $\psi(z)$ we can improve the efficiency of digital circuit test. Hovever, the process practically eludes formalizing and is based on the trial-and-error method. The principle of the method is estimation of signature testability for any new polynomial $\psi(z)$ and selection of polynomial such that the test efficiency be maximum.

The second argument of the function (16.1) is X_1, i.e. the first test pattern in the input test pattern. The argument effect on the signature value depends to a large extent on the mode of test generator operation. Thus for an incomplete period of a pseudorandom sequence, X_1 will vary with the input test sequence which can eventually improve test efficiency the same as the polynomial change. When implementing probabilistic testing most important is the selection of initial code X_1 that establishes the required dimensionality for uniform distribution of pseudorandom test patterns. In this case, only a small portion of the period of pseudorandom M-sequence of high order is used. Therefore its properties are described by aperiodic statistical characteristics.

The effect of these characteristics on the value of X_1 is great. To find out the relationship, we shall analyse the M-sequence characteristics for three forms of initial code X_1:

a) $X_1=100...0$;

b) X_1 obtained by using a perfect random sequence;

c) X_1 equal to another M-sequence whose high degree of generating polynomial is *int $\log_2 [deg\psi(z)]$*.

To specify the initial state X_1 for the pseudorandom test sequence generator based on a high-order generating polynomial $\psi(z)$, the second variant, i.e. the perfect random sequence, suits better. At the same time, the third variant is more preferable than the first one that is widely used for pseudorandom sequences with low-order generating polynomials.

The below theorem demonstrates that $X_1=100..0$ is suitable for the polynomials with small values of $deg\psi(z)$ used for exhaustive testing implementation.

Theorem 16.1. The signature of output sequence for an *m*-input combinational circuit obtained by the signature analyzer and the pseudorandom test generator with high degrees of generating polynomials being *m* and the test seguence length determined by 2^m-1, for the first test pattern $100...0$ is related to the signature with the initial pattern X_1 as follows

$$S(F/100...0) = V^l S(F/X_1) \qquad (16.3)$$

where the matrix

$$V = \begin{vmatrix} \alpha_1 & \alpha_2 & \alpha_3 & ... & \alpha_{m-1} & \alpha_m \\ 1 & 0 & 0 & ... & 0 & 0 \\ 0 & 1 & 0 & ... & 0 & 0 \\ 0 & 0 & 1 & ... & 0 & 0 \\ ... & ... & ... & ... & ... & ... \\ 0 & 0 & 0 & ... & 1 & 0 \end{vmatrix} \qquad (16.4)$$

and the coefficients $\alpha_i \in \{0,1\}$, $i=\overline{1,m}$, are determined in accordance with the generating polynomial $\phi(z)$ of signature analyzer, and l is the value of shift between the codes X_1 and $100...0$ in the pseudorandom sequence.

Proof. Let the output sequence of a combinational circuit consist of v ones symbols, with a one being also formed for the first test pattern X_1. then by the decisive property of signature analysis (formula 16.1), consisting in the fact that the signature of a bitwise modulo-2 sum of sequences under test is the bitwise sum of their signatures, we can write that $S(F/X_1) = S(y_1/X_1) \oplus S(y_2/X_1) \oplus ... \oplus S(y_v/X_1)$, where $\{y_j\}, j=\overline{1,v}$ is the binary sequence of a single one zeros. Accordingly, $\{y_j\}$ is the sequence comprising the unity value solely for the input test pattern X_1. We can similarly write that $S(F/100...0) = S(y_1/100..0) \oplus S(y_2/100..0) \oplus .... \oplus S(y_v/100..0)$. The value $S(y_1/100..0)$ differs from $S(y_1/X_1)$ solely by performing l steps more to calculate it in the signature analyzer. For l time steps the analyser processes zero input signals, i.e. behaves as an M-sequence generator for which $S(y_1/100..0)$ and $S(y_1/X_1)$ are the states related by

$$S(y_1/100...0) = V^l S(y_1/X_1),$$

where V is the binary matrix (16.4) that describes its behavior, and l is the distance (number of time steps) between these states. In view of the latter relation we can write.

$$S(F/1000...0) = V^l S(y_1/X_1) \oplus V^l S(y_2/X_1) \oplus ... \oplus V^l S(y_v/X_1) =$$
$$= V^l S(F/X_1),$$

This theorem can be generalized as follows

$$S(F/X_1) = V^{-l} S(F/100...0) = V^{L-l} S(F/100...0) =$$
$$= V^\lambda S(F/100...0), \quad (16.5)$$

where λ is the shift between the codes $100...0$ and X_1. An important corollary of Theorem 16.1 is the fact that the value of initial test pattern does not affect the validity of exhaustive testing for the circuit (Fig.16.2). This statement stems from the following condition: if an actual signature $S_a(F/100..0)$ mismatches a reference signature $S_r(F/100..0)$ for $X_1=100...0$, these signatures will differ for other values of X_1 since by multiplying the right and the left sides of inequality

$$S_a(F/100...0) \neq S_r(F/100...0).$$

by V^λ and in view of equality (16.5), we obtain

$$S_a(F/X_1) \neq S_r(F/X_1).$$

Therefore, we can use any pattern as an initial test pattern X_I to implement exhaustive VLSI self-testing.

In practice, code *100...0* is normally taken as an initial test pattern.

The length L of the test sequence is determined by the required validity of circuit test. For exhaustive test implementation, L is always 2^m-1, where m is the high degree of polynomial $\psi(z)$ used for constructing the test generator.

In the above discussion we assumed that signature analysis efficiency is the efficiency of only analysing circuit, with other factors being ignored. As the preceding examples proved, however, the efficiency of compact testing depends on many factors, including the compression circuit structure, which is described by a generating polynomial $\phi(z)$. The high degree of the polynomial is found by the inequality

$$deg\,\varphi(z) = Int\,\log_2 L. \qquad (16.6)$$

With the condition of (16.6) being met, the analysing circuit will satisfy some properties stemming from the coding theory. In particular, it would provide for detecting all single and double errors in a sequence under test. When exhaustive testing is implemented, $deg\,\phi(z) \geq deg\,\psi(z) = m$. A change in the form of generating polynomial $\phi(z)$ causes the sequences under test to be rearranged to form new sets with its own signature value each. This is a very important property of a signature analyzer that eventually improves the circuit test efficiency.

Consider the effect of changing the form of generating polynomial $\phi(z)$ on the validity of testing the circuit of Fig.16.1. Note that for the polynomial $\phi(z)=1 \oplus z^3 \oplus z^4$ the entire set of stuck-at faults is undetectable. At the same time, substituting $\phi(z)=1 \oplus z^3 \oplus z^4$ for $\phi(z)=1 \oplus z \oplus z^4$, we reveal that actual signature values obtained for the signature analyzer mismatch the reference ones as Table 16.4 shows. Analysis of the results in Table 16.4 shows that all single atuck-at faults for 2AND and 2OR gates are signatures testable. This can be attributed to the fact that none of the actual signatures matches the reference one whence all stuck-at faults are detectable.

The above example proves that by changing the form of generating polynomial $\phi(z)$ we can noticeably improve the efficiency of digital circuit testing. We should note, however, that searching for an optimum polynomial $\phi(z)$ or $\psi(z)$ cannot be practically formalized and is normally based on a simple search. Besides, selection of an optimum $\phi(z)$ clasely relates to $\psi(z)$. Therefore, a collection of generating polynomials $\phi(z)$ and $\psi(z)$ is selected in the specific structure for compact testing of a combinational circuit described by the Boolean expression $F=F(x_1,...,x_m)$.

Table 16.4

2OR gate		2AND gate	
Fault	Signature	Fault	Signature
$f_1 \equiv 0$	0 0 0 0	$f_2 \equiv 0$	0 0 0 0
$f_1 \equiv 1$	0 0 0 0	$f_2 \equiv 1$	0 0 0 0
$x_1 \equiv 0$	0 0 0 1	$x_3 \equiv 0$	0 0 0 0
$x_1 \equiv 1$	0 0 0 0	$x_3 \equiv 1$	0 0 1 0
$x_2 \equiv 0$	1 0 0 0	$x_4 \equiv 0$	0 0 0 0
$x_2 \equiv 1$	0 0 0 0	$x_4 \equiv 1$	0 1 0 0

To estimate the effect of signature analyzer initial setting A_0 on the final signature value, we shall prove the following theorem.

Theorem 16.2. For the initial state of analyzer $A_0 \neq 000...0$, signature $S(F/A_0)$ is a bitwise modulo-2 sum of signature $S(F/000...0)$ for $A_0=000..0$ and the vector $V^L A_0$, i.e.

$$S(F/A_0) = S(F/000...0) \oplus V^L A_0, \qquad (16.7)$$

where the matrix

$$V = \begin{vmatrix} \alpha_1 & \alpha_2 & \alpha_3 & ... & \alpha_{m-1} & \alpha_m \\ 1 & 0 & 0 & ... & 0 & 0 \\ 0 & 1 & 0 & ... & 0 & 0 \\ 0 & 0 & 1 & ... & 0 & 0 \\ ... & ... & ... & ... & ... & ... \\ 0 & 0 & 0 & ... & 1 & 0 \end{vmatrix}$$

and the coefficients $\alpha_i \in \{0,1\}$, $i=\overline{1,m}$, are determined in accordance with the generating polynomial $\phi(z)$ of signature analyzer, and L is the length of sequence under test.

Proof. Compression of a tested sequence of L symbols results in signature $S(F/000...0)$ for $A_0=000...0$. With the zero initial setting A_0 and the zero sequence under test, signature analyzer behaves as a normal pseudorandom sequence generator satisfying the condition $A^L = V^L A_0$, where A_0 is the state of generator in L time steps of operation. By using the signature analysis property that the signature of a bitwise modulo-2 sum of the tested sequences is the bitwise sum of signatures, we can write

$$S(F/A_0) = S(F/000...0) \oplus V^L A_0.$$

From Theorem 16.2 it follows that the initial setting of a signature analyzer does not affect the signature analysis validity. This follows the fact that if the actual signature $S_a(F/000...0)$ differs from the reference $S_r(F/000...0)$ for $A_0 = 000...0$, the said signatures will also differ for other values of A_0. From the inequality

$$S_a(F/000...0) \neq S_r(F/000...0)$$

it follows that

$$S_a(F/000...0) \, V^L A_0 \neq S_r(F/000...0) \, V^L A_0,$$

whence, in view of equality (16.7), we obtain

$$S_a(F/A_0) \neq S_r(F/A_0).$$

Thus the above analysis proves that generating polynomials for the test generator and the signature analyzer are the most significant signature arguments affecting the efficiency of signature analysis. The efficiency of signature analysis is less affected by the length of tested sequence L and is practically unaffected by X_1 and A_0. Taking account of the above facts we shall further use $X_1 = 100...0$ and $A_0 = 000...0$ as initial settings. The above properties of signature analysis and the relationship between signature and its arguments will be used in the following to analyse digital circuits for signature testability.

16.4. Generalized Condition for Signature Testability

Using the definition 16.1, we introduce the concept of a reference and actual signature of a single-output combinational circuit described by the Boolean function F.

Definition 16.2. The reference signature $S_r=S_r(F)$ of a single-output combinational circuit described by a Boolean expression $F=F(x_1,...,x_m)$ is the signature obtained by definition 16.1 for a fault-free state of the circuit.

Definition 16.3. An actual signature $S_a=S_a(F)$ of a single-output combinational circuit described by a Boolean expression $F=F(x_1,...,x_m)$ is the signature obtained by definition 16.1 for the case when the combinational circuit is either faulty or fault-free.

An underlying signature analysis property is the one based on the linearity of modulo-2 add operation. The property of 16.1 is its generalizaton for the Boolean function.

Property 16.1. Signature $S(F)$ of a Boolean function equal to the modulo-2 sum of function $F_1 \oplus F_2 \oplus ... \oplus F_w$ is defined as a bitwise modulo-2 sum of signatures $S(F_j)$, $j=\overline{1,w}$.

Using the above signature definitions and properties, we can define the concept of signature testability. For the start we divide the set of all possible faults into two subsets: detectable and undetectable. Leaving aside the latter subset, we shall investigate only detectable faults for testability. A detectable fault is taken to mean the fault associated with the test that can detect it by distortion of the digital circuit output response. Thus the presence of one or more detectable faults causes the circuit output response to be distorted. We shall formulate the generalized signature testability condition for detectable faults.

Definition 16.4. A tested fault of a single-output combinational circuit is signature testable if the value of actual signature $S_a(F)$ of a Boolean function

F describing the circuit differs from its reference value $S_r(F)$, i.e. when the inequality

$$S_a(F) \neq S_r(F) \qquad (16.8)$$

is satisfied.

The above definition is a generalized signature testability condition for any detectable fault class for the circuit which lead to one of its faulty modification. Specific features of a definite fault class allow us to refine the signature testability definition for the class. Let us consider the concept of signature testability for certain fault classes.

Chapter 17

EVALUATION OF SIGNATURES

17.1. Analytical Evaluation of Boolean Function Signatures

Now we shall examine different ways of evaluating a Boolean function signature for different combinations of its arguments (7.1). For the purpose we shall use sufficiently general assumptions which take account of the absence of signature testability dependence on specific values of the first test pattern X_1 and the initial state of signature analyzer A_0. Below are discussed the assumptions.

1. Evaluation of signature $S(F)$ for function F which is a constituent of unity depending on m variables.

2. Generating polynomials $\psi(z)$ and $\phi(z)$ for the test generator and signature analyzer take on an arbitrary ratio in the class of primitive polynomials.

3. The first test pattern $X_1=100...0$.

4. The initial state of signature analyzer $A_0=000...0$.

5. The test sequence length L satisfies the inequality $L>2^m-1$ where $m<deg\psi(z)$. Note that, in the general case, $deg\ \psi(z)=deg\phi(z)$. By way of example let us consider generation of a signature for a Boolean function $F=\bar{x}_1 x_2 \bar{x}_3 x_4$ whose value $S(F)$ will be defined by simulation. We shall take the polynomials $\psi(z)=1\oplus z\oplus z^4$ and $\phi(z)=1\oplus z^2\oplus z^5$ as generating ones and the length of test sequence $L=15$. The block-diagram of compact test implementation for a combinational circuit described by $F=\bar{x}_1 x_2 \bar{x}_3 x_4$ is given in Fig.17.1 whereas the timing chart of test generator and signature analyzer states is given in Table 17.1.

By analysing the signature generation chart, we can notice that the analyzer state remained unchanged for the first six cycles. And only in the seventh cycle the binary pattern *0101* corresponding to the constituent of unity $\bar{x}_1 x_2 \bar{x}_3 x_4$ has been applied to the input of circuit under test, causing the analyzer state to change to *10000*.

Starting with the seventh cycle, the analyzer behaved as normal M-sequence generator for the next eight cycles. This allows to reduce signature determination to the problem of evaluating a replica of M-sequence described by generating polynomial $\phi(z)=1\oplus z^2 \oplus z^5$ that has been shifted by eight cycles relative to the initial code *10000*.

In the general case, the problem of signature determination for an arbitrary generating polynomial will involve, under the assumptions made, calculation of the l th code of M-sequence generator with the initial code *100...0*. The M-sequence generator structure is described by polynomial $\phi(z)$ used for building the signature analyzer. The value of l is determined by the shift between the initial state X_1 of the test generator and the pattern for which $F=1$. The procedure of finding the value of l consists of two subproblems: 1) finding the coefficients to associate the two above codes of the test sequence produced by the test generator described by $\psi(z)$; 2) finding the value of l on the basis of obtained coefficients.

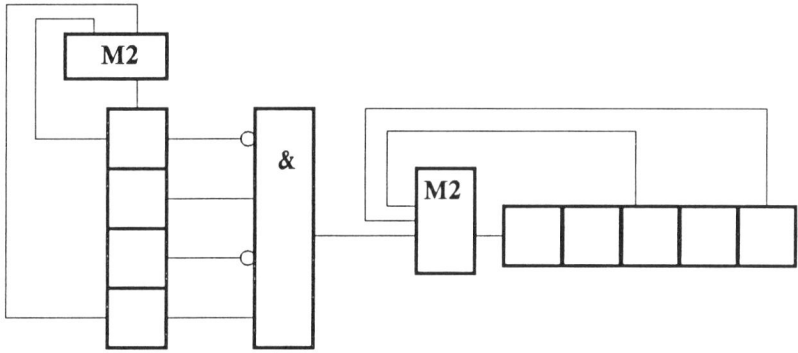

Fig. 17.1. The block-diagram of compact test implementation

Let us consider the possible way of solving the stated subproblems. Suppose that $deg\psi(z)=m$.

The first subproblem is to find the set of coefficients $\{\delta_i(l)\}$, $i=\overline{1,m}$, assuming the value *0* or *1*, which associate a binary vector $\gamma_1\gamma_2\gamma_3...\gamma_m$ defined by function

$$F = x_1^{\gamma_1} x_2^{\gamma_2} x_3^{\gamma_3} ... x_m^{\gamma_m}$$

with the vector *000...01*. The values $\gamma_i \in \{0,1\}$, $i=\overline{0,m}$, and $\gamma_i=1$ for $x_i^{\gamma_i} = x_i$ whereas $\gamma_i=0$ for $x_i^{\gamma_i} = \overline{x_i}$. The subproblem can be solved as follows. Based on vector $\gamma_1\gamma_2\gamma_3...\gamma_m$ we can form a matrix

$$W = \begin{vmatrix} \gamma_m & \beta_1 & \beta_2 & \beta_3 & \cdots & \beta_{m-1} \\ \gamma_{m-1} & \gamma_m & \beta_1 & \beta_2 & \cdots & \beta_{m-2} \\ \gamma_{m-2} & \gamma_{m-1} & \gamma_m & \beta_1 & \cdots & \beta_{m-3} \\ \cdots & \cdots & \cdots & \cdots & \cdots & \cdots \\ \gamma_2 & \gamma_3 & \gamma_4 & \gamma_5 & \cdots & \beta_1 \\ \gamma_1 & \gamma_2 & \gamma_3 & \gamma_4 & \cdots & \gamma_m \end{vmatrix} \qquad (17.1)$$

Table 17.1

k	x1 x2 x3 x4	F	a1 a2 a3 a4 a5
1	1 0 0 0	0	0 0 0 0 0
2	1 1 0 0	0	0 0 0 0 0
3	1 1 1 0	0	0 0 0 0 0
4	1 1 1 1	0	0 0 0 0 0
5	0 1 1 1	0	0 0 0 0 0
6	1 0 1 1	0	0 0 0 0 0
7	0 1 0 1	1	1 0 0 0 0
8	1 0 1 0	0	0 1 0 0 0
9	1 1 0 1	0	1 0 1 0 0
10	0 1 1 0	0	0 1 0 1 0
11	0 0 1 1	0	1 0 1 0 1
12	1 0 0 1	0	1 1 0 1 0
13	0 1 0 0	0	1 1 1 0 1
14	0 0 1 0	0	0 1 1 1 0
15	0 0 0 1	0	1 0 1 1 1

where $\beta_j \in \{0,1\}$, $j= \overline{1,m-1}$, are calculated as follows

$$\beta_1 = \gamma_1 \oplus \sum_{i=1}^{m-1}{}^{\oplus} \alpha_i \gamma_{i+1},$$

$$\beta_2 = \gamma_2 \oplus \sum_{i=1}^{m-2}{}^{\oplus} \alpha_i \gamma_{i+2} \oplus \sum_{j=m-1}^{m-1}{}^{\oplus} \alpha_i \beta_{i-m+2},$$

$$\beta_3 = \gamma_3 \oplus \sum_{i=1}^{m-3}{}^{\oplus} \alpha_i \gamma_{i+3} \oplus \sum_{i=m-2}^{m-1}{}^{\oplus} \alpha_i \beta_{i-m+3}, \qquad (17.2)$$

$$\cdots\cdots\cdots\cdots\cdots\cdots\cdots\cdots\cdots$$

$$\beta_{m-1} = \gamma_{m-1} \oplus \sum_{i=1}^{1}{}^{\oplus} \alpha_i \gamma_{i+m-1} \oplus \sum_{i=2}^{m-1}{}^{\oplus} \alpha_i \beta_{i-1},$$

where $\alpha_i \in \{0,1\}$, $i=\overline{1,m}$, are the coefficients of polynomial $\psi(z)$.

Based on the coefficient matrix W, we build the system of linear equations

$$\gamma_m \delta_1(l) \oplus \beta_1 \delta_2(l) \oplus \beta_2 \delta_3(l) \cdots \oplus \beta_{m-1} \delta_m(l) = 1,$$
$$\gamma_{m-1} \delta_1(l) \oplus \gamma_m \delta_2(l) \oplus \beta_1 \delta_3(l) \cdots \oplus \beta_{m-2} \delta_m(l) = 0,$$
$$\gamma_{m-2} \delta_1(l) \oplus \gamma_{m-1} \delta_2(l) \oplus \gamma_m \delta_3(l) \cdots \oplus \beta_{m-3} \delta_m(l) = 0, \qquad (17.3)$$
$$\cdots\cdots\cdots\cdots\cdots\cdots\cdots\cdots\cdots$$
$$\gamma_2 \delta_1(l) \oplus \gamma_3 \delta_2(l) \oplus \gamma_4 \delta_3(l) \cdots \oplus \beta_1 \delta_m(l) = 0,$$
$$\gamma_1 \delta_1(l) \oplus \gamma_2 \delta_2(l) \oplus \gamma_3 \delta_3(l) \cdots \oplus \gamma_m \delta_m(l) = 0$$

The values of sought-for coefficients $\delta_i(l)$, $i=\overline{1,m}$, are the solution to the above system.

By way of example we shall find coefficients $\delta_i(l)$, $i=\overline{1,4}$, for function $F = \overline{x}_1 x_2 \overline{x}_3 x_4$ that has been used for experimental evaluation of signature $S(F) = 10111$ (Refer to Table 17.1).

Based on $F = \overline{x}_1 x_2 \overline{x}_3 x_4$ we can state that $\gamma_1 \gamma_2 \gamma_3 \gamma_4 = 0101$. Numeric values of coefficients $\beta_j \in \{0,1\}$, $j=\overline{1,3}$, can be found by the expression (17.2).

$$\beta_1 = \gamma_1 \oplus \alpha_1 \gamma_2 \oplus \alpha_2 \gamma_3 \oplus \alpha_3 \gamma_4 = 0 \oplus 1\ 1 \oplus 0\ 0 \oplus 0\ 1 = 1,$$
$$\beta_2 = \gamma_2 \oplus \alpha_1 \gamma_3 \oplus \alpha_2 \gamma_4 \oplus \alpha_3 \beta_1 = 1 \oplus 1\ 0 \oplus 0\ 1 \oplus 0\ 1 = 1,$$
$$\beta_3 = \gamma_3 \oplus \alpha_1 \gamma_4 \oplus \alpha_2 \beta_1 \oplus \alpha_3 \beta_2 = 0 \oplus 1\ 1 \oplus 0\ 1 \oplus 0\ 1 = 1.$$

Then the coefficient matrix W (17.1) assumes the form

$$W = \begin{vmatrix} 1 & 1 & 1 & 1 \\ 0 & 1 & 1 & 1 \\ 1 & 0 & 1 & 1 \\ 0 & 1 & 0 & 1 \end{vmatrix}.$$

The system of linear equations (17.3) may be written as

$$\delta_1(l) \oplus \delta_2(l) \oplus \delta_3(l) \oplus \delta_4(l) = 1,$$
$$\delta_2(l) \oplus \delta_3(l) \oplus \delta_4(l) = 0,$$
$$\delta_1(l) \qquad \oplus \delta_3(l) \oplus \delta_4(l) = 0,$$
$$\delta_2(l) \qquad \oplus \delta_4(l) = 0.$$

The solution, to the above system are coefficients $\delta_1(l)=1$, $\delta_2(l)=1$, $\delta_3(l)=0$ and $\delta_4(l)=1$.

The second subproblem in finding the value of l is its calculation from the coefficients $\delta_i(l)$ which determins the relation between codes $\gamma_1\gamma_2\gamma_3\ldots\gamma_m$ and $000\ldots01$. Using the periodicity property of the test sequence of period $L=2^m-1$, we can demonstrate that $\delta_i(l)=\delta_i(l-L)$, $i=\overline{1,m}$, and $\delta_i(l-L)$ are coefficients that relate the codes $000\ldots01$ and $\gamma_1\gamma_2\gamma_3\ldots\gamma_m$.

For coefficients $\delta_i(-l_1)\in\{0,1\}$, $i=\overline{1,m}$, the problem of finding the value of l_1 equal to the delay of $100\ldots0$ with respect to $\gamma_1\gamma_2\gamma_3\ldots\gamma_m$ has been solved in view of $l_1=l-L+1$, since the code $100\ldots0$ is adjacent to $000\ldots01$, we use the technique of determining the value of l_1 to find l. For the purpose we may use the division of polynomial

$$\delta_1(-l_1)z^1 \oplus \delta_2(-l_1)z^2 \oplus \delta_3(-l_1)z^3 \oplus \delta_4(-l_1)z^4$$

by polynomial $\psi^{-1}(z)$ that is the reciprocal of $\psi(z)$. Division lasts to leaving a monomial z^{l_1} where the value of l_1 is determined by the power.

By way of example, let us find the value l_1 for an M-sequence generated by polynomial $\psi(z)=1\oplus z\oplus z^4$ with $\delta_1(l-L+1)=\delta_1(l)=1$, $\delta_2(l-L+1)=\delta_2(l)=1$, $\delta_3(l-L+1)=\delta_3(l)=0$ and $\delta_4(l-L+1)=\delta_4(l)=1$. In this case, the polynomial for the dividend has the form $z\oplus z^2\oplus z^4$ and the divisor is determined by polynomial $\psi^{-1}(z)=1\oplus z^3\oplus z^4$ that is the reciprocal of $\psi(z)=1\oplus z\oplus z^4$. Division proceeds as follows.

Evaluation Signatures

$$
\begin{array}{r}
z \oplus z^2 \oplus z^4 \lfloor 1 \oplus z^3 \oplus z^4 \\
z \oplus z^4 \oplus z^5 \overline{} \\
\overline{} z \oplus z^2 \\
 z^2 \oplus z^5 \\
 z^2 \oplus z^5 \oplus z^6 \\
\overline{} \\
z^6
\end{array}
$$

With $l-L+1=-6$, the shift between segments $\gamma_1\gamma_2\gamma_3...\gamma_m$ and $000...1$ in the test sequence is determined as $l=L-7=15-7=8$.

To estimate signature $S(F)$, we apply the classical relation that describes the behavior of M-sequence generator

$$A(k+l) = V^l A(k), \quad (17.4)$$

where $A(k)=100...0$ and V is the matrix

$$V = \begin{vmatrix} \alpha_1 & \alpha_2 & \alpha_3 & \cdots & \alpha_{m-1} & \alpha_m \\ 1 & 0 & 0 & \cdots & 0 & 0 \\ 0 & 1 & 0 & \cdots & 0 & 0 \\ \cdots & \cdots & \cdots & \cdots & \cdots & \cdots \\ 0 & 0 & 0 & \cdots & 1 & 0 \end{vmatrix},$$

where coefficients $\alpha_i \in \{0,1\}$, $i=\overline{1,m}$, are determined by the form of generating polynomial $\phi(z)$ that describes the signature analyzer. The state vector $A(k+1)$ of M-sequence generator is the desired signature $S(F)$.

For the above given example, $\phi(z)=1 \oplus z^2 \oplus z^5$ and $l=8$, therefore V^8 will be determined as follows

$$V^8 = \begin{vmatrix} 0 & 1 & 0 & 0 & 1 \\ 1 & 0 & 0 & 0 & 0 \\ 0 & 1 & 0 & 0 & 0 \\ 0 & 0 & 1 & 0 & 0 \\ 0 & 0 & 0 & 1 & 0 \end{vmatrix}^8 = \begin{vmatrix} 1 & 1 & 1 & 1 & 0 \\ 0 & 1 & 1 & 1 & 1 \\ 1 & 0 & 0 & 1 & 1 \\ 1 & 1 & 1 & 0 & 1 \\ 1 & 1 & 0 & 1 & 0 \end{vmatrix}.$$

According to expression (17.4), the value of signature that corresponds to the experimentally found value (Refer to Table 17.1).

$$S(F) = \begin{vmatrix} 1 & 1 & 1 & 1 & 0 \\ 0 & 1 & 1 & 1 & 1 \\ 1 & 0 & 0 & 1 & 1 \\ 1 & 1 & 1 & 0 & 1 \\ 1 & 1 & 0 & 1 & 0 \end{vmatrix} \begin{vmatrix} 1 \\ 0 \\ 0 \\ 0 \\ 0 \end{vmatrix} = \begin{vmatrix} 1 \\ 0 \\ 1 \\ 1 \\ 1 \end{vmatrix}.$$

Thus calculation of a signature for an arbitrary polynomial consists in sequential solution of the three subproblems. The time consumed by solution of each subproblem determines the total time required for signature calculation. The second subproblem is more difficult to solve because of the need to determine the shift l between the M-sequence segments. The complexity of subproblem is practically equivalent to calculation of a signature. Therefore we can use as an alternative the algorithm for high speed determination of shifted M-sequence replica to estimate l. In their turn, high-speed procedures of calculating coefficients $\{\delta_i(l)\}$, $i=\overline{1,m}$, decrease significantly the time required for solving the first subproblem.

The discussed analytical determination of signature $S(F)$ can be generalized to an arbitrary form of Boolean function. In this case, function F is transformed to the full disjunctive normal form. Then the signatures for all constituents of the disjunctive form are determined. The resulting signature is determined as a bitwise modulo-2 sum of signatures for the constituents of unity.

The necessity for transforming the initial function makes analytical determination of signatures for arbitrary generation polynomials ever more complex. Therefore more interest arouse simpler calculation algorithms for special cases of generating polynomial combinations.

17.2. Calculation of Signatures for Identical Generating Polynomials

Examine the possibility of using identical generating polynomials both for test generator and signature analyser. A peculiarity of this study is the use of a sole generating polynomial with the other signature arguments remained unchanged.

Thus the first test pattern $X_1=100...0$, the initial state of signature analyzer $A_0=000...0$, and the test sequence length $L=2^m-1$, where $m= deg\phi(z)=deg\psi(z)$. Besides, $\phi(z)=\psi(z)$ and the signature is calculated for the function F which is a constituent of unity dependent on m variables.

Examine the practical procedure of generating a signature for a Boolean function $F=\bar{x}_1x_2\bar{x}_3x_4$ for which we shall experimentally find the value of $S(F)$. Select the polynomial $\phi(x)=1\oplus z \oplus z^4$ as a generating polynomial for the test generator and the signature analyzer. Compact test implementation for the circuit described by the Boolean function $F=\bar{x}_1x_2\bar{x}_3x_4$ is practically identical to that of Fig.17.1. The only difference is the circuit of signature analyzer that has been defined by polynomial $\phi(x)=1\oplus z \oplus z^4$ Table 17.2 shows the timing chart for the states of test generator and the signature analyzer.

Table 17.2

k	x1 x2 x3 x4	F	a1 a2 a3 a4
1	1 0 0 0	0	0 0 0 0
2	1 1 0 0	0	0 0 0 0
3	1 1 1 0	0	0 0 0 0
4	1 1 1 1	0	0 0 0 0
5	0 1 1 1	0	0 0 0 0
6	1 0 1 1	0	0 0 0 0
7	0 1 0 1	1	1 0 0 0
8	1 0 1 0	0	1 1 0 0
9	1 1 0 1	0	1 1 1 0
10	0 1 1 0	0	1 1 1 1
11	0 0 1 1	0	0 1 1 1
12	1 0 0 1	0	1 0 1 1
13	0 1 0 0	0	0 1 0 1
14	0 0 1 0	0	1 0 1 0
15	0 0 0 1	0	1 1 0 1

One can see from the chart that the analyzer state remains unchanged for the first six cycles. Only in the seventh cycle when the input of circuit under test is applied with pattern *0101* associated with the constituent of unity $\bar{x}_1 x_2 \bar{x}_3 x_4$, the zero state changes for *1000*. For the successive cycles, the analyzer behaves as an M-sequence generator described by polynomial $\phi(x)=1 \oplus z \oplus z^4$. Signature calculation is greatly simplified because of the fact that the generating polynomials used for the test generator and the signature analyzer are identical. In particular, there is no need of finding the value of shift *l* between the M-sequence segments thereby signature calculation becomes less time-consuming.

With allowance made for the selected compact testing scheme peculiarity, let us consider signature calculation procedure. The first subproblem to be solved, as for the general case, is to find the set of coefficients relating a binary vector $\gamma_1 \gamma_2 \gamma_3 ... \gamma_m$ defined by the function

$$F = x_1^{\gamma_1} x_2^{\gamma_2} x_3^{\gamma_3} ... x_m^{\gamma_m}$$

with the vector *100...0*.

On the basis of $\gamma_1 \gamma_2 \gamma_3 ... \gamma_m$ we costruct the matrix *W* (17.1)

To formalize its construction one may use the recurrence

$$\gamma(k) = \begin{vmatrix} 0 & 1 & 0 & ... & 0 & 0 \\ 0 & 0 & 1 & ... & 0 & 0 \\ ... & ... & ... & ... & ... & ... \\ 0 & 0 & 0 & ... & 0 & 1 \\ \alpha_m & \alpha_1 & \alpha_2 & ... & \alpha_{m-2} & \alpha_{m-1} \end{vmatrix} \gamma(k-1) \qquad (17.5)$$

where $\gamma(k)$ is the row of matrix (17.1), $k = \overline{0, m-1}$, and $\gamma(0) = \gamma_1 \gamma_2 \gamma_3 ... \gamma_m$. By successively applying the relation (17.5), we obtain $\gamma(1) = \gamma_2 \gamma_3 \gamma_4 ... \gamma_m \beta_1$ where

$$\beta_1 = \alpha_m \gamma_1 \oplus \alpha_1 \gamma_2 \oplus \alpha_2 \gamma_3 \oplus ... \oplus \alpha_{m-1} \gamma_m$$

$\gamma(2) = \gamma_3 \gamma_4 \gamma_5 ... \gamma_m \beta_1 \beta_2$ where

$$\beta_2 = \alpha_m \gamma_2 \oplus \alpha_1 \gamma_3 \oplus \alpha_2 \gamma_4 \oplus ... \oplus \alpha_{m-2} \gamma_m \oplus \alpha_{m-1} \gamma_1$$

and etc.

The recurrence (17.5) is the matrix representation of calculating formulas (17.2) where the coefficients $\alpha_i \in \{0,1\}$, $i=\overline{1,m}$ are determined by the generating polynomial $\phi(z)$. Applying the matrix W obtained by (17.5), we can write similar to (17.3) that

$$\begin{vmatrix} 1 \\ 0 \\ 0 \\ \cdots \\ 0 \\ 0 \end{vmatrix} = W \begin{vmatrix} \delta_1(l) \\ \delta_2(l) \\ \delta_3(l) \\ \cdots \\ \delta_{m-1}(l) \\ \delta_m(l) \end{vmatrix}.$$

Let us find coefficients $\{\delta_i(l)\}$, $i=\overline{1,m}$, from the latter expression. Then we obtain

$$\begin{vmatrix} \delta_1(l) \\ \delta_2(l) \\ \delta_3(l) \\ \cdots \\ \delta_{m-1}(l) \\ \delta_m(l) \end{vmatrix} = W^{-1} \begin{vmatrix} 1 \\ 0 \\ 0 \\ \cdots \\ 0 \\ 0 \end{vmatrix}. \quad (17.6)$$

Therefore, the solution of the first subproblem; that is finding coefficients $\{\delta_i(l)\}$, $i=\overline{1,m}$, consists in application of the relation (17.6). As a result the expression for coefficients $\delta_i(l+1)$ assumes the following form

$$\begin{vmatrix} \delta_1(l+1) \\ \delta_2(l+1) \\ \delta_3(l+1) \\ \cdots \\ \delta_{m-1}(l+1) \\ \delta_m(l+1) \end{vmatrix} = W^{-1} \begin{vmatrix} 0 \\ 0 \\ 0 \\ \cdots \\ 0 \\ 1 \end{vmatrix}. \quad (17.7)$$

Next by using the relation

$$\begin{vmatrix} \delta_1(l+1) \\ \delta_2(l+1) \\ \delta_3(l+1) \\ \cdots \\ \delta_m(l+1) \end{vmatrix} - \begin{vmatrix} \alpha_1 & \alpha_2 & \alpha_3 & \cdots & \alpha_{m-1} & \alpha_m \\ \alpha_2 & \alpha_3 & \alpha_4 & \cdots & \alpha_m & 0 \\ \alpha_3 & \alpha_4 & \alpha_5 & \cdots & 0 & 0 \\ \cdots & \cdots & \cdots & \cdots & \cdots & \cdots \\ \alpha_m & 0 & 0 & \cdots & 0 & 0 \end{vmatrix} \begin{vmatrix} a_1(l) \\ a_2(l) \\ a_3(l) \\ \cdots \\ a_m(l) \end{vmatrix},$$

for $a_1(0)a_2(0)a_3(0)\ldots a_m(0)=100\ldots 0$, we obtain the value of signature $S(F)=a_1(l)a_2(l)a_3(l)\ldots a_m(l)$ as

$$S(F) = \begin{vmatrix} \alpha_1 & \alpha_2 & \alpha_3 & \cdots & \alpha_{m-1} & \alpha_m \\ \alpha_2 & \alpha_3 & \alpha_4 & \cdots & \alpha_m & 0 \\ \alpha_3 & \alpha_4 & \alpha_5 & \cdots & 0 & 0 \\ \cdots & \cdots & \cdots & \cdots & \cdots & \cdots \\ \alpha_m & 0 & 0 & \cdots & 0 & 0 \end{vmatrix}^{-1} \begin{vmatrix} \delta_1(l+1) \\ \delta_2(l+1) \\ \delta_3(l+1) \\ \cdots \\ \delta_m(l+1) \end{vmatrix} \qquad (17.8)$$

Substituting the values of coefficients $\{\delta_i(l+1)\}$, $i=\overline{1,m}$, calculated by (17.6), into the latter expression, we can finally obtain

$$S(F) = Q^{-1} W^{-1} \begin{vmatrix} 0 \\ 0 \\ 0 \\ \cdots \\ 0 \\ 1 \end{vmatrix},$$

where W^{-1} is the inverse of matrix W (17.1); Q is determined by coefficients $\alpha_i \in \{0,1\}$, $i=\overline{1,m}$, of the generating polynomial $\phi(z)$ and has the form

$$Q = \begin{vmatrix} \alpha_1 & \alpha_2 & \alpha_3 & \cdots & \alpha_{m-1} & \alpha_m \\ \alpha_2 & \alpha_3 & \alpha_4 & \cdots & \alpha_m & 0 \\ \alpha_3 & \alpha_4 & \alpha_5 & \cdots & 0 & 0 \\ \cdots & \cdots & \cdots & \cdots & \cdots & \cdots \\ \alpha_m & 0 & 0 & \cdots & 0 & 0 \end{vmatrix}.$$

For the example of Table 17.2, the matrix W has the form

$$W = \begin{vmatrix} 1 & 1 & 1 & 1 \\ 0 & 1 & 1 & 1 \\ 1 & 0 & 1 & 1 \\ 0 & 1 & 0 & 1 \end{vmatrix},$$

and the matrix Q has the form

$$Q = \begin{vmatrix} \alpha_1 & \alpha_2 & \alpha_3 & \alpha_4 \\ \alpha_2 & \alpha_3 & \alpha_4 & 0 \\ \alpha_3 & \alpha_4 & 0 & 0 \\ \alpha_4 & 0 & 0 & 0 \end{vmatrix} = \begin{vmatrix} 1 & 0 & 0 & 1 \\ 0 & 0 & 1 & 0 \\ 0 & 1 & 0 & 0 \\ 1 & 0 & 0 & 0 \end{vmatrix}.$$

Hence the value of signature is calculated as follows:

$$S(F) = \begin{vmatrix} 1 & 1 & 1 & 1 \\ 0 & 1 & 1 & 1 \\ 1 & 0 & 1 & 1 \\ 0 & 1 & 0 & 1 \end{vmatrix}^{-1} \begin{vmatrix} 1 & 0 & 0 & 1 \\ 0 & 0 & 1 & 0 \\ 0 & 1 & 0 & 0 \\ 1 & 0 & 0 & 0 \end{vmatrix}^{-1} \begin{vmatrix} 0 \\ 0 \\ 0 \\ 1 \end{vmatrix} =$$

$$= \begin{vmatrix} 1 & 1 & 0 & 0 \\ 1 & 0 & 1 & 0 \\ 0 & 1 & 0 & 1 \\ 1 & 0 & 1 & 1 \end{vmatrix} \begin{vmatrix} 0 & 0 & 0 & 1 \\ 0 & 0 & 1 & 0 \\ 0 & 1 & 0 & 0 \\ 1 & 0 & 0 & 1 \end{vmatrix} \begin{vmatrix} 0 \\ 0 \\ 0 \\ 1 \end{vmatrix} = \begin{vmatrix} 1 \\ 1 \\ 0 \\ 1 \end{vmatrix}.$$

Signature $S(F)$ can be determined similarly for other generating polynomials $\phi(z)$ for function F which is a constituent of unity whose *rank* is $deg\phi(z)$.

17.3. Finding Signatures for Reciprocal Polynomials

As is already known, the signature is a function dependent on many variables and determined by relation (16.1). Among the signature argument least significant are the first test pattern X_1 of test generator, the initial state of analyzer A_0, and the test sequence length L. Using the recommended values of arguments $X_1=100...0$, $A_0=000...0$ and $L=2^m-1$, where m is the number of variables of Boolean function F, we shall examine the signature as a function of three variables having the form

$$S = S\,[\psi(z), \varphi(z), F(x_1 \ldots x_m)] \qquad (17.9)$$

Consider the dependence of S on the form of function $F=F(x_1,\ldots,x_m)$ that describes the circuit for polynomial $\psi(z)$ which is reciprocal of $\phi(z)$, i.e. for $\psi(z) = \phi^{-1}(z)$. In this case, as we shall demonstrate below, the value of S depends on only the form of function F. Let us consider the relation $S=S(F)$ for

$$F = (\beta_1 + x_1^{\alpha_1})(\beta_2 + x_2^{\alpha_2})(\beta_3 + x_3^{\alpha_3}) \ldots (\beta_m + x_m^{\alpha_m}). \qquad (17.10)$$

where $\beta_i \in \{0,1\}$, $\alpha_i \in \{0,1\}$, $i=\overline{1,m}$; $x_i^{\alpha_i} = x_i$, for $\alpha_i=1$, and $x_i^{\alpha_i} = \overline{x_i}$, for $\alpha_i=0$. For the purpose we give some statements.

First of all we shall examine the function of (17.10) for $\beta_1+\beta_2+\beta_3+\ldots+\beta_m=0$. Here the function F is a constituent of unity, i.e. the logical product of all arguments or their negations.

The following theorem holds for the function.

Theorem 17.1. Signature $S(F)=a_1a_2a_3\ldots a_m$ of a Boolean function $F = x_1^{\alpha_1} x_2^{\alpha_2} x_3^{\alpha_3} \ldots x_m^{\alpha_m}$ that agrees with the definition (16.1) for $X_1=100\ldots0$, $A_0=000\ldots0$, $L=2^m-1$, and reciprocal polynomials $\psi(z)$ and $\phi(z)$ whose high degrees meet the equality $deg\phi(z)=deg\psi(z)=m$ may be determined by

$$S(F) = \alpha_m \alpha_{m-1} \alpha_{m-2} \ldots \alpha_1 \qquad (17.11)$$

Proof. In the test mode, the circuit that implements $F = x_1^{\alpha_1} x_2^{\alpha_2} x_3^{\alpha_3} \ldots x_m^{\alpha_m}$ is applied with 2^m-1 binary patterns $x_1x_2x_3\ldots x_m$ from the output of the pseudorandom test sequence (M-sequence) generator described by the primitive polynomial $\psi(z)$. Among the set of 2^m-1 input patterns $x_1x_2x_3\ldots x_m$ there may be all possible nonzero combinations as follows from the property of M-sequence.

Function $F = x_1^{\alpha_1} x_2^{\alpha_2} x_3^{\alpha_3} \ldots x_m^{\alpha_m}$ assumes a one on only one of 2^m-1 binary patterns. The form of the pattern is determined by the solution of $x_1^{\alpha_1} x_2^{\alpha_2} x_3^{\alpha_3} \ldots x_m^{\alpha_m} = 1$ that equals $\alpha_1\alpha_2\alpha_3\ldots\alpha_m$. An exception is the case when

$F=\bar{x}_1\bar{x}_2\bar{x}_3...\bar{x}_m$ for which a zero sequence is produced on the circuit output in the test mode.

Alteration of binary patterns produced by the M-sequence generator is uniquely determined by polynomial $\psi(z)$. The first of them is *100...0* for any $\psi(z)$. The pattern $\alpha_1\alpha_2\alpha_3...\alpha_m \neq 000...0$ for $F=1$ will be produced in $l \in \{1,2,3...,2^m-1\}$ cycles of generator operation where l is the function of generating polynomial $\psi(z)$. Thus we may conclude that storage elements of generator that produced M-sequence $x_1(l),x_2(l),x_3(l),...,x_m(l)=x_1(l),x_1(l-1),x_1(l-2),...,x_1(l-m+1)$ will contain the pattern $\alpha_1\alpha_2\alpha_3...\alpha_m$ in l cycles. In the l th cycle, the signature analyzer is applied with $F=1$ since it is only possible for $x_1x_2x_3...x_m=\alpha_1\alpha_2\alpha_3...\alpha_m$. Similarly, for the zero initial code $A_0=000...0$, the state of signature analyzer elements can be represented by the code $a_1(l)a_2(l)a_3(l)...a_m(l)=100...0$. Since $\phi(z)$ produces an M-sequence which is the reverse of the one produced by $\psi(z)$, we can state that

$$a_1(l) a_2(l) a_3(l) ... a_m(l) = a_1(l) a_1(l+1) a_1(l+2) ... a_1(l+m-1). \quad (17.12)$$

Starting with the l th cycle, the signature analyzer functions as an M-sequence generator for the successive $L+1-l$ cycles. Assuming that the l th cycle is the first for the analyzer and in view of the expression (17.12), we obtain that the signature value is determined by

$$a_1a_2a_3...a_m=a_1(L+1-l)a_1(L+1-l+1)a_1(L+1-l+2)...a_1(L+1-l+m-1).$$

M-sequences produced by reciprocal (inverse) polynomials are mutually inverse sequences. For such sequences, in view of the match between initial codes, the relation

$$x_1(l)=a_1(L+1-l-m+1), l \in \{1,2,3,...,2^m-1\},$$

holds true. From the relation it follows that $a_1a_2a_3...a_m=a_1(L+1-l)a_1(L+1-l+1)a_1(L+1-l+2)...a_1(L+1-l+m-1)=x_1(l-m+1)x_1(l-m+2)x_1(l-m+3)...x_1(l)= \alpha_m\alpha_{m-1}\alpha_{m-2}...\alpha_1$.

According to the above theorem, signature $S(F)$ is *1100* at the output of the circuit described by $F(x_1,x_2,x_3,x_4)=\bar{x}_1\bar{x}_2x_3x_4$.

Reciprocal polynomials $\psi(z)$ and $\phi(z)$ whose high degree is 4 are used as generating polynomials for the test generator and signature analyzer. An example of such polynomials is $\psi(z)=1 \oplus z \oplus z^4$ and $\phi(z)=1 \oplus z^3 \oplus z^4$. The initial code of analyzer is $A_0=0000$, the first test pattern $X_1=1000$ and the test sequence length is $L=15$.

For the case of F being a conjunction which *rank* is less that m, we deduce the theorem and the following two lemmas required to prove it.

Lemma 17.1. Conjunction of *rank* $n<m$ may be represented as a logical sum of constituents of unity of *rank* m whose number is 2^{m-n}.

Lemma 17.2. The logical sum of the constituents of unity is their modulo-2 sum.

Theorem 17.2. Signature $S(F)=a_1 a_2 a_3 ... a_m$ of a Boolean function $F = (\beta_1 + x_1^{\alpha_1})(\beta_2 + x_2^{\alpha_2})(\beta_3 + x_3^{\alpha_3}) ... (\beta_m + x_m^{\alpha_m})$, where $\beta_i \in \{0,1\}$, $a_i \in \{0,1\}$, $i=\overline{1,m}$; $x_i^{\alpha_i} = x_i$, for $\alpha_i=1$, and $x_i^{\alpha_i} = \overline{x_i}$, for $\alpha_i=0$, equals to

a) $000...0\,\alpha_{m-j+1}0...0 = 000...010...0$ for $\beta_j=1$, $j \in \{1,2,3,...,m\}$ and $\beta_1+\beta_2+\beta_3+...+\beta_{j-1}+\beta_{j+1}+...+\beta_m = 0$

b) $000...0$ for other cases.

Proof. For the case a), the function F is an elementary conjunction of the form $x_1^{\alpha_1} x_2^{\alpha_2} x_3^{\alpha_3} ... x_{j-1}^{\alpha_{j-1}} x_{j+1}^{\alpha_{j+1}} ... x_m^{\alpha_m}$, which, according to Lemma 17.1, is transformed to the sum of constituents of unity, i.e

$$F = x_1^{\alpha_1} x_2^{\alpha_2} x_3^{\alpha_3} ... x_{j-1}^{\alpha_{j-1}} x_{j+1}^{\alpha_{j+1}} ... x_m^{\alpha_m} =$$
$$x_1^{\alpha_1} x_2^{\alpha_2} x_3^{\alpha_3} ... x_{j-1}^{\alpha_{j-1}} x_j x_{j+1}^{\alpha_{j+1}} ... x_m^{\alpha_m} + x_1^{\alpha_1} x_2^{\alpha_2} x_3^{\alpha_3} ... x_{j-1}^{\alpha_{j-1}} \overline{x_j} x_{j+1}^{\alpha_{j+1}} ... x_m^{\alpha_m}.$$

In view of Lemma 17.2, the function F is transformed to $F = F_1 \oplus F_2$, i.e

$$F = x_1^{\alpha_1} x_2^{\alpha_2} x_3^{\alpha_3} ... x_{j-1}^{\alpha_{j-1}} x_j x_{j+1}^{\alpha_{j+1}} ... x_m^{\alpha_m} \oplus$$
$$x_1^{\alpha_1} x_2^{\alpha_2} x_3^{\alpha_3} ... x_{j-1}^{\alpha_{j-1}} \overline{x_j} x_{j+1}^{\alpha_{j+1}} ... x_m^{\alpha_m}.$$

By using the property of 16.1, we obtain $S(F) = S(F_1) \oplus S(F_2)$ where F_1 and F_2 are the constituents of unity for which

$$S(F_1) = \alpha_m \alpha_{m-1} \alpha_{m-2} ... \alpha_{m-j+2}\,1\,\alpha_{m-j}...\alpha_1$$

and

$$S(F_2) = \alpha_m \alpha_{m-1} \alpha_{m-2} \cdots \alpha_{m-j+2} 0 \alpha_{m-j} \cdots \alpha_1.$$

Then we shall finally have

$$S(F) = \alpha_m \alpha_{m-1} \alpha_{m-2} \cdots \alpha_{m-j+2} 1 \alpha_{m-j} \cdots \alpha_1 \oplus$$
$$\alpha_m \alpha_{m-1} \alpha_{m-2} \cdots \alpha_{m-j+2} 0 \alpha_{m-j} \cdots \alpha_1 = 000\ldots010\ldots0,$$

where the unity symbol is associated with the $(m-j+1)$ th position.

For the case b), the function is an elementary conjunction of *rank* $n \leq m-2$, therefore according to Lemma 17.1, it can be written as a modulo-2 sum of 2^{m-n} constituents of unity. According to the property (16.1), the signature of F is determined as the modulo-2 sum of signatures with identical symbols or an even number of ones in the like bits. Thus we finally obtain $S(F)=000\ldots0$.

For the above theorem the following corollary holds true.

Corollary 17.1. Signature $S(F)$ of function $F=0$ and $F=1$ is determined by $S(F)=000\ldots0$.

Using the outcome of the proved theorem and its corollaries, we shall prove the following theorem.

Theorem 17.3. Signature $S(\overline{F})$ of F negation is the signature $S(F)$ of the same function F, i.e.

$$S(\overline{F}) = S(F). \qquad (17.13)$$

Proof. By definition of logical add function, we may write $F + \overline{F} = 1$. According to Lemma 17.1 the latter relation is transformed to $F + \overline{F} = F \oplus \overline{F} \oplus F\overline{F} = F \oplus \overline{F} = 1$. Then, by using the property 16.1, we obtain $S(F=1)=S(F) \oplus S(\overline{F})$. Hence, in view of corollary 17.1, it follows that $S(F)=S(\overline{F})$.

Examine the relation $S=S(F)$ for an elementary disjunction

$$F = \beta_1 x_1^{\alpha_1} + \beta_2 x_2^{\alpha_2} + \beta_3 x_3^{\alpha_3} + \ldots + \beta_m x_m^{\alpha_m}, \qquad (17.14)$$

where $\beta_i \in \{0,1\}$, $\alpha_i \in \{0,1\}$, $i=\overline{1,m}$; $x_i^{\alpha_i} = x_i$, for $\alpha_i=1$, and $x_i^{\alpha_i} = \overline{x_i}$, for $\alpha_i=0$,

For the purpose let us prove the theorem.

Theorem 17.4. Signature $S(F)=a_1a_2a_3...a_m$ of an elementary disjunction $F = \beta_1 x_1^{\alpha 1} + \beta_2 x_2^{\alpha 2} + \beta_3 x_3^{\alpha 3} + ... + \beta_m x_m^{\alpha m}$ equals to

a) $\overline{\alpha}_m \overline{\alpha}_{m-1} \overline{\alpha}_{m-2} ... \overline{\alpha}_1$ for $\beta_1 \beta_2 \beta_3 ... \beta_m = 1$;

b) $000...0\ \alpha_{m-j+1}0...0 = 000...010...0$ for $\beta_j = 0$, where $j \in \{1,2,3,...,m\}$ and $\beta_1 \beta_2 \beta_3 ... \beta_{j-1} \beta_{j+1} ... \beta_m = 1$;

c) $000...0$ for other cases.

Proof. Let us represent an initial disjunction

$$F = \beta_1 x_1^{\alpha 1} + \beta_2 x_2^{\alpha 2} + \beta_3 x_3^{\alpha 3} + ... + \beta_m x_m^{\alpha m}$$

as conjunction

$$F = \overline{\beta_1 x_1^{\alpha 1}\ \beta_2 x_2^{\alpha 2}\ \beta_3 x_3^{\alpha 3}\ ...\ \beta_m x_m^{\alpha m}}.$$

Taking account of the output of Theorem 17.3, we determine the value of signature for F. For the purpose we will first look at the case a) for which

$$\overline{F} = \beta_1 x_1^{\overline{\alpha} 1} \beta_2 x_2^{\overline{\alpha} 2} \beta_3 x_3^{\overline{\alpha} 3} ... \beta_m x_m^{\overline{\alpha} m}.$$

Then by theorem 17.1 we obtain $S(F) = \overline{\alpha}_m \overline{\alpha}_{m-1} \overline{\alpha}_{m-2} ... \overline{\alpha}_1$. Cases b) and c) can be proved in a similar way.

Using the outcome of the above theorems and their corollaries, we may formalize calculation of signature values. Thus for a combinational circuit described by F, where F is represented, for example, in the full disjunctive normal form (FDNF), we can readily find its signature $S(F)$ by theorem 17.1.

By way of an example, let us consider a circuit described by the function $F = \overline{x}_1 x_2 x_3 x_4 + x_1 \overline{x}_2 x_3 \overline{x}_4 + \overline{x}_1 x_2 x_3 \overline{x}_4$. In the general form the function can be written as the sum $F = F_1 + F_2 + F_3$ where $F_1 = \overline{x}_1 x_2 x_3 x_4$, $F_2 = x_1 \overline{x}_2 x_3 \overline{x}_4$ and $F_3 = \overline{x}_1 x_2 x_3 \overline{x}_4$. On the basis of theorem 17.1 we may demonstrate that $S(F_1) = 1110$ since $F_1 = \overline{x}_1 x_2 x_3 x_4 = x_1^0 x_2^1 x_3^1 x_4^1$. In the same way we may write the values $S(F_2) = 0101$ and $S(F_3) = 0110$. Then, in view of $F = F_1 + F_2 + F_3 = F_1 \oplus F_2 \oplus F_3$ and by using the corollary to Theorem 17,1, we obtain $S(F)$ as a bitwise modulo-2 sum of signatures $S(F_1)$, $S(F_2)$ and $S(F_3)$. As a result $S(F) = 1110 \oplus 0101 \oplus 0110 = 1101$. The value of signature $S(F) = 1101$ obtained by calculation matches the one obtained experimentally.

The examined function F can be represented in the full disjunctive normal form or in any arbitrary form. In this case the following algorithm can be used:

1. Boolean function F written in an arbitrary form is transformed to the full disjunctive normal form.

2. According to Theorem 17.1 signatures for all constituents of unity involved in representation of F in FDNF are found.

3. The resulting signature of F is determined as a bitwise modulo-2 sum of signatures for the constituents of unity.

Using the above algorithm as well as Theorems 17.1, 17.2 and their corollaries, it is possible to analytically determine the values of signatures thereby excluding the labor-consuming simulation procedure.

A disadvantage of an analytical way of signature calculation is the rigidity of structure used for compact test implementation which may be attributed to the necessity of using reciprocol polynomials.

The most likely field of application for the above results is the analysis of signature-testability of digital circuit based on them. Consider signature-testability analysis techniques for specific digital circuit design.

CHAPTER 18

SIGNATURE TESTABILITY OF VLSI CHIPS

18.1. Signature Testability of a Circuit

For a single stuck-at fault, which is the most widespread model of a physical fault, we can formulate the signature testability concept as a theorem.

Theorem 18.1. The condition for signature testability of a single stuck-at fault $\chi \in \{\equiv 0, \equiv 1\}$ occurred at a node $g=g(x_1,...,x_m)$ of a combinational circuit which implements the Boolean function $F(x_1,...,x_m)=gG_1+\bar{g}G_2+G_3$ where $G_1=G_1(x_1,...,x_m)$, $G_2=G_2(x_1,...,x_m)$ and $G_3=G_3(x_1,...,x_m)$ with $G_1+G_2 \neq 0$ and $G_3 \neq 1$, is the following set of inequalities:

$$S(gG_1\overline{G}_3) \oplus S(gG_2\overline{G}_3) \neq 0,$$
$$S(\bar{g}G_1\overline{G}_3) \oplus S(\bar{g}G_2\overline{G}_3) \neq 0. \quad (18.1)$$

Proof. Define the reference signature value for function $F=gG_1+\bar{g}G_2+G_3$ by bringing the boolean expression $gG_1+\bar{g}G_2+G_3$ into the form $gG_1 \oplus \bar{g}G_2 \oplus G_3 \oplus gG_1G_2 \oplus \bar{g}G_2G_3$. Then by using the property (16.1) of the boolean function signature we obtain that $S_e(F)=S_e(gG_1+\bar{g}G_2+G_3)=S_e(gG_1 \oplus \bar{g}G_2 \oplus G_3 \oplus gG_1G_2 \oplus \bar{g}G_2G_3)=S(gG_1) \oplus S(\bar{g}G_2) \oplus S(G_3) \oplus S(gG_1G_3) \oplus S(\bar{g}G_2G_3)$.

If a single stuck-at fault $g \equiv 0$ occurs the boolean function expression

$$F = F(x_1,...,x_m) = gG_1 + \bar{g}G_2 + G_3 \quad (18.2)$$

will take the form $F'=G_2+G_3$, hence $S_p(F')=S_p[F(g\equiv 0)]=S_p(G_2+G_3)=S_p(G_2 \oplus G_3 \oplus G_2G_3)=S(G_2) \oplus S(G_3) \oplus S(G_2G_3)$. The condition for signature testability of fault $g \equiv 0$ as defined by 16.4 is the difference between $S_p(F')$ and $S_e(F)$, i.e. the fulfilment of condition (16.8). Substituting $S_p(F')$ and $S_e(F)$ into this inequality, we obtain $S(gG_1) \oplus S(\bar{g}G_2) \oplus S(G_3) \oplus S(gG_1G_3) \oplus S(\bar{g}G_2G_3) \oplus S(G_2) \oplus S(G_3) \oplus S(G_2G_3) \neq 0$. By applying transformations

$S(gG_1) \oplus S(gG_1G_3) = S(gG_1 \oplus gG_1G_3) = S(gG_1(1+G_3)) = S(gG_1\overline{G}_3)$ we shall finally get

$$S(gG_1\overline{G}_3) \oplus S(gG_2\overline{G}_3) \neq 0.$$

The latter inequality is the condition for signature testability of $g \equiv 0$ at the node $g = g(x_1,...,x_m)$ of the combinational circuit implementing the Boolean function (18.2). Similarly, we can demonstrate that inequality

$$S(\overline{g}G_1\overline{G}_3) \oplus S(\overline{g}G_2\overline{G}_3) \neq 0.$$

is the condition for testability of $g \equiv 1$.

Thus the system of inequalities

$$S(gG_1\overline{G}_3) \oplus S(gG_2\overline{G}_3) \neq 0,$$
$$S(\overline{g}G_1\overline{G}_3) \oplus S(\overline{g}G_2\overline{G}_3) \neq 0.$$

is a condition for signature testability of single stuck-at faults at a node of a combinational circuit described by $g = g(x_1,...,x_m)$.

We can similarly define the signature testability for another class of faults described by the inverse fault model. For the purpose we shall prove the following theorem.

Theorem 18.2. The condition for signature testability of an inverse fault $(g \equiv \overline{g})$ occurred at a node $g = g(x_1,...,x_m)$ of a single-input combinational circuit that implements the Boolean function $F(x_1,...,x_m) = gG_1 + \overline{g}G_2 + G_3$ where $G_1 = G_1(x_1,...,x_m)$, $G_2 = G_2(x_1,...,x_m)$ and $G_3 = G_3(x_1,...,x_m)$, with $G_1 + G_2 \neq 0$ and $G_3 \neq 1$ is the fullfilment of inequality:

$$S(G_1\overline{G}_3) \oplus S(G_2\overline{G}_3) \neq 0. \quad (18.3)$$

Proof. As the theorem 18.1 proved, the reference signature value for $F = gG_1 + \overline{g}G_2 + G_3$ can be defined as $S_e(F) = S_e(gG_1 + \overline{g}G_2 + G_3) = S(gG_1) \oplus S(\overline{g}G_2) \oplus S(G_3) \oplus S(gG_1G_3) \oplus S(g G_2G_3)$. If an inverse fault $g \equiv \overline{g}$ occurs, the Boolean function expression takes the form $F' = \overline{g}G_1 + gG_2 + G_3$ and the value of actual signature is $S_p(F') = S(F = \overline{g}G_1 + gG_2 + G_3) = S(\overline{g}G_1) \oplus S(gG_2) \oplus S(G_3) \oplus S(\overline{g}G_1G_3) \oplus S(gG_2G_3)$. As defined by 16.4, the condition for the signature testability of an inverse fault is the diference between $S_p(F')$ and $S_e(F)$ i.e. the fullfilment of inequality (16.8). Substituting $S_p(F')$ and $S_e(F)$ into the inequality, we obtain $S(gG_1) \oplus S(\bar{g}G_2) \oplus S(G_3) \oplus S(gG_1G_3) \oplus S(\bar{g}G_2G_3) \oplus S(\bar{g}G_1) \oplus S(gG_2) \oplus S(G_3) \oplus S(\bar{g}G_1G_3) \oplus S(gG_2G_3) \neq 0$. Rearranging the latter inequality, we shall finally get the condition for signature testability of an inverse fault as an inequality $S(G_1\bar{G}_3) \oplus S(G_2\bar{G}_3) \neq 0$.

Consider now the condition for signature testability of a bridging fault. For the purpose we examine the case of a bridging between inputs where variables x_i and x_j, $i,j \in \{1,2,...,m\}$, $i \neq j$ are produced. Let us prove the following theorem.

Theorem 18.3. The condition for signature testability of a bridging fault between two inputs to a single-output combinational circuit where x_i and x_j, $i,j \in \{1,2,...,m\}$, $i \neq j$, are produced and the combinational circuit is described by the boolean expression $F=F(x_1,...,x_m)=x_ix_jG_1+ x_i\bar{x}_jG_2+ \bar{x}_ix_jG_3+ \bar{x}_i\bar{x}_jG_4+G_5$, is the fulfilment of the following inequality

$$S(\bar{x}_iG_4\bar{G}_5) \oplus S(\bar{x}_jG_4\bar{G}_5) \oplus S(x_i\bar{x}_jG_2\bar{G}_5) \oplus S(\bar{x}_ix_jG_3\bar{G}_5) \neq 0 \quad (18.4)$$

for positive logic or

$$S(x_iG_4\bar{G}_5) \oplus S(x_jG_4\bar{G}_5) \oplus S(x_i\bar{x}_jG_2\bar{G}_5) \oplus S(\bar{x}_ix_jG_3\bar{G}_5) \neq 0 \quad (18.5)$$

for negative logic.

Proof. The expression $F=F(x_1,...,x_m)$, which describes the behavior of a single-output combinational circuit, may be presented as $F=F(x_1,...,x_m) = x_ix_jG_1+x_i\bar{x}_jG_2+ \bar{x}_ix_jG_3+ \bar{x}_i\bar{x}_jG_4 +G_5$ where $G_1+ G_2+ G_3+ G_4 \neq 0$ and $G_5 \neq 1$, with $\partial G_\nu/\partial x_i = \partial G_\nu/\partial x_j = 0$, $\nu = \overline{1,5}$. For the given Boolean function, the reference signature value appears as

$$S_e(F) = S(x_ix_jG_1) \oplus S(x_i\bar{x}_jG_2) \oplus S(\bar{x}_ix_jG_3) \oplus S(\bar{x}_i\bar{x}_jG_4) \oplus S(G_5) \oplus \\ S(x_ix_jG_1G_5) \oplus S(x_i\bar{x}_jG_2G_5) \oplus S(\bar{x}_ix_jG_3G_5) \oplus S(\bar{x}_i\bar{x}_jG_4G_5) \quad (18.6)$$

The effect of a short fault depends on the technology of the circuit. For the circuit with positive logic, a short fault is equivalent to inserting a dummy two-input AND gate. Thus conjunction x_ix_j will be generated at nodes x_i and x_j. With due regard for the condition, the Boolean function appears as $F' = x_ix_jG_1 + \overline{x_ix_j}G_4 + G_5$ and the actual signature

$$S_p(F') = S_p(x_ix_jG_1 + \overline{x_ix_j}G_4 + G_5) =$$

$$= S(x_i x_j G_1) \oplus S(\overline{x_i x_j} G_4) \oplus S(G_5) \oplus S(x_i x_j G_1 G_5) \oplus S(\overline{x_i x_j} G_4 G_5).$$

From the definition 16.4 we obtain that the condition for signature testability of a fault under test is the inequality

$$S(x_i x_j G_1) \oplus S(x_i \overline{x}_j G_2) \oplus S(\overline{x}_i x_j G_3) \oplus S(\overline{x}_i \overline{x}_j G_4) \oplus S(G_5) \oplus$$
$$S(x_i x_j G_1 G_5) \oplus S(x_i \overline{x}_j G_2 G_5) \oplus S(\overline{x}_i x_j G_3 G_5) \oplus S(\overline{x}_i \overline{x}_j G_4 G_5) \oplus$$
$$S(x_i x_j G_1) \oplus S(\overline{x_i x_j} G_4) \oplus S(G_5) \oplus S(x_i x_j G_1 G_5) \oplus S(\overline{x_i x_j} G_1 G_5) \neq 0.$$

Having been transformed, the inequality appears as

$$S(\overline{x}_i G_4 \overline{G}_5) \oplus S(\overline{x}_j G_4 \overline{G}_5) \oplus S(x_i \overline{x}_j G_2 \overline{G}_5) \oplus S(\overline{x}_i x_j G_3 \overline{G}_5) \neq 0.$$

For the gates with negative logic a short fault is equivalent to inserting a dummy two-input OR gate. Then the Boolean function takes on the form $F' = (x_i + x_j)G_1 + \overline{(x_i + x_j)}G_4 + G_5$, and its associated signature will be defined by the expression

$$S_p(F') = S(x_i G_1) \oplus S(x_j G_1) \oplus S(x_i x_j G_1) \oplus S(\overline{x}_i \overline{x}_j G_4) \oplus S(G_5) \oplus$$
$$S(x_i G_1 G_5) \oplus S(x_j G_1 G_5) \oplus S(x_i x_j G_1 G_5) \oplus S(\overline{x}_i \overline{x}_j G_1 G_5).$$

Having performed the same transformations as for the positive logic, we obtain that the inequality

$$S(x_i G_4 \overline{G}_5) \oplus S(x_j G_4 \overline{G}_5) \oplus S(x_i \overline{x}_j G_2 \overline{G}_5) \oplus S(\overline{x}_i x_j G_3 \overline{G}_5) \neq 0.$$

is the condition for signature testability of a short fault.

The condition for signature testability of any other short-type fault can be determined in the same way. Similar to (18.4) and (18.5) the condition may be represented as an inequality. This involves the necessity of computing the signatures for Boolean functions. The like problem arises at testing other fault types. Below it will be examined for various circuit structures against some fault types.

18.2. Signature Testability of Faults at Primary Nodes of Combinational Circuit

Preparatory to analyzing signature testability of two-stage CC, we will first examine signature testability of primary nodes in an arbitrary CC applied with input variables $x_1 x_2 x_3,...,x_m$ where the values of function $F(x_1...x_n)$ are generated. For the purpose we shall distinguish two subsets M_1 and M_2, where M_1 is the subset of single stuck-at faults at input nodes to CC and M_2 is the subset of faults at its output node, in the set M of single stuck-at faults of an arbitrary CC.

Let us first examine the faults of M_1 for signature testability. For the purpose we shall put the Boolean expression F, which describes the behavior of an arbitrary CC in the disjunctive normal form and write it as

$$F = x_i G_1 + \bar{x}_i G_2 + G_3, \quad (18.7)$$

where $\partial G_1 / \partial x_i = \partial G_2 / \partial x_i = \partial G_3 / \partial x_i = 0$, $G_1 + G_2 \neq 0$, $G_3 \neq 1$; and x_i is a variable generated at the i th input node of CC.

Now we prove the following theorem for single stuck-at faults at the i th node of CC.

Theorem 18.4. The signature $S_p[F(x_i = \chi)]$, $\chi \in [\equiv 0, \equiv 1]$, $i \in [1,2,3,...,n]$ for an output node of CC desribed by the Boolean function $F(x_1,...,x_n)$ and having the stuck-at fault at the i th input node assumes the value of $000...0\, a_{n-i+1} 0...0$, $a_{n-i+1} \in \{0,1\}$.

Proof. Consider first the stuck-at fault $x_i \equiv 0$. Then function F expressed as (18.7) takes on the form $F = G_2 + G_3$ where G_2 and G_3 is the disjunction of conjunctions whose rank is less than $n-1$ as follows from the relation $\partial G_2 / \partial x_i = \partial G_3 / \partial x_i = 0$. The condition results from the fact that $G_1, G_2, G_1 G_2$ are disjunctions of conjunctions without cofactor x_i. By theorem 17.2 the signature of conjunction whose rank is $n-1$ will appear as $000...1...0$ where the position of one in a single signature bit is determined by index i of the missing factor x_i. For conjunctions whose rank is less than $n-1$, the signature becomes $000...0$. Considering that the rank of product of conjunctions without factor x_i will not exceed $n-1$ and the linearity property of modulo-2 addition, we finally obtain

$$S_p[F(x_i \equiv 0)] = a_1 a_2 a_3 ... a_{n-i+1} ... a_n = 000 .. a_{n-i+1} ... 0$$

where $a_{n-i+1} \in \{0,1\}$.

For the stuck-at fault $x_i \equiv 1$, F will take on the form $F = G_1 + G_3$ where G_1 the same as G_2 and G_3, is the disjunction of conjunctions whose rank is less than $n-1$. Therefore, as in the case of $F = G_2 + G_3$ we obtain

$$S_p[F(x_i \equiv 1)] = 000..a_{n-i+1}...0.$$

The key outcome of the proved theorem is the fact that any single stuck-at fault occurred at the i th node of CC produces the signature of the form $S_p[F(x_i \equiv \chi)] = 000...a_{n-i+1}...0$ where $a_{n-i+1} \in \{0,1\}$. Based on the actual signature value obtained at the output node of CC with a fault from set M_1 we can formulate the condition for its detection. A fault $x_i = \chi$ of set M_1 will be detectable by signature analysis if the condition $S_p[F(x_i \neq \chi)] \neq S_e[F]$ is met. The inequality holds for the case when $S_e[F] \neq 000...a_{n-i+1}...0$. Therefore the M_1 faults are signature testable when the condition

$$S_e[F] \neq 000...a_{n-i+1}...0, \quad a_{n-i+1} \in [0,1], \quad i \in [1,2,3,...,n] \quad (18.8)$$

is met.

Let us examine the faults of M_2 for testability. For the purpose we shall give the theorem based on the corollary 17.1.

Theorem 18.5. Signature $S_p[F(x_i \neq \chi)]$, $\chi \in \{\equiv 0, \equiv 1\}$ on an output node described by the Boolean function $F(x_1,...,x_n)$ in a combinational circuit with a stuck-at fault $F = \chi$ on the node assumes the value $000...0$.

Based on the theorem, we can deduce that any fault of M_2 will be signature testable if the inequality

$$S_e[F] \neq 000...0. \quad (18.9)$$

is met.

Hence it appears that single stuck-at faults on the primary nodes in a CC implementing the Boolean function $F(x_1,...x_n)$ are signature testable when relation (18.8) enclosing the condition of (18.9) is met.

18.3. Signature Testability of Two-Level Combinational Circuits

Examine single stuck-at faults of a two-level combinational circuit employing multiple-input AND gates with inversing inputs and an OR gate for signature testability. The combinational circuit is described by the Boolean function $F(x_1,...,x_m)$ represented in the disjunctive normal form. We shall represent the entire set of stuck-at faults M as four subsets M_1, M_2, M_3 and M_4 where M_1 and M_2 are the subsets of faults on the primary nodes of CC; M_3 is the subset of faults on the input node of AND gates and M_4 is the subset of faults on their output nodes.

The condition for signature testability of faults on the primary nodes of a two-level CC, as in the case of an arbitrary CC, is the inequality (18.8). To examine fault subsets M_3 and M_4 for testability, we shall represent the original Boolean function $F(x_1,...,x_m)$ as

$$F = f_j + G_4 = x_i^* f_j^* + G_4, \quad j \in [1,2,3,...,r\,], \quad i \in [1,2,3,...,n\,]. \quad (18.10)$$

where $G_4 = f_1 + f_2 + ... + f_{j-1} + f_{j+1} ... f_r$; x_i^* is the input variable of the j th AND gate; f_j^* is an elementary conjunction for which condition $\partial f_j^* / \partial x_i = 0$ is met.

We will first analyse the fault subset M_3. Prove the following theorem.

Theorem 18.6. A single stuck-at fault $\chi \in [\equiv 0, \equiv 1\,]$ on the input node of the j th AND gate in a two level CC described by the Boolean function $F = f_j + G_4 = x_i^* f_j^* + G_4$, is signature testable when the inequalities

$$S(f_j) \oplus S(f_j G_4) \neq 0,$$
$$S(\overline{x_i^* f_j^*}) \oplus S(\overline{x_i^* f_j^*}\, G_4) \neq 0,$$

where x_i^* is the variable generated at an input to the j th AND gate are met.

<u>Proof.</u> Consider the fault $x_i^* \equiv 0$. For the case function $F(x_1,...,x_n)$ takes on the form $F = G_4$. Then by definition 16.4 we obtain that $S_p[F(x_i^* \equiv 0)] = S(G_4)$. Otherwise for $x_i^* \equiv 1$ we shall have $F = f_j^* + G_4$ and whence $S_p[F(x_i^* \equiv 1)] = S(f_j^*) \oplus S(G_4) \oplus S(f_j^* G_4)$. The discussed faults $x_i^* \equiv 0$. and

$x_i^* \equiv 1$ of are signature testable when the actual signature differs from the reference one, i.e. when inequality $S_p[F(x_i^*=\chi)] \neq S_e(F)$ where $S_e[F] = S(x_i^*f_j^*) \oplus S(G_4) \oplus S(x_i^*f_j^*G_4) = S(f_j) \oplus S(G_4) \oplus S(f_jG_4)$ is met. Substituting the values of $S_e[F]$ and $S_p[f(x_i^*=\chi)]$ into the above inequality, we obtain the system

$$S(f_j) \oplus S(f_jG_4) \neq 0, \quad (18.11)$$

$$S(\overline{x_i^*f_j^*}) \oplus S(\overline{x_i^*f_j^*}G_4) \neq 0, \quad (18.12)$$

where testability of fault $x_i^* \equiv 0$ is defined by (18.11), and that of $x_i^* \equiv 1$ is defined by (18.12).

Considering the fault subset M_4, it is worth noting that the condition for detecting a single stuck-at fault $f_j \equiv 0$ on the jth gate of the two-level CC is the inequality (18.11) as follows from the equivalent effect of fault $\equiv 0$ on the inputs and output of AND gate. It is readily proved that the testability condition for $f_j \equiv 1$ is defined by inequality (18.9).

Hence the entire set M of single stuck-at faults in a two-level CC implementing the Boolean function $F(x_1,...,x_n)=f_1+f_2+...+f_r$ represented in DNF is signature testable under the following conditions:

$$S_e[F] \neq 000..a_{n-1+i}...0; \quad a_{n-i+1} \in [0,1], \quad i \in [1,2,3,...n]$$
$$S(f_j) \oplus S(f_jG_4) \neq 0; G_4 = f_1 + f_2 + ... + f_{j-1} + f_{j+1} + ... + f_r, \quad (18.13)$$
$$S(\overline{x_i^*f_j^*}) \oplus S(\overline{x_i^*f_j^*}G_4) \neq 0; \quad x_i^*f_j^* = f_j,$$

where the output variable $f_j \in \{0,1\}$ of the AND gate is an elementary conjunction of input variables for CC and x_i^* is the input variable for AND gate, which assumes the value of either x_i or $\overline{x_i}$.

For the Boolean function F that has been represented in a canonical form the second inequality in (18.13) is transformed into a simpler form $S(f_j) \neq 0$. The third inequality can be simplified in a similar way. Thus for $\overline{x_i^*f_j^*}G_4 = 0$ we deduce that $S(\overline{x_i^*f_j^*})$ must be other than zero and for $\overline{x_i^*f_j^*}G_4 \neq 0$ the situation may arise when the fault occurred causes no change in the truth table of the

Boolean function $F = x_i^* f_j + G_4$. This is possible for redundant implementation of the Boolean function.

Eliminating the faults that change the original Boolean function implementation from M, we obtain the condition for signature testability of DNF implementing CC as

$$S [F] \neq 000..a_{n-i+1}...0; \quad a_{n-i+1} \in [0,1], \quad i \in [1,2,3,...,n],$$
$$S (f_j) \neq 000..a_{n-l+1}...0; \quad a_{n-l+1} \in [0,1], \quad l \in [1,2,3,...,n], \quad (18.14)$$

where the second inequality combines conditions $S(f_j) \neq 0$ and $S(x_i^* f_j^*) \neq 0$ $(x_i^* f_j^* = f_j)$.. Consider signature testability of a two-level CC that implements the Boolean function

$$F(x_1 x_2 x_3 x_4) = x_1 \overline{x_2} x_3 \overline{x_4} + x_1 x_2 \overline{x_3} \, \overline{x_4} + \overline{x_1} x_2 \overline{x_3} x_4..$$

Using the theorem 17.1 we obtain that

$$S(f_1) = 0101, \quad S(f_2) = 0011, \quad S(f_3) = 1010$$

and $S(F)=1100$. Analysing the obtained signature values against system (18.14), we can conclude that any fault in M for a two-level CC which inplements the original function is signature testable. In fact when any single stuck-at fault occurs the actual signature of the output node in CC will differ from the reference value $S_e[F]=1100$, which is to say that they are signature testable.

Hence by analysing the values of reference signatures for a two-level CC, we may estimate their signature testability without evaluating the signatures for single stuck-at faults thereby simplifying circuit analysis.

18.4. Analytic Approach to Evaluating Signatures and Signature Testability of faults by PSA.

Consider the probability for a $\equiv 0$ or $\equiv 1$ fault to occur at node g of a digital circuit. For the purpose we can represent the function f_i for node g as

Consider the probability for a $\equiv 0$ or $\equiv 1$ fault to occur at node g of a digital circuit. For the purpose we can represent the function f_i for node g as

$$f_i = gG_1^i + \bar{g}G_2^i + G_3^i, \quad i = \overline{1,m} \quad (18.15)$$

where

$$G_1^i = G_1^i(x_1, x_2, \ldots, x_n), \quad G_2^i = G_2^i(x_1, x_2, \ldots, x_n)$$

and

$$G_3^i = G_3^i(x_1, x_2, \ldots, x_n),$$

with $G_1^i + G_2^i \neq 0$ and $G_3^i \neq 1$; $g=g(x_1, x_2, \ldots, x_n)$ is the function describing an arbitrary node in the circuit.

Consider the procedure of testing a digital circuit by a PSA with internal modulo-2 adders. Its behavior is described by the matrix set of equations and the block-diagram is shown in Fig.9.3.

Since modulo-2 addition has the linearity property, the signature produced of sequences z_i described by polynomials $q_i(x)$, $i=\overline{1,m}$, equals that obtained on the similar SSA design at compression of a sequence described by polynomial

$$Q(x) = \sum_{j=1}^{m} q_j(x) x^{j-1} \quad (18.16)$$

It is evident that the value of signature obtained for the PSA by compressing z_i data on its i th line can be found by the formula

$$S_{PSAi} = V^{i-1} S_{SSAi} \quad (18.17)$$

where S_{SSAi} is the signature obtained at z_i data compression on the SSA whose feedback structure is identical to that of PSA with internal adders. The behavior of such SSA is described by the same matrix equations (9.3) as the PSA in question, but the input vector appears as $Z(k)=(z(k),0,0,\ldots,0)'$. Hence the signature for the PSA in question can be obtained by the formula

$$S_{PSA} = \sum_{i=1}^{m} {}^{\oplus} V^{i-1} S_{SSAi} \quad (18.18)$$

We will first consider generation of signature by the PSA with data being "compressed" on only two outputs from a digital circuit implementing f_i and f_j. For the sake of definiteness, we assume $i<j$. Define the reference signature for functions $f_i = gG_1^i + \bar{g}G_2^i + G_3^i$ and $f_j = gG_1^j + \bar{g}G_2^j + G_3^j$. It is evident that these expressions can be represented as

$$f_i = gG_1^i \oplus \bar{g}G_2^i \oplus G_3^i \oplus gG_1^i G_3^i \oplus \bar{g}G_2^i G_3^i,$$
$$f_j = gG_1^j \oplus \bar{g}G_2^j \oplus G_3^j \oplus gG_1^j G_3^j \oplus \bar{g}G_2^j G_3^j.$$

When a digital circuit is tested exhaustively the resulting signature depends on the kind of test pattern generator (the count and order of input binary patterns applied), characteristic polynomial and the initial state of PSA as well as the structure of connecting analyzer inputs to outputs of the circuit under test and the function implemented by the circuit. With the test generator, test experiment sheme, polynomial and PSA original state being specified, the resulting signature is the function of only Boolean expression F.

By formula (18.18) the PSA signature resulting from data compression on only the j th inputs will appear as

$$S_{PSAij} = S_{PSAi} \oplus S_{PSAj} = V^{i-1} S_{SSA}(f_i) \oplus V^{j-1} S_{SSA}(f_j).$$

By applying decomposition (18.15) and excluding the SSA index for compactness, we obtain

$$S_{PSAij} = V^{i-1} [S(gG_1^i) \oplus S(\bar{g}G_2^i) \oplus S(G_3^i) \oplus S(gG_1^i G_3^i) \oplus S(\bar{g}G_2^i G_3^i)] \oplus$$
$$V^{j-1} [S(gG_1^j) \oplus S(\bar{g}G_2^j) \oplus S(G_3^j) \oplus S(gG_1^j G_3^j) \oplus S(\bar{g}G_2^j G_3^j)].$$

When a fault $g \equiv 0$ occurs, the expression (18.15) takes on the form $f_i^* = G_2^i + G_3^i$, $i = \overline{1,m}$ and the resulting PSA signature for the two input in question will appear as

$$S^*_{PSAij} = V^{i-1}[S(G^i_2) \oplus S(G^i_3) \oplus S(G^i_2 G^i_3)] \oplus$$
$$V^{j-1}[S(G^j_2) \oplus S(G^j_3) \oplus S(G^j_2 G^j_3)].$$

The condition for signature testability of $g \equiv 0$ is the difference between error function signature and the reference one, i.e. the condition

$$S_{PSA} \oplus S^*_{PSA} \neq 0.$$

must be met.

Substituting the obtained values of reference signature and faulty circuit signature into the latter expression, and using the linearity property of modulo-2 addition, we obtain

$$S(gG^i_1 \overline{G^i_3}) \oplus S(gG^i_2 \overline{G^i_3}) \oplus V^{j-i}[S(gG^j_1 \overline{G^j_3}) \oplus S(gG^j_2 \overline{G^j_3})] \neq 0.$$

The conditions for signature testabilbty of $g \equiv 1$ on the i th and j th inputs to the PSA in question can be obtained similarly:

$$S(\overline{g}G^i_1 \overline{G^i_3}) \oplus S(\overline{g}G^i_2 \overline{G^i_3}) \oplus V^{j-i}[S(\overline{g}G^j_1 \overline{G^j_3}) \oplus S(\overline{g}G^j_2 \overline{G^j_3})] \neq 0.$$

Generalization of the resulting relations on all analyzer's inputs brings the following system of inequalities, which is the condition for signature testability of single stuck-at faults for a PSA with internal modulo-2 adders

$$\sum_{i=1}^{m} \oplus V^{i-1}[S(gG^i_1 \overline{G^i_3}) \oplus S(gG^i_2 \overline{G^i_3})] \neq 0,$$
$$\sum_{i=1}^{m} \oplus V^{i-1}[S(\overline{g}G^i_1 \overline{G^i_3}) \oplus S(\overline{g}G^i_2 \overline{G^i_3})] \neq 0.$$
(18.19)

If a fault is untestable by an identical SSA at some outputs from the circuit under test the associated terms under the summation sign will be zero.

It is easy to notice that the system (18.19) degenerates into testability conditions for SSA subject to a single PSA input.

Let us consider testing of the circuit by another equally popular analyzer, PSA with external modulo-2 adders or the BILBO-design PSA (see Fig.12.2). The system of equations describing its behavior appears as

$$A_B(k) = V_B A_B(k-1) \oplus Z_B(k),$$
$$Y_B(k) = A_B(k), \quad k = \overline{1,l}, \quad (18.20)$$

where the form of matrix V_B is us expressed by (9.4).

The PSA in question is described by the same primitive polynomial $\varphi(x) = x^m \oplus \alpha_1 x^{m-1} \oplus \alpha_2 x^{m-2} \oplus \ldots \oplus \alpha_{m-1} x^1 \oplus 1$ as the earlier discussed one. Hence there is a matrix T to associate the states of PSA in question with those of PSA with internal adders by the formula $A(k)=TA_B(k)$.

Substituting this formula into system (18.20), we obtain

$$A(k) = TV_B T^{-1} A(k-1) \oplus T Z_B(k),$$
$$Y_B(k) = T^{-1} A(k),$$

where $TV_B T^{-1}=V.$

The matrix

$$T = \begin{vmatrix} 1 & \alpha_1 & \alpha_2 & \ldots & \alpha_{m-2} & \alpha_{m-1} \\ 0 & 1 & \alpha_1 & \ldots & \alpha_{m-3} & \alpha_{m-2} \\ 0 & 0 & 1 & \ldots & \alpha_{m-4} & \alpha_{m-3} \\ 0 & 0 & 0 & \ldots & \alpha_{m-5} & \alpha_{m-4} \\ \ldots & \ldots & \ldots & \ldots & \ldots & \ldots \\ 0 & 0 & 0 & \ldots & 1 & \alpha_1 \\ 0 & 0 & 0 & \ldots & 0 & 1 \end{vmatrix}$$

satisfies the equation $TV_B T^{-1}=V.$

Thus the relation (18.20) and the form of matrix T imply that the siqnature obtained in the BILBO-design PSA relates to the signature produced by the PSA with internal adders as

$$S_{PSA}^B = T^{-1} S_{PSA} \quad (18.21)$$

if the inputs to the PSA with internal adders are applied with the following binary sequences:

$$\text{input} \quad 1: \quad \alpha_{m-1}z_m \oplus \alpha_{m-2}z_{m-1} \oplus \cdots \alpha_1 z_2 \oplus z_1;$$
$$\text{input} \quad 2: \quad \alpha_{m-2}z_m \oplus \alpha_{m-3}z_{m-1} \oplus \cdots z_2;$$
$$\cdots \quad \cdots \quad \cdots \quad \cdots \quad \cdots \quad \cdots$$
$$\text{input} \quad m-1: \quad \alpha_1 z_m \oplus z_{m-1};$$
$$\text{input} \quad m: \quad z_m.$$

From the relations obtained as well as expressions (18.16) and (18.20) we may deduce that for a binary sequence z_j compressed in the BILBO-design PSA on only the j th line the signature is defined by the formula

$$S^B_{PSAj} = T^{-1} \sum_{i=1}^{j} {}^\oplus \alpha_{j-1} V^{i-1} S_{SSAj} \quad (18.22)$$

where $\alpha_0 = 1$ and S_{SSAj} is the signature obtained in the SSA whose feedback loops are completely identical to those of PSA with internal adders.

From the linearity property of modulo-2 addition we can derive the formula to calculate signatures obtained in the BILBO-desigh PSA:

$$S^B_{PSA} = \sum_{j=1}^{m} {}^\oplus S^B_{PSAj} = T^{-1} \sum_{j=1}^{m} {}^\oplus (\sum_{i=1}^{j} {}^\oplus \alpha_{j-i} V^{i-1}) S_{SSAj}, \quad (18.23)$$

where $\alpha_0 = 1$.

Substituting $V = TV_B T^{-1}$ into the expression (18.23) subject to (18.21), we get the relationship

$$S^B_{PSA} = \sum_{j=1}^{m} {}^\oplus (\sum_{i=1}^{j} {}^\oplus \alpha_{j-i} V_B^{i-1}) S^B_{SSAj}, \quad (18.24)$$

where $a_0 = 1$ and S^B_{SSAj} is the signature produced by compressing the sequence z_j, $j = \overline{1,m}$, in the SSA whose feedback loops are completely identical to those of the BILBO-design PSA and the behavior is described by the set of equations (18.20) were the input vector degraded into $Z^B(k) = (z^B(k),0,0,...,0)'$.

By performing computations similar to those for the PSA with internal adders, we obtain the conditions for signature testability of stuck-at faults by the BILBO-design PSA:

$$\sum_{j=1}^{m} {}^{\oplus}((\sum_{i=1}^{j} {}^{\oplus}\alpha_{j-i}V^{i-1})(S(gG_1^i\overline{G_3^i}) \oplus S(gG_2^i\overline{G_3^i}))) \neq 0;$$

$$\sum_{j=1}^{m} {}^{\oplus}((\sum_{i=1}^{j} {}^{\oplus}\alpha_{j-i}V^{i-1})(S(\overline{g}G_1^i\overline{G_3^i}) \oplus S(\overline{g}G_2^i\overline{G_3^i}))) \neq 0.$$
(18.25)

It is obvious that in the above relations the values of signature and matrix V may be used as S and V both for SSAs with internal and external adders.

The same as for a PSA with internal adders, the system (18.25) degrades into the signature-testability conditions for SSA subject to one of its inputs.

Therefor the relation (18.19) or (18.25) and the analytical approach to signature evaluation allow any node in the digital circuit under test to be checked for signature testability.

Appendix 1.

Primitive polynomials (Table A1.1) have minimal number of nonzero coefficients for all value of $m \leq 30$, where m is the degree of the polynomial. The Table A1.1 contains the numbers of coefficients, which equel to *1* for the polynomial.

As an example, for $m=2$ in the Table A1.1 we have

$$0\ 3\ 4\ 7\ 12$$

which correspond to the polynomial

$$\varphi(x) = 1 \oplus x^3 \oplus x^4 \oplus x^7 \oplus x^{12}$$

Table A1.1

m	$\phi(x)$	m	$\phi(x)$
1	0 1	18	0 7 18
2	0 1 2	19	0 1 5 6 19
3	0 1 3	20	0 3 20
4	0 1 4	21	0 2 21
5	0 2 5	22	0 1 22
6	0 1 6	23	0 5 23
7	0 1 7	24	0 1 3 4 24
8	0 1 5 6 8	25	0 3 25
9	0 4 9	26	0 1 7 26
10	0 3 10	27	0 1 7 27
11	0 2 11	28	0 3 28
12	0 3 4 7 12	29	0 2 29
13	0 1 3 4 13	30	0 1 15 16 30
14	0 1 11 12 14	31	0 3 31
15	0 1 15	32	0 1 27 28 32
16	0 2 3 5 16	33	0 13 33
17	0 3 17	34	0 1 14 15 34

m	$\phi(x)$	m	$\phi(x)$
35	0 2 35	66	0 1 9 10 66
36	0 11 36	67	0 1 9 10 67
37	0 2 10 12 37	68	0 9 68
38	0 1 5 6 38	69	0 2 27 29 69
39	0 4 39	70	0 1 15 16 70
40	0 2 19 21 40	71	0 6 71
41	0 3 41	72	0 6 47 53 72
42	0 1 22 23 42	73	0 25 73
43	0 1 5 6 43	74	0 1 15 16 74
44	0 1 26 27 44	75	0 1 10 11 75
45	0 1 3 4 45	76	0 1 35 36 76
46	0 1 20 21 46	77	0 1 30 31 77
47	0 5 47	78	0 1 19 20 78
48	0 1 27 28 48	79	0 9 79
49	0 9 49	80	0 1 37 38 80
50	0 1 26 27 50	81	0 4 81
51	0 1 15 16 51	82	0 3 35 38 82
52	0 3 52	83	0 1 45 46 83
53	0 1 15 16 53	84	0 13 84
54	0 1 36 37 54	85	0 1 27 28 85
55	0 24 55	86	0 1 12 13 86
56	0 1 21 22 56	87	0 12 87
57	0 7 57	88	0 1 71 72 88
58	0 19 58	89	0 38 89
59	0 1 21 22 59	90	0 1 18 19 90
60	0 1 60	91	0 1 83 84 91
61	0 1 15 16 61	92	0 1 12 13 92
62	0 1 56 57 62	93	0 2 93
63	0 1 63	94	0 21 94
64	0 1 3 4 64	95	0 11 95
65	0 18 65	96	0 2 47 49 96

m	φ(x)	m	φ(x)
97	0 6 97	128	0 2 27 29 128
98	0 11 98	129	0 5 129
99	0 2 45 47 99	130	0 3 130
100	0 37 100	131	0 1 47 48 131
101	0 1 6 7 101	132	0 29 132
102	0 1 76 77 102	133	0 1 51 52 133
103	0 9 103	134	0 11 135
104	0 1 10 11 104	135	0 11 135
105	0 16 105	136	0 1 125 126 136
106	0 15 106	137	0 21 137
107	0 2 63 65 107	138	0 1 7 8 138
108	0 31 108	139	0 3 5 8 139
109	0 1 6 7 109	140	0 29 140
110	0 1 12 13 110	141	0 1 31 32 141
111	0 10 111	142	0 21 142
112	0 2 43 45 112	143	0 1 20 21 143
113	0 9 113	144	0 1 69 70 144
114	0 1 81 82 114	145	0 52 145
115	0 1 14 15 115	146	0 1 59 60 147
116	0 1 70 71 116	147	0 1 37 38 147
117	0 2 18 20 117	148	0 27 148
118	0 33 118	149	0 1 109 110 149
119	0 8 119	150	0 53 150
120	0 7 111 118 120	151	0 3 151
121	0 18 121	152	0 1 65 66 152
122	0 1 59 60 122	153	0 1 153
123	0 2 123	154	0 2 127 129 154
124	0 37 124	155	0 1 31 32 155
125	0 1 107 108 125	156	0 1 115 116 156
126	0 1 36 37 126	157	0 1 26 27 157
127	0 1 127	158	0 1 26 27 158

m	φ(x)	m	φ(x)
159	0 31 159	190	0 1 2 47 190
160	0 1 18 19 160	191	0 9 191
161	0 18 161	192	0 1 3 112 192
162	0 1 87 88 162	193	0 15 193
163	0 1 59 60 163	194	0 87 194
164	0 1 13 14 164	195	0 1 2 37 195
165	0 1 30 31 165	196	0 1 2 101 196
166	0 1 38 39 166	197	0 1 2 21 197
167	0 6 167	198	0 65 198
168	0 2 15 17 168	199	0 34 199
169	0 34 169	200	0 1 2 163 200
170	0 1 3 42 171	201	0 14 201
171	0 1 3 42 171	202	0 55 202
172	0 7 172	203	0 1 2 45 203
173	0 1 2 10 173	204	0 1 2 86 204
174	0 13 174	205	0 1 2 21 205
175	0 6 175	206	0 1 2 147 206
176	0 1 2 43 176	207	0 43 207
177	0 8 177	208	0 1 2 83 208
178	0 87 178	209	0 6 209
179	0 1 2 4 179	210	0 1 2 31 210
180	0 1 2 52 180	211	0 1 2 165 211
181	0 1 2 89 181	212	0 105 212
182	0 1 2 121 182	213	0 1 2 62 213
183	0 56 183	214	0 1 2 87 214
184	0 1 3 41 184	215	0 23 215
185	0 24 185	216	0 1 2 107 216
186	0 1 2 53 186	217	0 45 217
187	0 1 2 20 187	218	0 11 218
188	0 1 2 186 188	219	0 1 2 65 219
189	0 1 2 49 189	220	0 1 3 53 220

m	$\phi(x)$	m	$\phi(x)$
221	0 1 2 18 221	252	0 67 252
222	0 1 2 73 222	253	0 1 2 33 253
223	0 33 223	254	0 1 2 7 254
224	0 1 2 159 224	255	0 52 255
225	0 32 225	256	0 1 3 16 256
226	0 1 2 57 226	257	0 83 257
227	0 1 2 21 227	258	0 83 258
228	0 1 2 58 228	259	0 1 2 254 259
229	0 1 2 21 229	260	0 1 3 74 260
230	0 1 2 25 230	261	0 1 2 74 261
231	0 26 231	262	0 1 2 252 262
232	0 1 2 23 232	263	0 93 263
233	0 74 233	264	0 1 2 169 264
234	0 31 234	265	0 42 265
235	0 1 2 45 235	266	0 47 266
236	0 5 236	267	0 1 2 29 267
237	0 1 2 163 237	268	0 25 268
238	0 1 2 5 238	269	0 1 2 117 269
239	0 36 239	270	0 53 270
240	0 1 3 49 240	271	0 58 271
241	0 70 241	272	0 1 3 56 272
242	0 1 4 81 242	273	0 23 273
243	0 1 2 17 243	274	0 67 274
244	0 1 2 96 244	275	0 1 2 28 275
245	0 1 2 37 245	276	0 1 2 28 276
246	0 1 2 11 246	277	0 1 5 254 277
247	0 82 247	278	0 5 278
248	0 1 2 243 248	279	0 5 279
249	0 86 249	280	0 1 3 146 280
250	0 103 250	281	0 93 281
251	0 1 2 45 251	282	0 35 282

m	$\phi(x)$
283	0 1 2 200 283
284	0 119 284
285	0 1 2 77 285
286	0 69 286
287	0 71 287
288	0 1 10 11 288
289	0 21 289
290	0 2 3 5 290
291	0 1 2 76 291
292	0 97 292
293	0 1 3 154 293
294	0 61 294
295	0 48 295
296	0 4 9 11 296
297	0 5 297
298	0 1 2 78 298
299	0 1 2 21 299
300	0 7 300

Appendix 2.

The Table A2.1 contains the number of coefficients α_j and β_i the primitive polynomial

$$\phi(x) = 1 \oplus \alpha_j x^j \oplus \beta_i x^i \oplus x^m$$

Table A2.1

m	j	i	m	j	i
8	1	2	16	3	9
8	1	3	16	5	10
8	2	5	16	4	10
12	3	8	19	3	4
12	4	5	19	3	5
13	1	2	19	3	6
13	1	4	19	1	8
13	3	4	19	3	8
13	3	5	19	6	8
13	5	6	19	6	9
13	1	7	19	4	10
13	4	7	19	7	10
13	2	8	19	2	12
13	3	8	19	3	12
13	1	10	19	4	13
14	1	3	19	1	14
14	3	4	19	4	14
14	2	7	24	1	21
14	5	7	26	3	4
14	5	8	27	1	3
16	6	7	30	7	8
16	2	9	32	1	5

m	j	i	m	j	i
34	3	5	70	5	11
37	3	4	72	10	17
38	1	10	74	3	4
40	10	11	75	6	13
42	1	5	76	5	6
43	5	7	77	4	10
44	3	4	78	8	19
45	1	6	80	10	13
46	4	7	82	3	4
48	3	5	83	1	4
50	3	9	85	1	2
51	5	6	86	3	5
53	1	2	88	5	13
54	2	7	90	7	18
56	7	9	91	5	7
59	4	10	92	4	15
61	2	4	96	14	21
62	1	6	99	4	6
64	5	7			
66	2	3			
67	6	8			
69	5	10			

References

In Russian

Agulnik, A.R., and Musaelyan, S.S. (1983) Construction of Nonlinear Binary Sequences, Izv. vuzov. Radioelektronika, 26, 4, 19-27.

Alekseev, A.I., Sheremetev, A.G., Tuzov, G.I. and Glasov, B.I. (1969) Theory and Application of Pseudorandom Signals, Nauka, Moscow.

Badulin, S.S., Barnaulov, V.A., Berdyshev, V.A. et al. (1981) Automated Design of Digital Devices. Radio i Svyaz, Moscow.

Petrenko, A.I., and othes (1979) Automated Design of Crystal-based VLSI. Radio i Svyaz.

Baran, E.D. (1982) Reliability of Testing of Binary Sequences by the Method of State Counting. Avtom. Vych. Tekh., 6, 66-70.

Barashko, A.S. (1990) Optimal Parallel Signature Analyzer. Avtom. Vych. Tekh., 1, 74-79.

Bucharajev, R.G., Zacharov, V.M. (1978) Controllable Random-Code Generators. Izd. Kazan. univ., Kazan.

Berstein, M.S. and Romankevich, A.M. (1974) Statistical Methods of Checking Logic Circuits. Kibernetika, 1, 52-7.

Bharucha-Reid, A.T. (1969). Elements of the Theory of Markov Processes and Their Application. Nauka, Moscow.

Gulajev, V.A. (1983). Technical Diagnostics of Control Systems. Naukova Dumka, Kiev.

Gill, A. (1974) Linear Sequential Machines, Nauka, Moscow.

Goryashko, A.P. (1984) Design of Easily Testable Discrete Devices: Ideas, methods, implementation, Avtom. i Telemech, no. 7, 5-35.

Dobris, G.V. (1974) A New Concept of Constructing Pseudorandom Number Generators Based on Shift Registers. In: Information and Measuring Devicis in Radioelectronics. Tez. dokl. konf., Riga, pp.109-111.

Kalosha, E.P., Kachan, I.V., Yarmolik, V.N. (1991) Investigation of Universal Unit BILBO, Avtom. i Telemech, no. 1, pp.105-112.

Kachan, I.V. (1990) Parallel Signature Analyzer for Self-Testing Digital Devices. Avt. i vych. tekch., rel.19, 91-95.

Kazmina, S.K. (1982) Compact Testing, Avtom. Telemech., 2, 137-89.

Kir'yanov, K.G. (1980) On the Theory of Signature Analysis. Communications equipment. Radio-izm. Tekh., 2(27), 1-46.

Knuth, D. (1977) The Art of Computer Programming, vol.2. Seminumerical Algorithms, Addison-wesley, Reading, M.A.

Kalosha, Ye.P., Katsnel'son, Ye.I., Yarmolik, V.N. (1989) On Credibility of Compact Testing Methods. Avt. i Telemeh. 9, 160-165.

Latypov, R.Kh. (1985) Signature Analysis Comparison Against a Trivial Compression Technigue at Liner Combinational Circuit fault detection, Avtom. i. Telemeh., 2, 165-167.

Lapin'sh, Ya.K, and Metra, I.A (1973) On a Method of Probability Dictribution Generator Design, Avt. i Telemeh., N 4, 32-35.

Litikov, I.P. (1983) Ring-wise Testing of Combinational Circuits. Avtom. Telemeh., 7, 145-153.

Litikov, I.P. (1990) Ring-wise Testing of Digital Units. Energoatomizdat., Moscow.

Litikov, I.P. (1984) Design of a Ring-wise Testing System for Devices with Memory. Avtom. Telemeh., 10, 158-165.

MacWilliams, F.J. and Sloane, N.I. (1976) Pseudorandom Sequences and Arrays. Proc. IEEE, 64, 1715-1729.

Novik, G.H. (1982) On the Signature Analysis Efficiency. Avtom. Telemeh., 5, 110-118.

Peterson, W.W. (1961) Error Correcting Codes, Wiley, New York.

Romankevich, A.M. (1978) On a Method for Constructing Nonlinear Generators of Pseudorandom Sequences. Kibernetika, 1, 136-137.

Romankevich, A.M. and Karachun, L.F. (1981) Towards the Estimation of Characteristics of Test Sequences in a Probability System of Technical Diagnostics. Upr. Sistemy Machiny, 3, 33-35.

Sagalovich, Yu.L. (1988) The Potential of a Fixed Signature Analyzer with Arbitrary Extension of the Test Sequence. Avt. Telemeh., 6, 144-147.

Sagalovich, Y.L., Yarmolik, V.N. (1990) Design of Signature Analyzer for a Two-level Combinational Circuit. Avtom. Telemeh., 12, 155-160.

Smirnov, N.I., Struchkov, A.A., and Sudovtsev, V.A. (1979) Fault Diagnosis in VLSI-based Radio Equipment. Zarub. radioel., 1, 53-60.

Sarwate, D.V. and Pursley, M.V. (1980) Cross Correlation Properties of Pseudorandom and Related Sequences. Proc. IEEE, 68, 593-619.

Sogomonyan, E.S., Slabakov, E.V. (1989) Self-Testing Devices and Fault-Tolerant Systems. Radio i Svyaz, Moscow.

Sklyarov, V.A., Novikov, S.V., Yarmolik, V.N. (1990) Automatic Computer Design, Vysch. Shkola, Minsk.

Tamamoto H. and Narita, Yu. (1982) A Method for Determining Optimal and Quasi-optimal Input Probabilities in Random Test Method Testing. Denski tsushin gakkai ronbunshi, 8, 1057-1064.

Totsenko, V.G. (1985) Algorithms for Technical Diagnosing of Discrete Units. Radio i Svyaz, Moscow.

Idzumi, T., and Kashivagi, H. (1982) Initiating M-sequences of High Order as the Source of Binary Random Number Generation. Kejsoku dzido seiche gankai robunsh, 18, 9, 925-935.

Faradzev, R.G. (1975) Linear Sequential Machines, Soviet Radio, Moscow.

Yarmolik, V.N. and Morozevich, A.N. (1977) A Design Method for Multichannel High-speed Pseudorandom Number Generators. Radio Elekt. Eng., Minsk, 7, 66-69.

Yarmolik, V.N., Leusenko, A.E., and Demidenko, S.N. (1981) Investigation of Linear Congruential Sequence Properties. In: Automation of Exploratory Investigation, Minsk, Issue 2, 118-123.

Yarmolik, V.N., Leusenko, A.E., and Morozevitch, A.N. (1982) Random Process Generation by Simple Digital Circuits. Izv. Vuzov. Priborostr., 25, 31-34.

Yarmolik, V.N. (1982). Pseudorandom Test Signal Sequence Generator. Manag. Mech. Avtom., 3, 57.

Yarmolik, V.N. (1983) Designing Generators of Pseudorandom Sequences of Test Signals. Avtom. Telemeh., 6, 155-162.

Yarmolik, V.N. (1983) A Test Signal Sequence Generator for Probabilistic Testing of Digital Devices. Elektron. Model., N5, 49-54.

Yarmolik, V.N. (1983). Design of High-speed Parallel Pseudorandom Number Generators. Izv. Vuzov. Priborostr., 6, 48-52.

Yarmolik, V.N. (1983) A Method for Determining Modulo-2 Adder Connection Layout for Shifted M-sequence Replica. Izv. Vuzov. Radioel., 11, 80-82.

Yarmolik, V.N. (1985) Hardware Test Pattern Generators, Izv. Vuzov. Priborostr., 7, 35-40.

Yarmolik, V.N. (1985) Design of Multi-channel Signature Analyzers. Avtom. Telemeh., 1, 127-132.

Yarmolik, V.N. (1985). On the Validity of Binary Data Sequence Testing by Signature Analyzer. Elektron. Model, 6, 49-57.

Yarmolik, V.N. (1985). Signature Analysis Application for Testing and Diagnosing Discrete Network Structures. Avt. i Vych. Tech., 4, 73-79.

Yarmilik, V.N. and Demidenko S.N. (1986) Generation and Application of Pseudorandom Sequences for Random Testing. Nauka i Technika, Minsk.

Yarmolik, V.N. (1986) M-sequence Shift and Add Property. Radiotech., 6, 53-58.

Yarmolik, V.N. (1986). Integrated LSI Self-test. Mikroelektronika, 15, 1, 70-76.

Yarmolik, V.N. (1986) Built-in VLSI Self-test. Microelektronika, 1, 70-76.

Yarmolik, V.N. (1987), Signature Evaluation for Discrete Network Structures. Avt. i Vych. Tech., 5, 77-81.

Yarmolik, V.N. (1987) Checking Discrete Network Elements for Signature Testability. Avt. i Vych. Tech., 6, 72-75.

Yarmolik, V.N., and Kalosha, E.P. (1988) Compact Testing of LSSD Structures with Storage Element Creating Faults. Izv. Vuzov. Priborostr., 12, 32-36.

Yarmolik, V.N. (1988) Estimation of Reference Signature Values. Elekt. model. 1988, 6, 69-71.

Yarmolik, V.N. and Kalosha, E.P. (1988) On the Validity of Signature Analysis. Izv. Vuzov. Priborostr., 5, 33-37.

Yarmolik, V.N. and Kalosha, E.P. (1988) Full Self-testing of LSIC with LSSD Structure. Mikroelek., 17, Rel. 2, 99-104.

Yarmolik, V.N. and Kalosha, E.P. (1989) A Method for Estimating Signatures in Diagnostics. Elektr. Model., 6, 50-54.

Yarmolik, V.N. and Kalosha, E.P. (1989) Ehaustive Self-testing CMOS VLSI, Mikroelek., 3, 282-285.

Yarmolik, V.N. and Kachan, I.V. (1989) Determination of Error Multiplicity Caused by Digital Circuit Faults. Avt. i Vych.Tech., Minsk, Rel.18, 101-106.

Yarmolik, V.N. and Kachan, I.V. (1989) Analysing the Efficiency of the BILBO Parallel Signature Analyzer. Avt. i Vych. Tech, 4, 81-86.

Yarmolik, V.N. (1989) Checking Digital Circuits for Signature Testability. Avtom, Telemeh., 10, 159-167.

Yarmolik, V.N. and Kachan, I.V. (1990) Checking Digital Circuits for Testability by Parallel Signature Analyzers. Avt. i Vych. Tech., 3, 89-95.

Yarmolik, V.N. and Kalosha, E.P. (1990) Method for Analytical Definition of Signatures in Self-testing LSSD Structures. Mikroelek. 19, 3, 246-251.

Yarmolik, V.N. and Kachan, I.V. (1990) Analytical Signature Determination for Parallel Signature Analyzers. Izv. Vuzov. Priborostr., 9, 33-38.

Yarmolik, V.N. (1991) Fast Algorithm of Signature Calculation for Network Descrete Structures, Avt. i Vych. Tech, 3, 79-83.

Yarmolik, V.N. (1992) Regular Sequences Signature Calculation, Avtom. Telemeh., 1, 146-155. , 3, 79-83.

Yarmolik, V.N. and Bukov, Y.V. (1992) Random Testing of Digital Structures, Avt. i Vych. Tech., 5, 57-58.

Yakovlev, V.V. and Fedorov, R.F. (1974) Stochastic Computers, Mashinistroyenie, Leningrad.

In English

Agrawal P., Agrawal V.D. (1975) Probabilistic Analisis of Ramdom Test Generation Method for Irredundant Combinational Logic Networks, IEEE Trans. Comput., 7, 691-695. (Vol.C-24).

Agrawal P. Agrawal V.D. (1976) On Monte Carlo Testing of Logic Tree Networks, IEEE Trans Comput., 6, 644-667. (Vol.C-25).

Agrawal V.D. (1978) When to Use Random Testing, IEEE Trans.Comput., 11, 1054-1055. (Vol.C-27).

Ando H. (1980) Testing VLSI With Random Access Scan, Proc. COMPCON, 50-52.

Arvillias F.C., Maritsas D.C. (1977) Combinatorial Logic Free Realization of High-Speed M-Sequence Generation, Electronics Letters 7, 500-502. (Vol.13).

Arvillias A.C., Maritsas D.C. (1979) Toggle-Registers Generating in Parallel k-th Decimations of M-Sequences x^r+x^i+1 Design Tables, IEEE Trans. Comput., 2. P.89-101. (Vol.C-28).

Bardell P.H., McAnney W.H., Savir J. (1987) Built-in test For VLSI Pseudorandom Techniques. Lohn Wiley & Sons.

Bardell P.H., Lapointe M. (1991) Production Experience with Built-In Self-Test in the IBM ES/9000 System, Proc. IEEE Test Conf., 28-36.

Barzilai Z., Coppersmith D., Rosenberg A.L. (1983) Exhaustive Generation of Bit Patterns With Applications to VLSI Self-Testing, IEEE Trans. Comput., 2, 190-194. (Vol.C-31).

Barzilai Z.J., Savir J., Markowsky G., Smith V.G. (1981) The Weighted Syndrom Sums Approach to VLSI Testing, IEEE Trans. Comput., 12, 996-1000. (Vol.C-30).

Bennetts R.G. (1984) Design of Testable Logic Circuits. Reading, MA: Addison-Wesley.

Bennetts R.G., Chen C.L. (1986) Linear Dependencies in Linear Feedback Shift Registers, IEEE Trans. Comput., 12, 1086-1088. (Vol.C-37).

Beucler F.P., Manner M.J. (1984) HILDO: The Highly Integrated Logic Device Observer, VLSI Design. June. 46-50. (Vol.5).

Boundary-Scan Architecture Standard Proposal, Version 2.0 JTAG. (1988) 31 March.

Chan J., Abraham J. (1990) A Study of Faulty Signatures Using a Matrix Formulation, Proc. IEEE Test Conf., 553-561.

Chen C.L. (1988) Exhaustive Test Pattern Generation Using Cyclic Codes, IEEE Trans. Comput., 2, 225-228. (Vol.C-37).

Chen T., Sudana G. (1992) A Self-Testing and Self-Repairing Structure for Ultra-Large Capacity Memories, Proc. IEEE Test Conf., 623-631.

Chin C.K., McCluskey E.J. (1987) Test Length for Pseudorandom Testing, IEEE Trans. Comput., 2, 252-256. (Vol.C-36).

Chickermane V., Lee J., Patel J. (1992) Design for Testability Using Architectural Descriptions, Proc. IEEE Test Conf., 752-761.

Damiani M., Olivo P., Favalli M., Ricco B. (1989) Analytical Model for the Aliasing Probability in Signature Analysis Testing, IEEE Trans. CAD., 11, 1133-1444. (Vol.8).

Eichelberger E.B., Williams T.W. (1977) A Logic Design Structure for LSI Testability, Proc. 14th Design Automation Conf., 462-468.

Eichelberger E.B., Wiliams T.W., Muehldorf E.I., Walther R.G. (1978) Logic Design Structure for Testing Internal Array, Degest of papers 3rd USA-Japan Comput Conf., 266-272.

Eichelberger E.B., Lindbloom E. (1983) Random-Pattern Coverage Enhancement and Diagnosis for LSSD Logic Self-Test, IBM J. Res. Develop., 3, 265-272. (Vol.27).

Eiki H., Inagaki K., Yajima S. (1980) Autonomous testing and its application to Testable Design of Logic Circuis, Digest of papers FTCS-10, 173-178.

El-Ziq Y.M. (1983) S^3: VLSI Self-testing Using Signature Analysis and Scan Path Techniques, Proc. ICCAD. 73-76.

Ens E., Miller D.M. (1983) Syndrom-testable Internally Unate Combinational Networks, Electron. Lett., 19, 637-639. (Vol.19).

Fasang P.P. (1980) BIDCO, Built-In Digital Circuit Observer, Proc. IEEE Test Conf., 261-266.

Fasang P.P. (1983) Microbit Brings Self-testing on Board Complex Microcomputers, Electronics, 10. March, 116-119.

Fasang P.P. (1982) Circuit Module Implements Practical Self-testing, Electronics, 19, 164-167.

Ferreira J., Pinto F., Matos J. (1992) A Boundary Scan Test Controller for Hierarchical BIST, Proc. IEEE Test Conf., 217-223.

Fredricssen H.A. (1975) Class of Nonlinear de Bruijn Cycles, J. of Combinatorial Theory, V19-A, 192-199.

Fradricssen H. (1972) Generation of the Ford Sequence of length 2^n, n Large, Journal of Combinatorial Theory, V.12-A, 153-154.

Frohwerk R.A. (1977) Signature Analysis: A New Digital Field Service Method, Hewlett-Packard J., 5, 2-8. (Vol.28).

Fujiwara H., Kinoshita K. (1978) Testing Logic Circuits with Compressed Data, Degest of papers FTCS-8, 108-113.

Funatsu S., Kawai M., Yamada A. (1989) Scan Design at Nec, IEEE Design & Test of Comput., 3, .50-57.

Funatsu S., Wakatsuri N., Arima T. (1975) Test Generation Systems in Japan, Didest of papers 12th Annu. Des. Autom. Conf., 114-122.

Funatsu S., Wakatsuki N., Yamada A. (1978) Designing Digital Circuits With Easily Testable Consideration Digest of papers IEEE Semiconduct Test Conf., 98-102.

Gelsinger P.P. (1987) Design and Test of the Intel 80386, IEEE Design and Test, June. 42-48.

Godfrey K.R. (1966) Three-level M-sequences, Electronics Letters, 7, 241-243. (Vol. 2).

Godoy H.C., Franclin G.B., Bottoff P.S. (1977) Automatic Checking of Logic Design Structure for Compliance with Testability Ground Rules, Proc. 14th Desin Automation Conf., 460-478.

Golomb S.W. (1980) On the Classification of Balanced Binary Sequences of Period 2^n-1, IEEE Trans. Inform. Theory., 6, 730-732. (Vol.IT-26).

Gutfreund K. (1983) Integrating the Approaches to Structures Design for Testability, VLSI Design, 6, 34-37, 40-42. (Vol4).

Hayes J.P. (1976) Check Sum Methods For Test Data Compression, Automat. Fault-Tolerant Comput., 1, 3-17.

Hayes J.P. (1976) Transition Count Testing of Combinational Logic Circuits, IEEE Trans. Comput., 6, 613-620. (Vol.C-25).

Harwood W., McDermott M. (1989) Testability Features of The MC68332 Modular Microlar Microcontroller, Proc. IEEE Test Conf., 616-623.

Hassan S.Z., Lu D.L. McCluskey E.J. (1983) Parallel Signature Analyzer - Detection Capability and Extensions, Digest of Papers COMPCON, . 440-445.

Hilla S. (1992) Boundary Scan Testing For Multichip Modules, Proc. IEEE Test Conf., 224-231.

Hortensiuos P.D., McLeod P.D., Miller D.M., Card H.C. (1989) Cellular Automata-based Pseudorandom Number Generators for Built-in Self-test, IEEE Trans. Cimput.-Aided Design Integrated Circuits Syst., 8, 842-859, (Vol.8).

Hsiao M.Y., Patel A.M., Pradhan D.K. (1977) Store Address Generator With On-Line Fault-Detection Capability, IEEE Trans. Comput., 11, 1144-1147. (Vol.C-26).

Hurd W.J. (1974) Efficient Generation of Statistically Good Pseudonoise by Linearly Interconnected Shift Registers, IEEE Trans. Comput., 2, 146-152. (Vol.C-23).

Hortensius P.D., McLeod R.D., Card H.C. (1989) Parallel Random Number Generation for VLSI Systems Using Cellular Automata, IEEE Trans. Comput., 10, 1466-1672. (Vol.C-38).

Iliman R.J. (1985) Self-Tested Data Flow Logic: A New Approach, IEEE Design Test Comput., 2, 50-58. (Vol.2).

Ivanov A., Agarwal V.K. (1989) An Analysis of the Probabilistic Behavior of Linear Feedback Signature Registers, IEEE Trans. CAD, 10, 1074-1088. (Vol.8).

Jarwala N., Yau C.W. (1989) A New Framework for Analyzing Test Generation and Diagnosis Algorithms for Wirign Interconnects, Proc. IEEE Test Conf., 63-70.

Karpovsky M.G., Nagvajara P. (1990) Optimal Robust Compression Of Test Responses, IEEE Trans. Comput., 1, 138-141. (Vol.C-39).

Karpovsky M.G. (1986) Finite Orthogonal Series in the Design of Digital Devices. John Willy & Sons.

Komonytsky D. (1982) LSI Self test Using Level Sensitive Scan Design and Signature Analysis, Proc. IEEE Test Conf., 414-424.

Komonytsly D. (1983) Synthesis of Techniques Creates Complete System Self-test, Electrohics, 10, March, 110-115.

Koenamann B., Mucha J., Zwiehoff G. (1979) Built-in Logic Block Observation Technique, Proc. IEEE Test Conf., 37-41.

Koenamann B., Mucha J., Zwiehoff G. (1980) Built-in Test for Complex Integrated Circuits, IEEE Journal Solid-state Circuits., 6, 315-321. (Vol.SC-15).

Lala P.K. (1985) Fault Tolerant and Fault Testable Hardware Design, Englewood Cliffs, NJ: Prentice-Hall.

Lambidonis D., Ivanov A., Agarwal V. (1991) Fast Signature Computation for Linear Compactors Proc. IEEE Test Conf., 808-817.

Landis D.L. (1989) A Self-Test Sestem Architecture for Reconfigurable LSI, Proc. IEEE Test Conf., 275-282.

Lawis T.G., Payne W.H. (1973) Generalized Feedback Shift Register Pseudorandom Number Algorithm, Journal of ACM, 3, 456-468. (Vol.20).

Le Blanc J.J. (1984) LOCST: A Built-in Self-test Technique, IEEE Design & Test, 11, 45-52.

Lempel A. (1970) On a Homomorphism of the de Brujn Graph and It's Applications to the Design of Feedback Shift Registers, IEEE Trans. Comput., 12, 1204-1208. (Vol.C-19).

Lipsky L., Seth S.C. (1989) Signal Probabilities in AND-OR Trees, IEEE Trans. Comput., 11, 1558-1563. (Vol.C-38).

Losq J. (1978) Efficiency of Random Compact Testing, IEEE Trans. Comput., 516-525. (Vol.C-27).

Malaiya Y.K., Yang S. (1984) The Coverage Problem for Random Testing, Proc. IEEE Test. Conf., 237-245.

Maierhofer J. Hierarchical Self-Test Concept Based on the JTAG Standard, Proc. IEEE Test Conf., 127-134.

McCanny J.V., White J.C. (1987) VLSI Technology and Design. Academic Press. London.

Mucha J. (1981) Hardware Techniques for Testing VLSI Circuits Based on Built-in Test, Proc. COMPCON, 366-369.

Muzio J.C., Miller D.M. (1982) Spectral Techniques for Fault Detection, Digest of papers FTCS-12, 297-302.

Moore G. (1980) VLSI: What Does the Future Hold? Electron. Aust., 42, 141.

McCluskey E.J. (1984) A Survey of Design for Testability Scan Techniques, VLSI Design., 12, 38-61. (Vol.5).

McCluskey E.J. (1984) Verification Testing - A Pseudoexhaustive Test Technique, IEEE Trans. Comput., 6, 541-546. (Vol.C-33).

McLeod G. (1992) BIST Techniques for ASIC Design, Proc. IEEE Test Conf., 496-505.

Miller D.M., Muzio J.C. (1983) Specitral Fault Signatures For Internalily Unate Combinational Networks, IEEE Trans Comput., 11, 1058-1064. (Vol.C-32).

Nicolaidis M. (1992) Transparent BIST for RAMs, Proc. IEEE Test Conf., 598-607.

Nicolaidis M. (1990) Efficient UBIST Implementation for Microprocessor Sequencing Parts, Proc. IEEE Test Conf., 316-326.

Olivo P., Damiani M., Ricco B. (1989) On the Design of Miltiple-input Shift-Registers for Signature Analysis Testing, Proc. IEEE Test Conf., 936.

Parker P.K. (1976) Compact Testing: Testing with Compressed Data, Digest of papers FTCS-6, 93-98.

Parker P.K., McCluskey E.J. (1975) Analysis of Logic Circuits with Fault Using Input Signal Probabilites, IEEE Trans Comput., 6, 574-578. (Vol.C-245).

Parker P.K. McCluskey E.J. (1975) Probabilistic Treatment of Ceneral Combinatorial Networks, IEEE Trans. Comput., 6, 668-670. (Vol.C-24).

Posse K. (1991) A Design-for-Testability Architecture for Multichip Modules, Proc. IEEE Test Conf., 113-121.

Powell T.J., Hwang F. (1988) Testability Features in the TMS370 Family of Microcomputers, Proc. IEEE Test Conf., 153-160.

Pradhan D.K. (1986) Fault-Toterant Computing. Volume 1. Englewood Cliffs, NJ: Prentice-Hall.

Pradhan V.V. (1987) Developing a Standard for Boundary-scan Implementation, Proc. IEEE Test Conf., 462-466.

Pratt A.R. (1970) Fast Pseudo-Random Number Generators for Computers The Radio and Electronic Engineering, vol. 40, 2, P.83-88.

Pries W., Thanailakis A., Card H.C. (1986) Group Properties of Cellular Automata and VLSI Applications, IEEE Trans. Comout., 12, 1013-1024. (Vol.C-35).

Reddy SM. (1977) A Note on Testing Logic Circuits by Transition Counting, IEEE Trans. Comput., 3, 313-314. (Vol.C-26).

Reddy S.M., Saluja K.K. Karpovsky M.G. (1985) Data Compression Technique for Built-in Self Test, Digest of papers FTCS-15, 299.

Russell G., Sayers I.L. (1989) Advanced Simulation and Test Methodologies for VLSI Design. Van Nostrand Reinhold Int. Co. Lid., 388.

Saluja K.K., Karpovsky M. (1983) Testing Computer Hardware through Data Compression in Space and Time, Proc. IEEE Test Conf., 83-88.

Sarma S. (1991) Buit-in-Self-Test Considerations in a High-Performance, General-Purpose Processor, Proc. IEEE Test Conf., 21-27.

Savir J., Ditlow G.S., Bardell P.H. (1984) Random Pattern Testability, IEEE Trans. Comput., 1, 79-90. (Vol.C-34).

References

Savir J. (1980) Syndrom-Testable Design of Combinational Circuits, IEEE Trans. Comput., 6, 442-451. (Vol.C-29).

Savir J. (1979) Syndrom-testable Design of Combinational Circuits, Digest of papers. FTCS-9, 137-140.

Savir J. Bardell P.H. (1983) On Random Pattern Test Length, 95-106.

Sas J., Cattoor F., De Man H. (1992) Strategies for Programmable Data Paths Based on Cellular Automata, Proc. IEEE Test Conf., 110-120.

Saxena N., Franco P., McCluskey E. (1991) Refind Bounds on Signature Analysis Aliasing for Random Testing, Proc. IEEE Test Conf., 818-827.

Schneider R.R. (1967) On the Necessity to Examine D Chains in Diagnostic Test Generation, IBM J. Res. Develop., 1, 114. (Vol.9).

Shedletsky J.J. (1977) Random Testing: Practically Versus Verified Effectiveness, Proc. FTCS-7, 157-179.

Siavoshi F. (1988) WTPGA: A Novel Weighted Test-Pattern Generation Approach for VLSI Built-In Self Test, Proc. IEEE Test Conf., 256-262.

Smith I.F. (1980) Measures of the Effectivenes of Fault Signature Analysis, IEEE Trans. Comput., 6, 510-514. (Vol.C-29).

Stanke W. (1973) Primitive Binary Polynomials, Mathematic of Computation, 124, 277-280. Vol.27).

Stewart J.H. (1978) Application of Scan/Set for Error Detection and Diagnostics, Digest of papers Semiconduct Test Conf., 152-158.

Stroele A. (1992) Self-Test Sceduling with Bounded Test Execution Time, Proc. IEEE Test Conf., 130-140.

Stroele A., Wunderlich H. (1990) Error Masking in Self-Testable Circuits, Proc. IEEE Test Conf., 544-552.

Stroud C. (1991) Buit-In Self-Test for Hight-Speed Data-Path Circuitry, Proc. IEEE Test Conf., 47-56.

Su C., Kime C. (1990) CAD of Pseudoexhaustive BIST for Semiregular Circuits, Proc. IEEE Test Conf., 680-689.

Susskind A.K. (1981) Testing by Verifying Walsh Coefficients, IEEE Trans. Comput., 2, 198-201. (Vol.C-30).

Tang D.T., (1984) Chen C.L. Iterative Exhaustive Pattern Generation for Logic Testing, IBM J. Res. Develop., 2, 212-219. (Vol.28).

Tang.D.T., (1984) Chen C.L. Logic Test Pattern Generation Using Linear Codes, IEEE Trans. Comput., 9, 845-849. (Vol.C-33).

Tang D.T., Woo L.S. (1983) Exhaustive Test Pattern Generation with constant Weigtht Vectors, IEEE Trans. Comput., 12, 1145-1150. (Vol.C-32).

Tausworthe R.C. (1965) Random Number Generated by Liner Recurrence Modulo Two, Mathem. Computation., 90, 201-209.

Teller-Giron R., David R. (1974) Random Fault-detection in logical networks, Digest of papers. Int. Symp. on Discrete System. Riga, 232-241.

Thorel P., David R., Pulou J., Rainard J.I. (1987) Design for Random Testability, Proc. IEEE Test Conf., 923-929.

Tootill J.P.R., Riobinson W.D., Eagle D.J. (1973) An Asymptotically Random Tausworthe Sequence, Journal of ACM, 3, 469-481. (Vol.20).

Wang L.T., McCluskey E.J. (1988) Curcuits for Pseudoexhaustive Test Pattern Generation, IEEE Trans. CAD, 10, 1068-1080. (Vol.7).

Wang L.T., McCluskey E.J. (1986) Condensed Linear Feedback Shift Register (LFSR) Testing-A Pseudoexhaustive Test Technique, IEEE Trans. Comput., 4, 367-370. (Vol.C-35).

Wang L.T., McCluskey E.J. (1986) A hybrid Design of Maximum Length Sequence Generators, Proc. IEEE Test Conf., 38-47.

Wang L.T., McCluskey E.J. (1988) Hybrid Designs Generating Maximum-Length Sequences, IEEE Trans. CAD, 1, 91-99. (Vol.7).

Wang L.T., McCluskey E.J. (1988) Lineare Feedback Shift Register Design Using Cyclic Codes, IEEE Trans. Comput., 10, 1302-1306. (Vol.C-37).

Wagner K.D. (1983) Design for Testability in the Amdahl 580, Proc. COMPCON, 384-388.

Wagner K.D., Chin C.K., McCluskey E.J. (1987) Pseudorandom Testing, IEEE Trans. Comput., 3, 332-343. (Vol.C-36).

Whetsel L. (1988) A View of the JTAG Port and Architecture, Proc. ATE & Instrum. Conf., 385-401.

Williams M.J.Y., Angel J.B. (1973) Enhancing Testability of Large Scale Integrated Circuits Via Test Points and Additional Logic, IEEE Trans. Comput., 1, 46-60. (Vol.C-22. Jan).

Williams T.W., Daehn W., Gruetzner M., Stark C.W. (1987) Aliasing Error in LFSR's, Proc. IEEE Test. Conf., 39-45.

Williams T.W., Daehn W., Gruetzner M., Starke C.W. (1988) Bounds and Analysis of Aliasing Errors in Linear Feedback Shift-registers, IEEE Trans CAD, 1, 75-83. (Vol.7).

Williams T.W. (1985) Test Length in a Self-Testing Environment, IEEE Design & Test of Comput., 2, 59-63. (Vol.2).

Willett M. (1976) Characteristic M-Sequences, Mathematics of Computation, 134, 306-311. (Vol.30).

Wolfram S. (1983) Statistical Mechanics of Cellular Automata, Rev. Modern Phys., 55, 601-644.

Wunderlich H.J., Hellebrand S. (1989) The Pseudo-Exhaustive Test of Sequential Circuits, Proc. IEEE Test Conf., 19-27.

Wunderlich H.J. (1988) Multiple Distributions for Biased Random Test Patterns, Proc. IEEE Test Conf., 236-244.

Virupakshia A.R., Redoly V.C. (1983) A Simple Random Test Procedure for Detection of Single Intermittent Fault in Combinational Circuits, IEEE Trans. Comput., 6, 594-597. (Vol.C-32).

Yarmolik V.N. (1990) Fault Diagnosis of Digital Circuits. John Willy & Sons.

Yarmolik V.N., (1988) Demidenko S.N. Generation and Application of Pseudorandom Sequences for Random Testing. John Willy & Sons.

Yarmolik V.N., Kalosha E.P. (1991) Analysing Digital Circuits for Signature Testability, Proc. 2nd Euro. Test Conf., 501.

Index

A
actual binary sequence, 233
addressable storage element, 18
architectural design, 3
atual signature, 286
automaton, 74

B
BILBO (Built-In Logic Block Observer), 6, 7, 8, 211, 212
binary polynomial division algorithm, 141
boolean differences, 3
boolean function, 29
bipolar technology, 1
bistochastic circulant, 168
bistochastic transition probability matrix, 167
boundary scan, 20,30
Boundary-Scan Technique, 20
built-in self test, 5

C
canonical sum, 87
canonical sum of products, 87
cellular automata, 73
comparision circuit, 106
complementary MOS (CMOS) technology, 1
compression in space and time, 206
compression technique, 141
Computer-Aided Design (CAD), 4
computer-aided design system, 3
congruential pseudorandom sequences, 38
controllability, 19,120
correlation function, 85
cyclic shift register, 135

D
d-algorithm, 3
DeBruijn sequences, 71
detection probability, 91, 92, 96, 99, 101

E
emitter-coupled logic (ECL), 1
error escape probability, 164
escape tetection, 118
Euler function, 40
exhaustive test generator, 123
exhaustive testing, 123
expected (error-free) output sequence, 233

F
fanout, 90
fault coverage, 96, 114
fault escape probability, 115
flip-flop, 6
ford sequences, 71
full disjunctive normal form (FDNF), 175

G
gate, 5
generating polynomial, 46, 48, 52
generator, 7, 8, 46

Gold sequences, 71

H

Hamming code, 208
hardware, 3
hardware overhead, 5, 17
HILDO, 28
Huffman model, 111

I

I^2 logic, 1
independent events, 92
Integrated circuit, 1, 2
integrator, 108
interchannel error aliasing, 214
irreducible primitive polynomial, 7
iterative algorithm, 136

J

JTAG architecture, 20

K

keyword, 7

L

Large-scale integrated circuit, 2
lay out design, 4
linear dependence, 66
LOCST technique, 25
logic design, 3
LSSD - level sensitive scan design, 13, 15, 17, 20

M

M-sequence, 39
manifestation probability, 91, 92
Markov chain, 98, 166, 170
Markov process, 98

maximal-length sequence, 34
Medium-scale integrated circuit, 2
Mersen number, 56
Microbit, 6, 36, 37
microelectronic, 1
microprocessor, 5
minimal detection, 105
modulo-2 arithmetic operation, 141
Moore's law, 1
multi-line compression schemes, 205
multifunctional signature analyzer, 156
multilevel sequences of maximal length, 75
multiplication, 142

N

Non-reconvergent fanout, 104

O

observability, 19, 120
Ones counting, 85
optimal probability, 102
orthogonal disjunctive normal form (ODNF), 171

P

parallel signature analyzer, 205
parallel signature analyzer, 214
parity checking, 208
primitive polynomial, 27, 28, 124
preweighted vectors, 132
probabilistic expression, 89
probability distribution for M-multiple errors, 221
probability distribution of error se-

quences, 171
programmable logic arrays (PLA), 5
propagation probability, 91, 92
PRSG -pseudorandom test sequence,
PRTG pseudorandom test pattern generator, 24, 27, 28, 30,31, 34, 36, 37
PSA - parallel signature analyzer, 7, 8, 25, 28, 30, 31, 34
PSA simulation, 270, 233
PSA with T-flip-flop, 260
pseudorandom number generator, 6
pseudorandom number sequences, 39
pseudorandom test patterns, 7, 115
pseudorandom test sequence generator, 7, 22, 212
pseudorandom testing, 111, 116

Q

quadratic signature analyzer, 152, 156

R

Random Access Scan Technique, 18, 19
random patterns, 97
random sequences, 83
random test length, 98
random testing, 83, 116
reduced Humming code, 208
redundant, 120
redundancy, 20
relaxation method, 106
related sequences, 65, 71
representative faults, 95

regular binary sequences, 187
resistor-transistor logic (RTL), 1
response analyzer, 7
reverse primitive polynomial, 42
reversible counter, 106
Ring VLSI Self-Test, 28

S

scan-path technique, 8
scan-technique, 8
self-test products, 6
self-testing, 5, 6, 23
self-testing microcomputer, 36
self-testing VLSI, 123
self-testing VLSI chips, 4, 7, 8
self-testing techniques, 5
sequences, 78
sequential circuit, 111
sequential PRPG, 53
servointegrator, 106
shift register, 6, 7, 8, 144, 145
signal probability, 93
signature, 7
signature analysis, 141, 6, 36
signature analyzer, 6, 22, 24, 25, 141, 144, 146
signature analysis efficiency, 163
signature testability, 273, 306, 307
small-scale integrated circuit, 2
software, 3
STAFAN (STAtistical Fault ANalysis), 90
Stanford Scan Path Design, 8
State count testing, 209
stochastic integrator, 107, 108, 109

storage elements, 8
stuck at faults, 94
stuck at 0 (≡ 0) fault, 94
stuck at 1 (≡ 1) fault, 94
structured design, 24
structural dependence, 66

T

TDI (test data input), 12, 15, 22, 31
TDO (test data output), 12, 15, 22, 31
testability design techniques, 3, 8,
test generator, 5, 77,
test generation, 3, 4
test length, 99, 118
test pattern generation, 7
test response analysis, 127
test sequence length, 96
test stimuli generation, 5, 6
test structures, 4
test stimuli, 27
transistor, 1, 2, 5
transistor-transistor logic (TTL), 1
transition counting, 178
trivial testing, 185

U

undetectable errors, 165
Ultrahigh-speed integrated circuit, 2
uniform distribution, 93
universal module, 7
universal module BILBO, 212
untestable fault, 119

V

Very Higth-speed integrated circuit, 2
Very Large-Scale Integrated circuit (VLSI), 2, 3, 9
VLSI self test design, 7

W

worst-case fault, 98